TERRAIN MODELLING IN SURVEYING AND CIVIL ENGINEERING

edited by

G Petrie

Professor of Topographic Science
University of Glasgow

and

T J M Kennie

Lecturer in Engineering Surveying
University of Surrey

Whittles Publishing

in association with
Thomas Telford Ltd

Published by
Whittles Publishing Services
Roseleigh House, Latheronwheel, Caithness, KW5 6DW

Thomas Telford Ltd.,
Thomas Telford House, 1 Heron Quay,
London E14 9XF

British Library Cataloguing in Publication Data

Terrain modelling in surveying and civil engineering.
  1. Civil engineering. Applications of surface modelling by
computer systems.
I. Petrie C. II. Kennie T.J.M.
624.0151636
ISBN 1–870325–30–3

Typeset by Thomson Press (I) Ltd., New Delhi
Printed by Bell and Bain Ltd., Glasgow

Contributors

D.R. Catlow	Formerly with Laser-Scan Limited, Cambridge Science Park, Milton Road, Cambridge CB4 4FY, UK. Currently GIS Senior Software Analyst, Unisys Europe, European GIS Centre, Amsterdam, The Netherlands
G.S. Craine	Director, MOSS Systems Limited, Barclays House, 51 Bishopric, Horsham, West Sussex RH12 1QJ, UK
A.M. Durrant	Consultant, Mined Logic International, 38 White Cross Road, Swaffham, Norfolk PE37 7QY, UK
M.W. Griffin	Software Projects Manager, EASAMS Limited, Lyon Way, Frimley, Camberley, Surrey GU16 5EX, UK
S.R. Gould	Senior Software Engineer, Ground Modelling Systems Limited, Rockingham Drive, Linford Wood, Milton Keynes MK14 6NG, UK
D.J. Gugan	Software Consultant, Laser-Scan Limited, Cambridge Science Park, Milton Road, Cambridge CB4 4FY, UK
T.J. Hartnall	Senior Software Engineer, Laser-Scan Limited, Cambridge Science Park, Milton Road, Cambridge CB4 4FY, UK
R. Hogan	President, HASP Inc., Loveland, Colorado 80537, USA
T.J.M. Kennie	Formerly Lecturer in Engineering Surveying, University of Surrey. Currently Group Training and Development Manager, Debenham, Tewson and Chinnocks, 44 Brook Street, London, W1A 4AG, UK
M.R. Ketteman	Formerly with HASP Europe. Currently Business Development Manager, CIML, Celtic House, Heritage Gate, Derby, DE1 1OX, UK
I. McAulay	Associate and Computer Specialist, Turnbull Jeffrey Partnership, Sandeman House, 55 High Street, Edinburgh EH1 1SR, UK
M.J. McCullagh	Lecturer in Geography, Department of Geography, University of Nottingham, University Park, Nottingham NG7 2RD, UK
R.A. McLaren	Director, Know Edge Limited, 33 Lockharton Avenue, Edinburgh EH14 1AY, UK
E. Malcolmson	Managing Director, Axis Software Systems Limited, Park House, 88–102 Kingsley Park Terrace, Northampton NN2 7HJ, UK
P.H. Milne	Senior Lecturer in Surveying, Department of Civil Engineering, University of Strathclyde, Rottenrow, Glasgow G1 1XW, UK
G. Petrie	Professor of Topographic Science, Department of Geography

and Topographic Science, University of Glasgow, Glasgow G12 8QQ, UK

J.W. Shearer Lecturer in Topographic Science, Department of Geography and Topographic Science, University of Glasgow, Glasgow G12 8QQ, UK

F. Steidler Wild Heerbrugg Limited, 9435 Heerbrugg, Switzerland

J. Strodachs Managing Director, Applications in CADD Ltd., 21 Britannia Street, Shepshed, Leicestershire LE12 9AE, UK

L.W. Thorpe Scicon Limited, 49 Berners Street, London WIP 4AQ, UK

M. Turnbull Partner and Landscape Architect, Turnbull Jeffrey Partnership, Sandeman House, 55 High Street, Edinburgh EH1 1SR, UK

H. Webb Applications Engineer, Intergraph (UK) Limited, Delta Business Park, Great Western Way, Swindon SN5 7XP, UK

Contents

Foreword

The origins of this book lie in two three-day courses on Terrain Modelling in Surveying and Civil Engineering organised by the two editors and given at the Universities of Surrey and Glasgow respectively in April and September 1987. As word spread about the existence of printed course notes, requests were received for copies. After discussion with Dr Whittles, it was decided that his new company would undertake the publication of an expanded version of the course notes reorganized into the form of a book and with several additional contributions covering topics or systems which had not been included at the outset.

In keeping with its origins, this volume is aimed at the professional practitioner interested in or involved with digital terrain modelling in surveying and civil engineering and in fields closely related to these disciplines. Thus it is not intended as a comprehensive textbook on the mathematical theory underlying surface modelling, nor is it a volume reporting the results of the latest academic research into automatic image correlation techniques for the production of digital elevation data using photogrammetric methods. Instead, it is intended in the first instance to provide an account of the current state of the art in the areas of data acquisition and systems and software. These are matters of high concern to practitioners, yet the information is very scattered and often difficult to obtain. These topics are covered in Parts A and C of this volume, together with a short introduction to the theoretical background of digital terrain modelling given in Part B, to form the first half of the book. The second half provides an account of the ways in which digital terrain modelling can be applied to civil engineering projects (covered in Part D) and to the closely related fields of landscape visualization, radio communications planning and defence applications (Part E). Finally the vital matter of the accuracy, procurement and testing of digital terrain modelling systems is covered in Part F. Of course, there is much more to be said about all of these subjects and it is hoped that the extensive lists references provided at the end of several chapters will allow the reader to explore each of these areas much further if required.

As usual with volumes having many individual contributors, the editors are extremely grateful to all of them for their cooperation, efforts and patience during the gestation period of the book. As will also become apparent to those reading the book, it would have been impossible to produce it without extensive help from numerous systems suppliers in the form of printed information, illustrations and answers to queries. Their assistance is most gratefully acknowledged, as is the permission given by Messrs Blackie & Son Ltd. to use a number of diagrams and illustrations published in previous volumes edited by us. The present publisher, Dr Keith Whittles, has been extremely helpful throughout the duration of this project and it is a pleasure to record both his excellent cooperation and his unswerving faith in this volume. The professionalism of Dr Linda Nash in acting as technical editor has also been very apparent to the present writers and must not pass without an appreciative record. Lastly, but certainly not least, the writing and editing of a book always impacts heavily on the families of the authors and editors. In our case, the understanding, forbearance and support afforded by our respective wives, Kari and Sue, has been of a very high order and has also helped greatly to turn the project into a reality.

G.P.

T.J.K.M.

Preface

The subject of terrain modelling has been one of continuous growth over the last 20 years. At its root has been the explosive development in computer technology which has taken place over this period, both in terms of computational power and in the widespread, almost universal availability of computers. Growth in the applications of terrain modelling has been especially rapid in recent years, largely as a result of spectacular advances in computer graphics and imaging technology, and parallel developments in the software used for display and visualization. At the same time, there has been a vast increase in the speed with which data about the terrain surface can be acquired, due to the development of electronic theodolites and distance measuring equipment for field survey, analytical and automated photogrammetric instrumentation for aerial surveys and digitally-based methods of extracting information from existing maps. As a result, digital terrain modeling (DTM) techniques are used routinely in many fields of activity—especially in civil engineering, landscape architecture and planning, aircraft simulators, radiocommunications planning, defence planning, and other related areas.

In spite of all this activity, there are few publications available on this subject area in the English language. Davies and McCullagh's book, *Display and Analysis of Spatial Data*, published in 1975, dealt with the overall area of spatial data in its widest sense, though not with digital terrain modelling in particular. Since then, a few proceedings have appeared reporting the results of specialist workshops (see references). Most recently, an excellent text by Lancaster and Salkauskas on the purely mathematical aspects of curve and surface fitting has been published, part of which has much relevance to terrain modelling. However, no book has appeared which is devoted specifically to digital terrain modelling and, in particular, to current methods of data acquisition, to the present state of development in DTM systems and software, and to the applications of terrain modelling in civil engineering and the associated subjects mentioned above. This book redresses the balance.

At the same time, it is necessary to declare what this book does not cover. In particular, it does not attempt to set out or explain the detailed mathematical basis of terrain modelling—though a short introduction is given. Also, it does not cover the subject of surface, subsurface and solid modelling as applied to the geosciences. There is of course an overlap in certain areas, but the data collection techniques, such as those associated with the drilling of geological boreholes and with geophysical surveys involving gravity, seismic and geomagnetic measurements, are very different to those used for the measurement and description of the terrain surface itself. Also, the methods used for the subsurface modelling of geological structures and their applications to underground mining, oil and gas reservoir simulation and exploitation, etc., are often very different to those used in terrain surface modelling. The subject of this present volume is large and varied enough to justify publication and to keep the editors and their contributing authors quite fully occupied.

It also seems appropriate to discuss here the relationship of digital terrain modelling to the other closely related fields of digital mapping, automated mapping/facilities management (AM/FM), and geographic and land information systems (GIS and LIS), all of which are concerned with spatial data about the terrain surface and the objects occurring on it. Digital mapping is concerned with computer-based methods of producing maps at medium to large scales and uses much the same technologies for data acquisition, processing, display and output as those employed in digital terrain modelling. However, as the name suggests, the final product in the context of surveying and civil engineering is almost invariably some form of topographic map, whereas in DTM, the emphasis is on the surface form of the terrain as represented by a set of elevation values at known locations and often represented graphically by perspective block diagrams, slope maps, etc. Thus many digital terrain models do not contain cultural information—on roads, buildings, boundaries, etc.— of the type contained in a standard topographic map. Of course, it must be recognized that there is no clearcut dividing line in this respect and, as will be seen in Parts C and D of this book, the production of digital maps and terrain models is closely integrated in certain of those systems concerned principally with the planning and design of civil engineering projects at large scales.

With regard to the fast-developing fields of GIS, LIS and AM/FM, at the moment these are less well integrated with digital terrain modelling. In the first place many of the current GIS, LIS and AM/FM systems are designed only to deal with 2D data and not the 3D data generated and handled by a terrain modelling system. Furthermore, in many cases, the basic locational information used by public utilities, local government authorities, planning organizations, etc., is derived from large-scale maps which contain only planimetric information, i.e., there is no elevation data to be extracted from such maps. However, in the long run, this situation seems certain to change, and a much closer integration of DTM into certain types of GIS, LIS and AM/FM systems can be expected to take place in the coming years. A major incentive to do so will come about as regional, state-wide and national digital terrain models are established. These will form massive databases to which the database management systems (DBM), query languages and analysis techniques which are integral to GIS, LIS and AM/FM systems can and will be applied with advantage. Undoubtedly these will add to what already appears to be an interesting and exciting field, and will result in new challenges for the future.

References

ASPRS-ACSM (1978) *Proceedings of the Digital Terrain Models (DTM) Symposium.* St Louis, Missouri, 567pp.

Davis, J.C. and McCullagh, M.J. (1975) *Display and Analysis of Spatial Data.* John Wiley and Sons, London, 383pp.

Kubik, K. and Roy, B.C. (1986) *Digital Terrain Model Workshop Proceedings.* Ohio State University, Columbus, Ohio, 227pp.

Lancaster, P. and Salkauskas, K. (1986) *Curve and Surface Fitting: An Introduction.* Academic Press, London, 280pp.

Toomey, M. (1984) *Digital Elevation Model Workshop Proceedings.* Edmonton, Alberta, 231 pp.

1 Introduction to terrain modelling—application fields and terminology

T.J.M. KENNIE and G. PETRIE

1.1 Introduction

Surface modelling is a general term which is used to describe the process of representing a physical or artificially created surface by means of a mathematical expression. Terrain modelling is one particular category of surface modelling which deals with the specific problems of representing the surface of the Earth. The techniques of terrain modelling are in widespread use and have been applied widely in the physical and earth sciences. In particular the technique has been applied to problems in the following fields.

(i) Topographic mapping. Digital representations of the terrain often form one of the main elements of the mapping process. However, unlike surface modelling where a unique mathematical expression can often be used to define the feature of interest, it is virtually impossible to define precisely the structure of the terrain by a single mathematical function. Consequently, many different approaches to the problem have been advocated. Both large-scale (site-specific) and small-scale (national) topographic modelling can be carried out. The former are of particular importance since they form the basis for modelling civil engineering projects.

(ii) Civil engineering. Terrain models are in widespread use in the field of civil engineering. Initially they were used almost exclusively for the determination of earthwork cut and fill volumes for highway engineering projects. Nowadays, however, the same principles are applied to both linear projects (such as roads and railways) and projects with a large areal extent (such as landscaping and estate development). Furthermore, the modelling techniques can also be used to create digital design models of proposed structures (such as roads, buildings, etc.). The ability to merge the design and terrain models also enables realistic visual impressions to be created of the environmental impact of civil engineering project.

(iii) Hydrographic/bathymetric mapping. Small-scale models of the sea bed using depths obtained from the echo sounders can also be produced. Compared with their topographic equivalents it is less easy to gain access to, or visualize the sea bed, and difficult also to identify, locate and measure distinct lines on the terrain (such as ridges, valleys and breaklines) which play such a vital role in modelling terrain surfaces. More recently digital models of the sea bed at large scales have also been created, typically as part of a marine site investigation for an offshore structure.

(iv) Geological and geophysical mapping. Geology has been a discipline where terrain modelling has found widespread application over a long period of time. Besides terrain surfaces, models of underground surfaces defining specific geological strata can also be created. Models of this type are often derived from sparse or scattered data obtained during drilling. The presence of faults is a special problem of geological applications, with many special techniques being devised by geologists to deal with the specific problems of modelling such features.

The closely related field of geophysical exploration and mapping is an area of activity where terrain modelling is used almost routinely and many contributions in terms of basic theory and techniques have been made by geophysicists.

(v) Mining engineering. The application of terrain modelling in this field is closely related to that of geology and geophysics. Again much of the data used is highly clustered, and poorly distributed. Consequently much work has been made by mining engineers with such data; in particular, the work done by Krige is highly regarded and has led to the term 'Kriging' coming into widespread use in all forms of modelling and interpolation. Another application in this area is concerned with digital modelling of coal stocks to give volume estimates.

(vi) Simulation and terrain visualization. Many people take the view that the most advanced and sophisticated forms of terrain modelling are in the fields of simulation and visualization, particularly for flight and radar simulation. In the former case, realistic representations of the Earth's surface, derived from terrain models, are combined with the requirement for real time and constantly changing simulations of the pilot's view of the ground. Finally, terrain modelling using fractals can be used to produce views of artificially created surfaces and is of particular relevance in the field of computer graphics and animation.

(vii) Military engineering. An understanding of the terrain is of vital importance to a military commander. It is therefore not surprising that substantial effort and large sums of money have been spent on research and development into the military applications of terrain models and other digital mapping products. Terrain models may be used, for example, to derive the visibility from a specific point on the terrain; this may then be used by the commander to determine the optimum positions for radar installations, ground-to-air missile launchers or communications equipment. If information about slope is also derived from the model, this may be used to plan suitable routes for tracked and untracked vehicles. Sophisticated missile guidance systems which make use of terrain contour matching and area correlation algorithms also make extensive use of terrain modelling data. Static simulations of the terrain for military battlefield planning or environmental impact analysis are also important.

1.2 Terminology

The concept of creating digital models of the terrain is a relatively recent development, and the introduction of the term Digital Terrain Model (DTM) is generally attributed to two American engineers working at the Massachusetts Institute of Technology during the late 1950s. The definition given by them (Miller and La Flamme 1958) was as follows:– "The digital terrain model (DTM) is simply a statistical representation of the continuous surface of the ground by a large number of selected points with known X, Y, Z coordinates in an arbitrary coordinate field". This early work of Miller and La Flamme was concerned specifically with the use of cross-sectional data to define the terrain. Since then, several other terms—e.g. Digital Elevation Model (DEM), Digital Height Model (DTM) and Digital Ground Model (DGM)—have been coined to describe this, and other closely related processes. Although in practice these terms are often presumed to be synonymous, in reality they often refer to quite distinct products. The following discussion is therefore provided in an attempt to simplify and standardize the use of these terms.

First of all, one can consider a selection of the definitions or meanings of the different words as encountered in various dictionaries or in the technical literature on the subject. Thus *ground* is defined as "the solid surface of the Earth"; "a solid base or foundation"; "a surface of earth" or "a portion of the Earth's surface". By contrast, *height* is given as "measurement from base to top"; "distance upwards"; and "elevation above the ground or a recognised level". In the case of *elevation*, "angular height above the horizon"; and "height above a given level, especially that of the sea" are two useful definitions. Finally *terrain* has the rather different concept or meaning of being "a tract of country considered with regard to its natural features"; or "an extent of ground, region, territory, etc."

From these definitions, some differences begin to manifest themselves between the different terms.

(i) In the case of the *Digital Elevation Model (DEM)*, the term elevation emphasizes the measurement of height above a datum and the absolute altitude or elevation of the points contained in the model. DEM is a term which is in particularly widespread use in the U.S.A., and generally refers to the creation of a regular array of elevations, normally in a square grid or hexagonal pattern, over the terrain. The DEMs are created by photogrammetric or cartographic methods, rather than field survey.

(ii) Digital Height Model (DHM): DHM is a less common term and has the same definition as DEM since the words elevation and height are normally regarded as synonymous. The term seems to have originated in Germany.

(iii) Digital Ground Model (DGM): seems to lay its emphasis on a digital model of the solid surface of the Earth. In this way, there is presumed to be some connection between the elements which are no longer considered discrete. This connection generally takes the form of an inherent interpolation function which may be used to generate any point on the ground surface. The term is in general usage in the UK, although its use has to some extent been superseded by *digital terrain model.*

(iv) Digital Terrain Model (DTM): is a more complex and all-embracing concept involving not only heights and elevations but other geographical elements and natural features such as rivers, ridge lines, etc. In the narrower sense, a DTM only represents terrain relief. However in its most general form, a DTM is considered by most people to include both planimetric and terrain relief data. Moreover, unlike the previous definitions, this representation may also include derived data about the terrain such as slope, aspect, visibility and so on.

(v) The term *Digital Terrain Elevation Data (DTED)*, as used by the US Defense Mapping Agency (DMA), describes essentially data produced by the same process although it specifically uses a grid-based method of data storage.

Reference

Miller, C., and LaFlamme, R.A. (1958) The digital terrain model—theory and applications. *Photogrammetric Engineering* **24** (3) 433–442.

Part A
Data Acquisition

2 Field data collection for terrain modelling

T.J.M. KENNIE

2.1 Introduction

The acquisition of accurate three-dimensional coordinates which represent the surface of the Earth is a vital stage in the process of terrain modelling. It is possible to form such models using a range of different techniques. The particular technique used will depend on factors such as the size of the area to be surveyed, the accuracy required of the data and the type of information which will eventually be extracted from the model. At small scales, for example for flight simulation or visualization, a model could be created by digitizing an existing contour map. Photogrammetry may also be used to produce models, generally by measuring points on a regular grid. Both of these techniques will be discussed in more detail subsequently in this book. At larger scales, however, for engineering and mining projects, it is now common practice to create models by ground surveying. The remainder of this chapter will therefore be restricted to a discussion of the instruments and procedures which are used to acquire terrain data using modern electronic equipment, together with methods of recording this data in the field.

2.2 Electronic tacheometers

Although techniques such as grid levelling and stadia tacheometry may be used to produce terrain models, their use has become much less common since the introduction of the electronic tacheometer. The term electronic tacheometer refers to an instrument which is able to measure electronically both angles and distances and to perform some limited computational tasks using an integral microprocessor, e.g., reducing a slope distance to the horizontal or calculating coordinates from bearings and distances. The instrument may also, but not necessarily, be able to store data electronically, either in an internal memory unit or, more commonly, in an external solid-state data recorder.

A further term which is in widespread use to describe instruments in this category is total station. Unfortunately, while the use of this term normally implies that the instrument is capable of electronic angle and distance measurement, it has also been applied to instruments which do not measure angles electronically. It is therefore necessary to use this term carefully if confusion is to be avoided.

While there exist individual design features which are unique to a particular manufacturer, it is generally possible to differentiate between the integrated and modular designs of electronic tacheometer. In view of the rate of development of

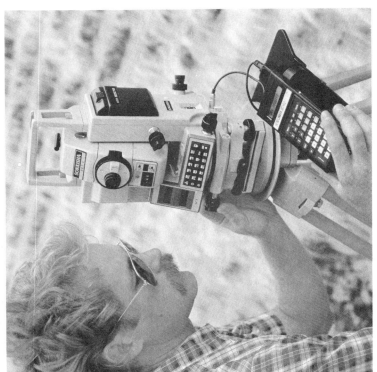

Figure 2.1 (a) Sokkisha Set 2 Integrated Electronic Tacheometer with SDR 2 dedicated data recorder. Courtesy Sokkisha (UK) Ltd. (b) Sokkisha SDR 2 'Electronic Field Book'. Courtesy Sokkisha (UK) Ltd.

Figure 2.2 Geodimeter 440 Integrated Electronic Tacheometer with Geodat 126 dedicated data recorder. Courtesy Geotronics Ltd.

microprocessor technology and the ever increasing range of instruments on the market, it would be inappropriate to attempt to describe the whole range of instruments in this field. Consequently, two examples which are representative of instruments in each category will be described in order to enable the general characteristics of each design to be more fully appreciated.

2.2.1 Integrated design

Many of the earlier examples of electronic tacheometers, such as the AGA Geodimeter 700 and the Zeiss (Oberkochen) Reg Elta, were of an integrated design. The primary feature of such a design is that the electronic theodolite and EDM instrument form a single integrated unit. One of the main advantages of such an

Table 2.1 Integrated electronic tacheometers

	Sokkisha SET 2	Sokkisha SET 3	Geodimeter 440
Angle measurement			
Standard deviation of mean pointing (face I and II)			
Horizontal	± 2″	± 5″	± 2″
Vertical	± 2″	± 5″	± 2″
Distance measurement			
Range (m) in average/ good conditions			
1 prism	1300/1600	1000/1300	2300/3000
3 prisms	2000/2500	1600/2100	3500/5500
Maximum	2600/3200	2200/2800	5000/7000
Standard deviation	± 3mm ± 2ppm	± 5mm ± 3ppm	± 5mm ± 5ppm

approach is the need to transport only a single unit and also to be able to dispense with auxiliary cables which are often required to link separate units. Furthermore, since the units are matched together during manufacture, and often use coaxial optics for telescope pointing and transmitting the EDM signal, instruments of this type rarely suffer from a lack of collimation of the telescope line of sight and the EDM signal. Two instruments which are typical of this design are the Sokkisha SET2/3 (Fig. 2.1a) and the Geodimeter 440 (Fig. 2.2). The general characteristics of these instruments are summarized below in Table 2.1.

The Sokkisha SET 2 and SET 3 are typical of the relatively low cost electronic tacheometers which are being manufactured by Japanese optoelectronics companies.

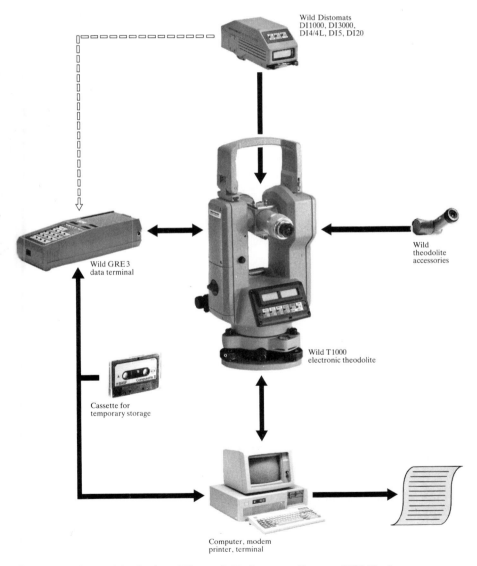

Figure 2.3 Wild Modular Design of Electronic Tacheometer. Courtesy Wild Heerbrugg.

Table 2.2 Modular electronic tacheometers

	Wild (Heerbrugg)		Kern (Aarau)	
	T-1000/DI-1000	T-2000/DI-5	E1/DM 550	E2/DM 503
Angle measurement				
Standard deviation of				
mean pointing				
(face I and II)				
Horizontal	± 3″	± 0.5″	± 2″	± 0.5″
Vertical	± 3″	± 0.5″	± 2″	± 0.5″
Distance measurement				
Range (m) in				
average conditions				
1 prism	500	2500	1800	2500
3 prisms	800	3500	3000	3500
Maximum	800	5000	3300	5000
Standard deviation	± 5mm ± 5ppm	± 3mm ± 2ppm	± 5mm ± 5ppm	± 3mm ± 2ppm

Other recent examples include the Topcon ET-1 and the Nikon DTM-5. Over 20
preprogrammed functions can be interrogated by using the integral keyboard. The
system can also be linked to a data recorder such as the Sokkisha SDR2 (Fig. 2.1b)

The Geodimeter 440 is a further example of an integrated instrument. It has several
unique features including an electronic level display, one-way speech communication
and a visible guidance system to assist the prism carrier to locate himself along the
beam of the EDM signal. It can also be linked to a data recorder.

Figure 2.4 Kern E2/DM 503 Modular
Electronic Tacheometer with DIF 41 data
recorder (the DIF 41 is a data recorder
based on the Hewlett Packard HP–41CV
calculator). Courtesy Kern (Aarau).

2.2.2 Modular design
In this case, the electronic theodolite and EDM instrument are separate units which can be operated independently. This arrangement is more flexible, since theodolites and EDM units with differing accuracy specifications can be combined. It may also be a more cost-effective solution, since the individual units may be replaced and upgraded as developments occur. The Wild (Heerbrugg) and Kern (Aarau) designs are typical of instuments in this category.

Wild (Heerbrugg) offer two modular systems based around the Wild T-2000 and T-1000 electronic theodolites. Several combinations of theodolite and EDM are available, but the most popular arrangements are the Wild T-1000/DI1000 (Fig. 2.3) and the Wild T-2000/D15. Both would normally be linked to a data recorder such as the Wild GRE3. The specifications of both systems are given in Table 2.2

Kern (Aarau) also offer a range of components which can be combined to form a modular electronic tacheometer. Two of the most common arrangements are given in Table 2.2, and Fig. 2.4 illustrates the Kern E2/DM 503 combination attached to a DIF 41 data recorder.

2.3 Data recorders

Traditionally the surveyor has recorded survey observations manually in a fieldbook. Standard fieldbook layouts have developed over the years, and these attempt to identify gross booking or reading errors. Some form of numbering and a field sketch are also normally required, particularly for recording survey detail. Conventional techniques such as these are quite adequate for many survey projects. However, with developments in digital mapping and terrain modelling in recent years, the traditional method of recording survey observations and subsequently transferring them via a computer keyboard has been found, in practice, to be a slow and error-prone activity. Consequently the need arose for a device which could electronically store survey field observations. Such a device is referred to as a data recorder, or alternatively a data collector or data logger.

When used in conjunction with electronic tacheometers, data recorders which receive values direct from the instrument display enable the surveyor to eliminate sources of gross error such as those caused by mis-booking or mistakes in keyboard data entry. Furthermore, they also enable substantial increases in survey productivity (in terms of points observed) to be realized. Bennett (1983) estimated that field productivity using electronic tacheometers and data recorders rose to about 800 points per day compared with about 300 points per day using conventional techniques.

2.3.1 Design of data recorders
The data recorder and its associated peripherals for data transfer (Fig. 2.5) form a crucial link between the field observations and the end product. The distinction can be made between recorders dedicated solely to the logging and processing of survey observations and the more general-purpose, hand-held computers which have been adapted to perform logging but which can also operate as stand-alone computers. Each design has its particular merits; ideally, however, users should ensure that the particular data recorder which they choose satisfies some, if not all of the following hardware and software requirements.

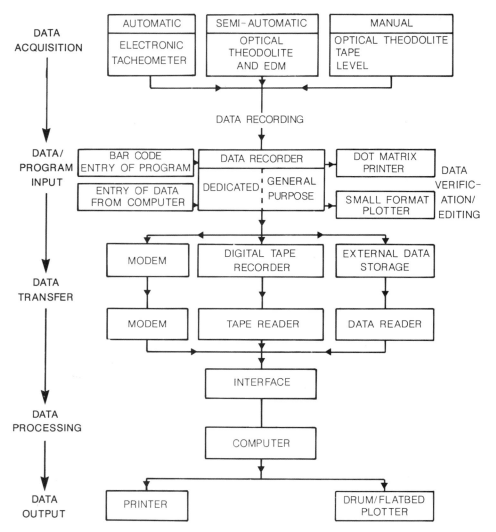

Figure 2.5 Data recorder and associated peripherals for data transfer.

2.3.1.1 *Hardware requirements*

(i) Physical. The unit should be easy to handle under field conditions. It should therefore be compact and lightweight. It may also be important to consider whether it can be mounted on to the instrument tripod without causing undue strain. If the recorder is to be used primarily in temperate climates it must be weatherproof. Similarly, if the working conditions are likely to be hot and dry, the unit should be insulated from the effects of extreme heat and dust.

(ii) Data storage/transmission. The storage capacity of the recorder should enable a reasonable number of observations to be recorded (at least 3–4 h of intense field observations). The retention period of the data should also be independent of whether the recorded is switched on or off and also ensure that data is retained for a

reasonable time period (typically three months). Data transmission from the re-corder to the host computer or peripheral (via modem or digital tape recorder) should be possible at reasonably high transmission rates without data being corrupted.

(iii) Screen/keyboard. The screen should be easy to read in direct sunlight. An LCD (liquid crystal display) is preferable to an LED (light-emitting diode) display, although the former may not operate below 0°C unless some form of screen heater is included in the design. Ideally the screen should enable several lines of data to be displayed. The keyboard should contain numeric and preferably alpha characters in a layout which is either standard (QWERTY) or easy to operate. The keys themselves should also be reasonably large, since the user may be wearing gloves during their operation.

2.3.1.2 *Software requirements*

If a dedicated data recorder or data logging software for a hand-held computer is purchased, it is unlikely that the user will have a significant level of control over the structure of the software (although some manufacturers allow the user to define the form of the data entry to a limited extent). In these circumstances, or if the user is considering developing software in-house, it is important that the software should satisfy the following basic requirements.

(i) Data acquisition. The software should be flexible and accommodate various forms of data entry (manual, semi-automatic and fully automatic). Equally important, the form of data entry should not impose major changes to standard survey practice. It should be easy to review the data. It may also be useful to be able to search and edit a particular parameter (for example the height of instrument). The program should contain procedures which avoid accidental deletion of stored data.

(ii) Prompts and error checking. The software prompts should be meaningful and guide the user to input the correct information. Also, the data entry should be checked for gross errors and validated for accuracy. For example, the software should error-trap an angle entry of 360° 12′ 67″. Similarly the software should be able to confirm, for example, whether the variation between the face left and face right pointing to the same target is within some preset tolerance.

(iii) Field computation. Ideally, the software should enable the user to carry out a limited amount of computation while in the field. Typically this may include the facility to determine the position of the instrument by angular or distance resection.

2.3.2 Dedicated surveying data recorders

The design of data recorders dedicated solely to recording and processing surveying data has altered dramatically in the past 15 years. The first designs, such as the Geodat 700, used punched paper tape as the recording medium. This approach suffered from many practical difficulties including breakage of the tape by mishandling or excessive humidity, a cumbersome and heavy recording mechanism and high power consump-tion. Although crude in design by present-day standards, this device was a significant development and provided the first effective method of recording field observations automatically.

   The next development, which overcame many of the problems associated with paper tape recorders, involved the use of magnetic tape as the recording medium. The Wild TC-1 used this system. A digital tape recorder, protected against damp and dust

Table 2.3 Dedicated surveying data recorders

Manufacturer	Model	Size (mm) (W × L × H)	Weight (kg)	Storage Capacity (kbytes)	Display (lines × characters)	Keyboard (No. of keys, type)	Comments
Nikon	DR-1	100 × 200 × 37	0.6	60/cartridge	4 × 16	40 A/N(S)	Uses removable data cartridges designed for DTM-1 Total Station
Pentax	DC-1	100 × 210 × 35	0.6		2 × 16	34 A/N(S)	Designed for PTS-10 Total Station
Sokkisha	SDR-2	89 × 152 × 38	0.45	32	1 × 16	33 A/N(S)	Designed for SET range of Total Stations
Topcon	FC-1	90 × 205 × 58	0.65	60	1 × 16	28 A/N(S)	Designed for ET-1 Total Station
Zeiss (W. Germany)	REC 200	90 × 191 × 55	0.6	up to 440 lines of data/cartridge	1 × 16	20 N	Use with MEM Data Storage Modules and Elta range of Total Stations
Zeiss (E. Germany)	Micronic 445L	85 × 170 × 35	0.49	32–48	1 × 16	20 N	Can be linked to separate storage modules
Geotronics	Geodat 126	90 × 206 × 56	0.5	500–1000 points	1 × 12	35 A/N(S)	Based on Hewlett-Packard HP-41 CV/CX calculator
Wild	GRE 3	82 × 222 × 66	0.8	32–128	3 × 8	18 multi-function N	Basic programs can be loaded from host computer
Kern	DIF 41	75 × 175 × 35	0.35	3.1	1 × 12	35 A/N(S)	Based on HP-41 series of calculators
Kern	Alphacord	98 × 218 × 54	0.98	64–16	2 × 16	42 A/N	CP/M operating system. Basic, Fortran or Pascal programs can be loaded from host computer
AMS	Datasafe II	100 × 222 × 52	0.86	64–256	2 × 16 or 4 × 16	42 A/N	Alphacord based on this design

N, numeric; A/N, Alphanumeric; (S), shift key.

Table 2.4 General-purpose hand-held computers

Manufacturer	Model	Size (mm) (W × L × H)	Weight (kg)	Storage capacity (kbytes)	Display (lines × characters)	Keyboard (no. of keys, type)	Comments
Husky	Hunter	216 × 156 × 32	1.2	144, 208 352, 496	8 × 40	50 A/N	Software for logging survey data available from Ground Modelling Systems and Optimal Portable Disk Drive now available
Zeiss (W. Germany)	REC 500	216 × 56 × 32	1.2	144	8 × 40	50 A/N	Identical to Husky Hunter
Epson	HX-20	290 × 215 × 44	1.7	32	4 × 20	61 A/N	Survey Software available (Longdin and Browning Ltd)
Immediate Business Systems	Field-Work Fifty	274 × 190 × 55	1.35	64–256	2 × 40	36 A/N	Bubble memory. No facilities for automated data acquisition
Microscribe	600	214 × 162 × 37	1.38	128, 320	8 × 40	47 A/N	
D&H Surveys	Psion Organiser		0.25	32	2 × 16	A/N	

by a sealed cover, enabled about 1200 points to be recorded on one digital tape. One of the main advantages cited by users of this technique is the relative ease of sending data of this type from a remote location to a central processing facility for plotting. Although the magnetic tape units are no longer manufactured, they are still used by several surveying organizations.

The most recent development has been the production of solid-state electronic data recorders. Most data recorders available at present are of this design. The distinction can be made between solid-state data storage modules which plug into an electronic tacheometer, and hand-held units with an integral keyboard. The latter are more common and range from devices based on electronic calculators (such as the Geodat 126), to more sophisticated devices such as the Kern Alphacord. Table 2.3 summarizes the features of some of the most common devices.

In general, however, all dedicated designs suffer from two weaknesses. First, users are often tied to a particular manufacturer's design since it can be difficult to interface recorders to different electronic tacheometers. Second, it may be difficult, if not impossible, to alter the software within the recorder.

2.3.3 General-purpose hand-held computers

A more flexible approach which has become particularly common in recent years is the use of general-purpose hand-held computers for data recording. Two different types of computers may be used: those which are fully weatherproof of rugged construction and suitable for outdoor use (such as the Husky Hunter), and those which are not designed for use outdoors (such as the Epson HX-20). Table 2.4 summarizes the features of a selected number of instruments which fall within the former category, which offers several advantages for survey work.

The first benefit is the ability to reprogram the form of data entry, for example, for specialist surveying operations or for other types of surveying equipment which dedicated recorders would find difficult to accommodate. The second advantage is the flexibility to create specialist programs to process data in the field. Finally, in view of the general-purpose nature of these devices it is also possible to use the computer for conventional surveying computational tasks and other office-based tasks such as word processing. Further details of this approach can be found in Fort (1987, 1988) and Anderson (1988), and Fort (1988) provides a more comprehensive summary of the specifications for a selection of both dedicated and general-purpose data recorders.

2.4 Field procedure

The potential increase in survey productivity which can be achieved using electronic tacheometers linked to data recorders has been mentioned previously. It is vital, however, to ensure that this increase in field productivity is translated into increased project productivity. To achieve this, it is essential to survey in a systematic manner, to produce a clear and unambiguous field sketch and to relate efficiently the recorded observations to the terrain. This final requirement is normally achieved by coding the observations and indicating whether the data relates to a single point feature, or to a series of linked points normally referred to as a string. Topographic details such as hedges, ditches or roads may therefore be considered as strings of points to be joined by a common line style. Two requirements relating to strings are that first, by definition, a string cannot consist of points with different codes, and second, where two or more strings join, it is normally necessary to create an additional point which has identical coordinates but a different code.

2.4.1 Data storage
The method of storing data varies considerably depending on the type of data recorder which is being used. It is possible, however, to distinguish between stored data which relates to the position of the object and information which describes the object being surveyed.

Various options exist for recording field observations ranging from recording raw observations (horizontal angle, slope distance and vertical angle), to storing three-dimensional coordinates. The former approach is preferred by many users since it eliminates the possibility of systematic errors (such as refractive index) inadvertently being applied to a series of measured distances.

Feature coding is the process of describing features in the data recorder by means of some numeric or alphanumeric description. The coding process is normally carried out as an integral part of the field data acquisition phase, although instances of coding being performed after the survey has been performed have also been reported. While full alphanumeric coding may eliminate the need for a filed sketch over small sites, for large projects it is vitally important that a numbered sketch be prepared in the field. Coding systems vary considerably and may reflect either the design of the data recorder being used or the in-house standards of a particular organization. Generally, however, they should enable the surveyor to describe certain characteristics of the feature or how it should be represented on the final map. Thus a coding system could be used to define the type of annotation to be attached to a point feature. Similarly, it may be used to describe whether points on a string should be joined by a straight or curved line.

2.4.2 Sampling Patterns
The selection of a suitable sampling pattern is of vital importance when acquiring data by ground surveying techniques. Two different strategies may be adopted. The first involves acquiring data on a pre-determined grid basis. Generally the grid interval will be fixed although like photogrammetric progressive sampling it may be varied according to the terrain characteristics. Unlike photogrammetric methods the grid will tend to be site specific, have an arbitrary orientation and will not be linked to the national grid. The time required to set out such a grid and the possibility of not recording changes in ground characteristics which occur between the grid nodes precludes the general application of this approach.

The second and more common approach involves a selective sampling strategy. In this case the height points to be recorded are chosen so as to best represent the undulations of the terrain. Thus points are chosen selectively in the form of string lines along changes of slope, along ridge lines and along breaklines such as the base of a stream. By subsequently generating a triangular mesh using either the Delaunay method or the radial sweep technique the terrain characteristics can be represented more effectively.

2.5 Conclusions

A wide range of instruments is currently available for the field capture of three-dimensional terrain data, ranging from integrated measuring and recording devices to modular systems based on several individual units. For the future, it seems likely that the major changes in field data collection will involve increases in the storage capacity, and possibly modifications to the method of acquiring data. In the first case there is a need for a storage system which enables several thousand measurements to be stored. This could be achieved either by using high capacity reusable modules for on-board recording or by using high capacity hand-held data recorders. On a practical level it is also essential that the battery system for both the electronic tacheometer and the data recorder provides sufficient power for at least 10 hours' continuous use. Finally, the concept of data collection remote from the instrument has also been suggested, e.g., the LASERFIX system suggested by Gorham (1988). There is some merit in this approach since the coding and field sketching process is most effectively carried out at the measurement point rather than at the instrument station. It is also attractive since the technical difficulties of transmitting data from the electronic tacheometer to the prism have largely been overcome. In practice, however, the problems of carrying a measuring prism with a data recorder attached may prove to be a serious disadvantage and may preclude the general acceptance of this approach to field data recording.

References and further reading

Anderson, H. (1988) The Psion Organiser II—more than a data logger. *Civil Engineering Surveyor* **VIII**, (7) (September), 6–7.

Bennett, M.L. (1985) Early experiences of automatic data capture. *Land and Minerals Surveying* **3**, 414–420.

Fort, M.J. (1987) The optimal way. *Civil Engineering Surveyor* **VII**, (8) 13–19.

Fort, M.J. (1988) Data recorders and field computers. *Civil Engineering Surveyor* **VIII**, (6) 1–8.

Gorham, B. (1988) Measurement of spatial position using laser beams. *Land and Minerals Surveying* **6**, (3) (March), 121–126.

Grant, M.A. (1981) The practicalities of field data recording. Paper F1, *Survey and Mapping '81*, 17pp.

Kennie, T.J.M. (1990) Electronic angle and distance measurement. Chapter 1 in *Engineering Surveying Technology*, eds. Kennie, T.J.M. and Petrie, G., Blackie and Son Ltd, 485 pp.

Schwendener, H.R. (1978) The measuring and recording systems of the Wild Tachymat TC1. *The ACSE Journal*, October, 13–17.

Smallshire, V.W. (1985) Use of a portable microcomputer as a survey data logger. *Civil Engineering Surveyor*, July/August, 14–19.

3 MOSS survey developments

G.S. CRAINE

3.1 Introduction

Due to the influence of computer technology, revolutionary changes have occurred within survey practice during the past 20 years. These changes have improved the existing methods of recording and processing survey information and, of more widespread significance, they have provided the opportunity for automating many of the manual tasks of contouring and volumetric analysis associated with site development and the preparation of land information systems.

3.2 Development of MOSS survey system

Electronic instruments, recorders, fast and cheap processors, plotters and interactive graphics terminals have facilitated the automation of many surveying processes.

The potential of these developments was appreciated in the early 1970s, and MOSS was one of the first systems that adopted these advances and attempted to redefine procedures and techniques to allow full advantage to be taken of the new equipment, thus providing greater automation.

Automation was not achieved simply by introducing new devices into the 'production line' but by re-assessing the processing procedures and all previous manual tasks with the aim of improving methods. The introduction of field-feature stringing methods integrated the tasks of field recording and plotting and allowed the rapid production of computerized drawings. The instruments developed for angle and distance measurement have totally revolutionized survey practice. No longer are multiple rounds of face left and right angles required nor is considerable time spent measuring traverse legs.

The benefit of the new developments and techniques has been to remove the laborious and time-consuming aspects of the work, such as measurement finesse, booking of readings, office calculations and plotting. The professional surveyor is left to concentrate on his primary objective of collecting the required information at the required accuracy while improving his efficiency.

The MOSS survey system was developed to assist this process of automation. During the evolutionary period there was much argument about the potential of new field methods and equipment but this is now accepted. The new instruments and graphic hardware are widely available at reasonable cost and everyone has the opportunity to take advantage of these techniques. However, the MOSS survey features have been considerably enhanced to capitalize on the latest instrument and computer hardware advances and to provide greater flexibility for recording, processing and storing information associated with surveys and digital mapping.

The new survey developments within MOSS have been aimed at the smaller survey organization and the system can be offered in modules as required. The system is initially available on Apollo workstations offering the latest state-of-the-art technology with high-quality graphics. These machines offer large memory and disk support providing unlimited capacity for model storage, together with guaranteed interactive response.

The existing MOSS survey facilities have been used successfully worldwide for several years on a range of projects from small site developments to major highway projects and mine excavations. Considerable practical experience has been obtained from users of the MOSS system during this period and the techniques provided by MOSS have strongly influenced attitudes towards the potential for digital techniques.

Automated digital modelling is now widely accepted and more emphasis is being placed on the automatic recording of greater detail with the aim of producing high-quality cartographic output, digital surface models and supporting information. The introduction of electronic instruments and data collectors of increased capacity indicates the future direction of developments and software must be planned to capitalize upon these advances.

As a result of these considerations, the survey system has been rewritten to provide a greater range of facilities. In addition, greater emphasis has been placed on accommodating individual surveyor's field practice.

3.2.1 Survey control

Traverse recording is available within MOSS, independent of the detail survey observations, and it provides for adjustment by the Bowditch method or Crandall's and Bird's least-squares methods. The traverse observations can now be recorded with the detailed feature observations and the total survey processed, together with any closure adjustment. New features are available to allow stations to be located by intersection and resection techniques, and the observations can be included with the traverse data. These new location techniques can use either stored stations or current traverse stations and are determined by least-squares methods. Fly stations and open traverses may be observed from the new traverse to relate local detail.

3.3 Field recording techniques

The introduction of field recording and feature-coding methods is probably the most contentious issue involving computerized survey systems. In the past, manual methods have required the collection of field point data which are numbered and associated with other points in a sketch to allow the preparation of the survey drawing. This procedure is possible with MOSS and all observations are stored in a complementary point string which can be accessed by interactive editing to create the desired features.

MOSS feature-coding methods provide complete flexibility for recording string and point features. In its most simplistic form, the staff man may be considered as a plotter pen tracing out the features. Obviously this is not as straightforward as digitizing, but it allows the creation of skeletal features. Strings or part strings may be recorded sequentially or by uniquely coding a series of randomly observed points on each feature.

Complete field stringing to produce a perfect survey without any editing is as unrealistic as recording only points for later interactive creation of the model. These are the two extremes of field data recording, and the best solution lies between the two and it will vary in different situations.

Maximum practical use of stringing is the key to efficient data recording. This will vary for different projects, such as rural and urban landscapes, and because of the unique nature of surveys, no two surveyors will record detail in the same way, not even in the same organization.

The recommended method is to produce a skeleton of the detail, similar to previous field sketches, recording all detail and as much line work as possible from each survey station. Interactive editing is used to connect the short strings and points to complete the features. The final model may require complex feature-coding, but it is unnecessary to introduce this in the field when simple initial-character feature-identifiers will suffice. The complex labels can be added in the office.

The aim of any surveyor is to balance the work required in the field and the office to produce the most efficient compromise for each survey. This will vary between surveyors so the instruments and computer system must be sufficiently flexible in allowing him to record, identify and assemble the final survey to produce the optimum result.

When reading papers and articles on computerized survey techniques, it often appears that all surveyors are using the latest electronic equipment. This is not the case, and the software should provide for a range of measurement types from chain and offset, stadia readings and geometric observations. Fortunately data collectors are becoming increasingly popular, storing the observations in the field either directly from the instrument or by manual input. Direct keyboard input of data is feasible but time-consuming when compared with field coding.

Observations record point data, but there is also a requirement to allow taped measurements, based on observations to record associated points. Many of these procedures are based on previous manual methods but have to be included within the software to accommodate different survey styles.

3.3.1 Feature recording

The techniques available for feature recording provide the surveyor with total flexibility in his method of undertaking the survey. Survey detail may be identified in the field as points and strings making maximum use of field coding or alternatively the survey detail may be recorded as points with unique identifiers, which can be used for interactive creation of the features. The preferred solution will lie between these extremes, depending upon the surveyor and the situation.

The ideal approach for greatest automation of the survey is to record a skeletal survey of the detail, comprising as many strings as possible. The string coding methods assist the recording of string data by allowing the recording of sequences of points with minimal feature coding, or random selection of string detail if the points are fully coded. Automatic sorting of the observations produces the required strings and, where necessary, completes the labelling. A new feature has been introduced to allow all observed points to be recorded in a complementary point string, and each point is identified by a uniquely specified number. The string can be displayed and used to create the feature strings interactively, quite independently of the field-feature coding. This alternative method can be used instead of field stringing, or it could be used in conjunction with the field sketch to relate points to field coded strings. Although the stringing technique is considered the preferred method, there is a practical limit to the amount of complete stringing and connections that can be coded in the field. Interactive editing will always be used to finalize the survey and the

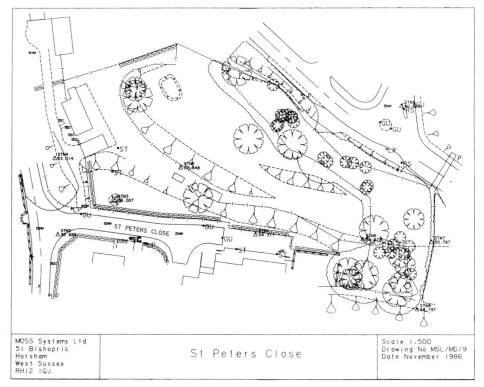

MOSS Systems Ltd
51 Bishopric
Horsham
West Sussex
RHI2 IQJ.

St Peters Close

Scale 1,500
Drawing No MSL/MD/9
Date November 1986

Figure 3.1 Detail survey.

amount of editing will be relative to the complexity of the survey content. Fig. 3.1 illustrates a typical survey.

3.3.2 Detailed recording

(i) Straight and curved strings. The majority of observations will relate to string features, which may consist of a series of straight and curved lines. Previously, only straight links have been allowed but the recording methods have been extended to allow the identification of points on curved sections of continuous strings. The technique provides perfect reproduction of plan features and reduces the number of field observations. However, this only ensures plan definition but additional points can be introduced during string processing to provide both plan and level definition on the curves, which is essential for future analysis of the digital surface. These points are introduced to a user-defined tolerance and are used in any subsequent triangulation, contouring, sectioning and volumetric analysis. Fig. 3.2 illustrates the application of these techniques to road channel lines.

The curve recording techniques have also been developed to allow identification of complex site features such as the hammerhead layout in the example illustrated. This may be recorded as continuous strings by using identifiers for discontinuity breaks and gaps within the feature.

(ii) Offsetting and paralleling of features. In addition to direct observations, facilities are available to allow associated strings or points to be offset from a current feature.

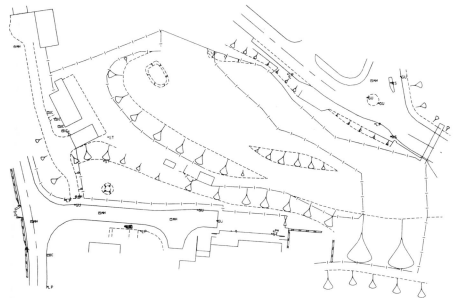

Figure 3.2 Curved string recording.

The obvious application is for offsetting a curve and adjacent footpath from an observed road channel line. The new feature may be offset horizontally and vertically at each point, or a simpler truly parallel string may be created. Individual features such as trees or street furniture may be recorded by this technique.

(iii) Recording of buildings. Buildings often require a series of observations to record all the detail. Several methods are provided to allow the surveyor total flexibility in the field for recording the detail in the most convenient manner.

Buildings may be recorded as a string, part strings or points from one or more survey stations, which can be edited later to produce the required feature. An alternative is to use chain and offset measurements, referred to temporary stations located at the building corners for fixing detail. This is an extremely useful method, especially where points are obstructed from the instrument station.

A third taping alternative may be used in conjunction with either of these methods, which allows a new point to be fixed in relation to an observed link by an extension and lateral offset. A whole building can be recorded in this way based on one fixed link. Levels may be continued between points by additional vertical offsets, or added later by ground levelling.

These techniques, together with interactive editing, allow the recording and creation of the most complex layouts. Additional facilities are available for automatic closing and squaring of the building strings.

(iv) Unique point string. A new feature has been introduced to allow the recording of an independent point string where each point has a unique numerical identifier. This string may be used to construct the survey detail, but the technique has a further purpose in that it allows observations to be referenced in the field. Points common to several features may be recalled by a unique number or by the relative position of the observation in the data set. A blank number without an observation simply repeats the previous observation for inclusion in another feature string.

(v) Recording of objects. Rectangular and circular objects describe common survey features, and special facilities have been included to simplify their location. The surveyor has the choice of recording the details as a real coordinate string or by position and orientation for location of a unique macro-symbol.

(vi) Flexibility. The aim of the new extensions to the field survey software has been to provide a very flexible field data collection method to allow the recording of detail for reproduction of high-quality cartographic drawings and the production of digital models.

The system permits a variety of recording techniques including the latest electronic theodolites, chain, and offset and taping methods and should be adaptable to every surveyor's unique style.

Advantage is taken of the range of data collectors available for all the common instruments to assist with field recording. The input coding provides for all observations and adjustments together with the full range of string and point feature indicators.

3.4 Survey/digitizing

Survey and digitizing are often considered as separate processes, but they have a common function in the preparation of digital information. In the development of the new survey facilities within MOSS they were considered to be alternative methods of collecting the same digital data. This is important for consistency of data recording and also when considering interactive editing features, which are clearly related to digitizing. The end result of these operations is the same, although carried out in different situations.

The process of digitizing converts an existing plan into digital form and has no regard for the method of preparation whether by ground or aerial survey. Obviously, to regard detail survey as field digitizing is a gross simplification of the process. The survey process identifies all detail and content as well as recording and ordering the information. The complexity of the survey will dictate the balance between field stringing and the requirement for interactive editing to produce the required digital model.

The survey process also allows the identification of the additional points on features required to produce a three-dimensional surface model. Often this information is missing from plans supplied for digitizing, which are no more than a plan representation.

3.5 Contouring, design and analysis

MOSS provides a unique contouring system based on the triangulation of the recorded digital information. All strings links are automatically included in the original triangulation which guarantees the maintenance of the angular surface features. The new curve features are generated as strings to within a close tolerance to ensure the integrity of the triangulation and derived contours. Fig. 3.3 illustrates the accuracy and acceptability of the resulting contours.

In addition to contouring, the digital model provides the basis for the detailed design of engineering, site development and landscaping projects. Comprehensive facilities are also available for producing equivalent digital designs and further analysis.

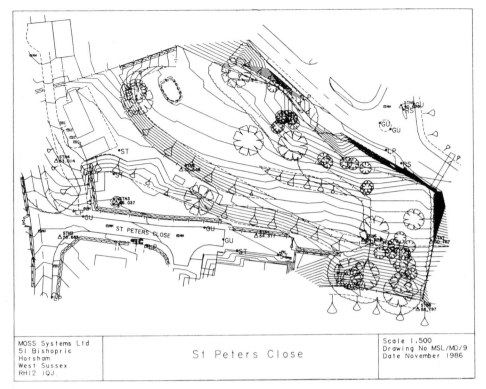

Figure 3.3 Land survey with derived contours.

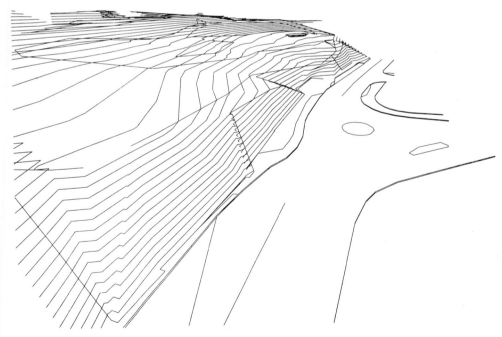

Figure 3.4 Original survey with derived contours.

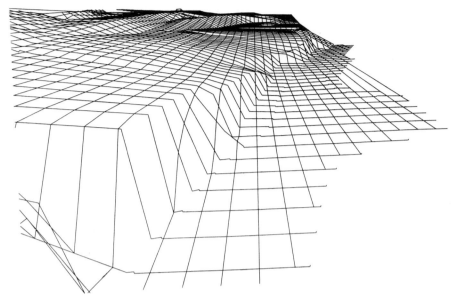

Figure 3.5 Original survey with derived grid.

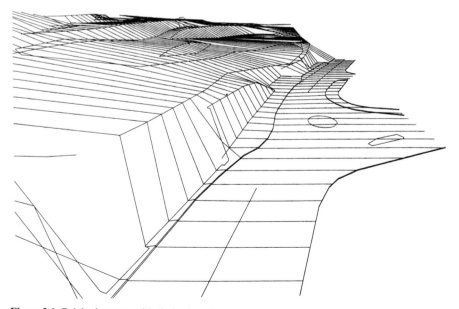

Figure 3.6 Original survey with derived sections.

The digital surface models accurately define all surfaces and volumetric analysis based on sections or triangular prisms produces accurate results for either simple stockpile or complex land reclamation and engineering projects.

The analysis routines also provide for sectioning and production of perspective views. Fig. 3.4, 3.5 and 3.6 illustrate perspectives of alternative surfaces representing detail generated from the survey model.

3.6 Graphic output

Comprehensive drawing facilities are provided for the generation of plan, section and perspective drawings. Standard detail interpretation plus line and symbol macro-facilities allow the rapid production of high-quality output. Feature interpretation can be linked to field-feature coding, and unique drawing macros simplify and automate the production of standard drawings. The interactive graphics facilities provide complete interaction and enhancement of the final drawings with additional linework and text annotation.

3.7 Conclusions

These latest extensions to the MOSS survey module provide increased flexibility for recording detail and they accommodate various survey styles. The survey module together with the supporting graphics features allows full automation of drawing production together with the added potential of a complementary digital surface model. These software developments, together with recent advances in computer and survey technology, now provide a unique environment that is most satisfying and rewarding for the surveyor.

4 Photogrammetric methods of data acquisition for terrain modelling

G. PETRIE

4.1 Introduction

The matter of acquiring the basic terrain elevation data needed to implement digital terrain modelling techniques is of course a fundamental one, well known to all those who attempt to use these techniques in practice. At the present stage of development, there is little digital elevation data available off the shelf in the UK except that derived from very small (e.g. 1:250 000) scale topographic maps, which has a low sampling density and relatively moderate accuracy in terms of the requirements of civil engineers and surveyors.

Since data acquisition is so important to all practitioners of terrain modelling, this immediately poses the question as to which *techniques* should be considered for use in the collection of elevation data. The three main methods which can be used to acquire the elevation data are:

Table 4.1 Sources of DTM data

Source of DTM data	Method used	DTM accuracy	Areal coverage	Typical applications
Ground survey	Total or semi-total station	Very high	Limited to specific sites	Small area site planning and design
Photogrammetric measurements	Stereoplotting machines (with or without correlators)	(i) High (if from spot heights) (ii) Lower (if from contours)	Large area projects, especially in rough terrain	(i) Large engng. projects-dams, reservoirs, roads, open cast mines. (ii) Nationwide, especially in association with orthophotos
Cartographic (existing topographic maps)	Either (i) manual digitizing; (ii) semi-automatic line following (iii) fully automatic raster scanning	Low—derived from contours on medium and small-scale topo maps	Nationwide at small scales	Aircraft simulators, landscape visualization, landform representation military battlefield simulation

(i) *ground survey methods* using electronic tacheometers, i.e., total or semi-total stations;

(ii) *photogrammetric methods* based on the use of stereoplotting machines; and

(iii) *graphics digitizing methods* by which the contours shown on existing topographic maps are converted to strings of digital coordinate data and the required elevations derived from them.

Table 4.1 presents the main characteristics of these three alternative techniques as applied to terrain modelling, in terms of the elevation accuracy required, the extent of the area which has to be modelled and the specific application for which the modelling process will be carried out.

4.2 Photogrammetric methods of terrain data collection

The positional and elevation accuracies which can be achieved using photo-grammetric methods, the scale of the final maps or terrain model, the possible contour interval, etc., are dependent on various interrelated factors, but chiefly:

(a) the *scale and resolution* of the aerial photography;

(b) the *flying height* at which the photography was taken;

(c) the *base: height ratio* (i.e., the geometry) of the overlapping photographs; and

(d) the *accuracy of the stereoplotting equipment* used for the measurements.

4.2.1 Photographic scales and mapping scales

Addressing first the matter of the *photographic scales* which may be encountered in terrain mapping and modelling, Fig. 4.1 shows how these are directly related to the flying height of the aircraft and to the focal length of the aerial camera used for the photographic mission. The possible range of scales has been plotted in Fig. 4.1 for

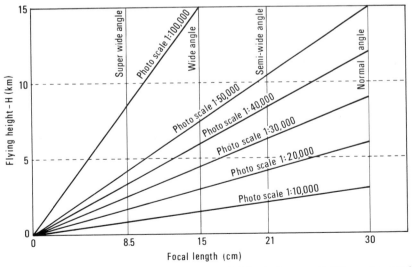

Figure 4.1 Aerial photographic scales from flying heights in the range 0–15 km using standard lenses of $f = 8.5$, 15, 21 and 30 cm respectively.

Table 4.2 Photographic scales, mapping scales and contour intervals

Photographic scale	Ground resolution @ 40 l p/mm	Flying height (m)		Mapping scale	Enlargement factor (photo: map)	Contour interval
1:3 000	0.075 m	450	$\longrightarrow$	1:500	6 ×	0.5 m
1:5 000	0.125 m	750	$\longrightarrow$	1:1 000	5 ×	1 m
1:10 000	0.25 m	1 500	$\longrightarrow$	1:2 500	4 ×	2 m
1:25 000	0.625 m	3 750	$\longrightarrow$	1:10 000	2.5 ×	5 m
1:50 000	1.25 m	7 500	$\longrightarrow$	1:50 000	1 ×	10 m
1:80 000	2.0 m	12 000	$\longrightarrow$	1:100 000	0.8 ×	20 m

those focal lengths (f) commonly used in aerial survey work: $f = 30$ cm (normal-angle); $f = 21$ cm (semi-wide-angle); $f = 15$ cm (wide-angle) and $f = 8.5$ cm (super-wide-angle), corresponding to angular coverages of 56°, 75°, 93°, and 120° respectively, when used with the standard format size of 23 × 23 cm.

It will be seen that the use of these cameras gives rise to a wide range of photographic scales from 1:3 000 to 1:5 000 at the lower end to 1:100 000 scale at the upper end of the spectrum. The actual choice of photographic scale which is made will usually depend on the required scale of mapping and the desired contour interval. Table 4.2 summarizes the relationship between these parameters for the case of standard (23 × 23 cm) format photography taken with a wide-angle ($f = 15$ cm) camera and a 60% forward overlap over the scale range 1:3 000 to 1:80 000.

These figures represent a rough yardstick of current practice by British air survey companies employing high-precision plotting machines on mapping contracts in the UK and abroad. It will be noticed that the ratio between photographic scale and map scale declines markedly as one goes from large-scale mapping for engineering purposes (where enlargement factors of 6 are normal) to small-scale topographic mapping (where slight enlargements or even reductions are common). This is due to the fact that, even on small-scale maps at 1:50 000 to 1:100 000 scales, it is necessary to detect, interpret and map features of rather small dimensions such as individual buildings, secondary roads, rivers and streams. If the scale and ground resolution of the photogrammetric imagery becomes too small, then the completeness of the map will suffer, leading to a large expense being incurred through the necessity for extensive field completion work by surveyors.

4.2.2 Accuracy of photogrammetrically-measured terrain elevations

The *accuracy* with which the spot heights or elevations for terrain modelling may be measured depends firstly on the relationship of the distance between successive photographs (the air base B) and the flying height H—the so-called base: height, $B:H$, ratio, which summarizes the geometrical characteristics of the photographs. Secondly, the heighting accuracy will also depend on the accuracy of the stereoplotting machines on which the measurements are carried out. Lastly, and perhaps most importantly, it depends on the flying height at which the photography was taken.

The first of these two factors, the *base:height (B:H) ratio*, is dependent on the focal length f and angular coverage of the taking camera, giving values of 0.3 with normal-angle ($f = 30$ cm and 60° coverage); 0.45 with semi-wide-angle ($f = 21$ cm and 75° coverage); 0.6 with wide-angle ($f = 15$ cm and 90° coverage) and 1.0 with super-wide-angle ($f = 8.5$ cm and 120° coverage) photographs. In general terms, the larger the value of the base: height ratio, the more accurate the measurements of terrain

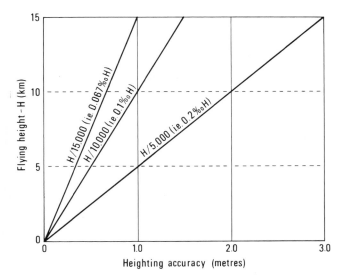

Figure 4.2 Heighting accuracy from flying heights in the range 0–15 km using stereoplotting machines with heighting accuracies of 1/5 000, 1/10 000 and 1/15 000 of the flying height (*H*) respectively.

elevation. Thus at first sight, super-wide-angle photography would appear to be optimized for height measurement. However, the negative side of using super-wide-angle photography is that it is associated with small photographic scale and poorer ground resolution from a given flying height. Thus wide-angle photography with $f = 15$ cm and a base:height ratio of 0.6 is most frequently used as a compromise between the requirements for good scale and resolution on the one hand and very accurate height measurements on the other.

For a particular base:height ratio, the heighting accuracy will be related directly to the flying height H. Taking the usual range of stereoplotting equipment available to air survey companies and national mapping agencies, the expected accuracy of spot heights using wide-angle photography will lie in the range 1/5 000 to 1/15 000 of the flying height *H*, depending on the type of instrument used (Fig. 4.2). From this it will be seen that the expected heighting accuracies of terrain elevations expressed as root mean square errors (RMSE) will lie in the range ± 0.1 to 3.0 m over the possible range of flying heights, up to a maximum of $H = 15$ km in the case of high-flying jet photographic aircraft. This accommodates the accuracy requirements of almost all terrain modelling carried out by surveyors and civil engineers.

4.2.3 Applicability of photogrammetric methods of acquiring terrain elevation data

From this discussion, it will be seen that photogrammetric methods can be used for digital terrain elevation data collection over a very wide range of scales and accuracies. Over relatively small areas where the required sampling density is very high and the specified height accuracy is very demanding, *ground survey* methods remain extremely important and highly applicable, e.g., for specific building sites. However, as soon as data has to be collected over extensive areas of terrain and especially if the terrain is rough, then photogrammetric methods come into their own. In the context of acquiring terrain elevation data over larger areas, photogrammetry is the dominant method of data acquisition in any situation where the information cannot be acquired from existing topographic maps. Even where such *contoured maps* do exist, the contour interval may be too wide or the accuracy of the contours shown on these maps may often be insufficient for terrain modelling purposes, especially for civil engineering design projects concerned with road or reservoir construction or the related projects undertaken by landscape architects. In these situations, photogrammetric methods of measurements will again be used in preference to the data produced from graphical contours.

This then poses a fundamental problem. All surveyors and many civil engineers are quite capable of carrying out the field survey of a relatively small site to produce the required elevation data. At the other end of the scale range, if appropriate, the task of manually digitizing existing contour sheets can also be undertaken in-house by surveyors, engineers, planners and landscape architects working in private firms, public utilities and local or national government agencies. However, in the case of photogrammetric data acquisition, the stereomeasuring equipment used is very complex and expensive, the associated procedures are very specialized, and highly trained, experienced personnel are required to operate the instrumentation. Thus resort must be made to the services offered by commercial air survey firms with their specialist equipment and expert staff.

4.2.4 Photogrammetric measurements of the terrain
The basic principle of a stereoplotting machine is that it recreates an exact 3D stereomodel of the terrain from the aerial photographs. The photogrammetrist then measures this stereomodel very accurately instead of going into the terrain to carry out the measurements using surveying instruments. The savings in time and cost of doing so are enormous.

The terrain models measured in the stereoplotting machine may be either optical or mechanical in nature. *Optical models* are formed by optical reprojection of the two aerial photographs making up a stereopair (Fig. 4.3). A measuring/tracing device is then used to measure and plot the features present in the model in all three dimensions *X*, *Y*, and *Z*. *Mechanical models* are formed by replacing the optically-projected rays

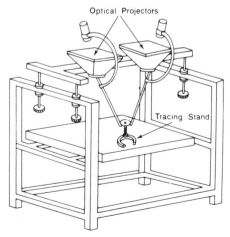

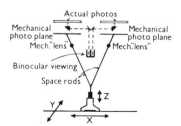

Figure 4.3 Optical projection stereoplotting machine.

Figure 4.4 Mechanical projection stereoplotting machine.

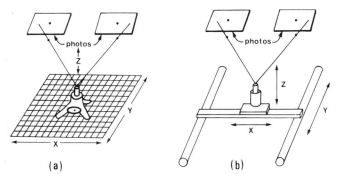

Figure 4.5 Measuring/digitizing arrangements. (a) Using tablet digitizer (X/Y) plus vertical column equipped with rotary or linear encoder (Z). (b) Using cross-slides (X/Y) and vertical column (Z) all equipped with rotary and linear encoders.

by their mechanical equivalents in the form of rods (Fig. 4.4). Optical models (in the so-called optical projection machines) are still common in North America, but the use of mechanical projection machines is almost universal in the UK and most countries in Western Europe. The former are simpler and less expensive but less flexible in the types of photography that can be accommodated. The latter are more complex and more expensive but also more accurate and more flexible.

All the three-dimensional (X, Y and Z) measurements made in a stereoplotting machine may be *digitized* by mounting the measuring mark on a suitable cross-slide system and encoding each of the axes individually (Fig. 4.5). Alternatively, a tablet digitizer may be mounted on the base surface of the machine to give positional (X, Y) information while the terrain heights (Z) are obtained in digital form by adding a linear or rotary encoder to the height measuring device. The former arrangement is preferred since the use of cross-slides can easily allow measuring resolutions or accuracies of 5–$10\,\mu$m, whereas the tablet digitizer is limited to 25–$50\,\mu$m at best.

4.2.5 Photogrammetric measurement patterns
The required terrain elevation information may be obtained in any one of several sampling patterns.

(i) Systematic (e.g., grid-based) sampling. A systematic pattern of *spot heights* may be measured in a regular geometric (square, rectangular, triangular) pattern (Fig. 4.6) as specified by the client. Such an approach is that favoured in any type of photogrammetric operation which is either fully or partially automated, where the locations of the required grid node points can be pre-programmed and driven to under computer control. Thus the grid can be preset to a specific interval, which has the consequence that the finer but perhaps significant terrain features will not be measured specifically.

(ii) Random (but specifically located) sampling. An alternative approach sometimes used by photogrammetrists is to measure heights selectively at significant points only—i.e., at the tops of hills, in hollows and along breaks of slope, ridge lines and streams. Thus all the points to be measured are identified by the photogrammetrist on the basis of his inspection and interpretation of the terrain features. Therefore all the measured points will be randomly located, i.e., an irregular network of points will result from the measurements. The consequence of this is that more thought must be

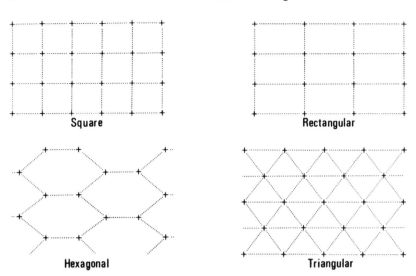

Figure 4.6 Systematic (grid-based) sampling patterns.

given to the structuring and arrangement of the measured data than if a purely grid-based approach to data acquisition had been adopted. In many ways, this irregular network approach is similar to that of the land surveyor who selects the significant points in the field, measuring their heights by optical or electronic tacheometric methods.

(iii) Composite sampling. Quite often, the approach to data acquisition taken by photogrammetrists will combine the elements of both of the above approaches, especially if a non-automated, operator-controlled type of stereoplotting machine is being used for the height measurements. The basic grid-measuring pattern will be supplemented by the measurements made at significant points in the terrain, e.g., on hilltops, along breaklines and streams as mentioned above in (ii). Furthermore, if a road line has already been selected, then additional points will be measured both along the centre line to give a longitudinal profile and at right angles to give a series of lateral cross-sections to provide accurate estimates of earthwork quantities (cut and fill) for alternative designs.

(iv) Measured contours. The final (alternative) approach is to systematically measure contours over the whole area of the stereomodel outputting the measurements in the form of strings of digital coordinate data. Again, this may be supplemented by spot heights measured along terrain break lines.

The decision as to which of these various patterns will be adopted in practice will often depend on the specific type of equipment available for the photogrammetric measurement. If an automated or semi-automated system is available, the use of a regular grid is virtually certain, which helps to account for the fact that the majority of the terrain modelling programs offered by the photogrammetric instrument manufacturers and system suppliers are grid-based. Also, the choice between measuring spot heights or contours will often be based on considerations of the accuracy of the height information required for a specific project. As already discussed, using a high-precision stereoplotting machine and wide-angle photography (with a $B:H$ ratio of 0.6), the expected spot height accuracy expressed in terms of root mean square error could lie between ± 0.1 and ± 0.2 m for a flying height H of 1 000 m; 0.5–1 m for a flying height of 5 000 m; 1–2 m for a flying height of 10 000 m, etc., corresponding to height accuracy measurements of 1/5 000 to 1/10 000 of the flying height.

With contouring, the expected accuracy will be less. The minimum possible contour interval will be 1/1 000 to 1/2 000 H which, for the flying height of 1 000 m, gives a minimum contour interval of 0.5–1 m. The expected or specified accuracy of such measured contours is that 90% of all the points lying on a specific contour shall lie within half the contour interval. Thus for the one-metre contour interval cited, 90% of the points should lie within ± 0.5 m of their correct height, so that the RMSE (67%) will be ± 0.3 m. So obviously, measured contour lines are rather less accurate than the spot heights measured from the same photography. On the other hand, such photogrammetrically measured contour lines will be much more accurate than any contour lines shown on existing maps which are derived by interpolation from measured spot heights—a point frequently forgotten by many users of digital terrain information.

4.2.6 Photogrammetric digitizing units

A variety of digitizing units can be attached to stereoplotting machines. Trying to sort out and evaluate their differing capabilities can be quite bewildering to the non-specialist. Hardware-based, firmware-based and software-based units are all used; a review is given in Petrie (1981). The *hard-wired and firmware-based units* are essentially

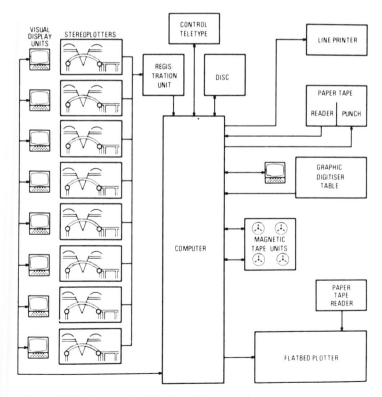

Figure 4.7 Hardware of the Hunting digital mapping system.

stand-alone electronic boxes which decode and count the signals coming from the linear or rotary encoders mounted on a stereoplotting machine and convert them to X, Y and Z coordinate values which are displayed and recorded, usually on magnetic tape, cassettes or cartridges. The *software-based units* are, as the name suggests, based on the use of an on-line computer which not only performs the data recording but can give prompts and assistance to the operator during the setting up of the stereomodel, as well as carrying out checks on the measured coordinate data. Some of these software-based units utilize a dedicated microcomputer attached on-line to the machine; others use a large time-sharing computer as in the system used by Hunting Surveys based on a large DEC PDP-11 minicomputer (Fig. 4.7). Essentially, however, all of these systems are carrying out the 'blind' digitizing technique also used widely for digitizing contours on existing maps. Checking must then be carried out using hard-copy plots generated at the end of a digitizing session. However, in a few cases, such as in the Kern DC-2B system (Fig. 4.8), a computer-driven plotting/drafting table has been used on-line which can plot out all the terrain elevation or contour data measured by the photogrammetrist for his inspection and possible correction during the actual measuring operation.

However, the direct attachment of *graphics workstations* to stereoplotting machines to allow direct checking and interactive editing of the measured data is a current trend. Some photogrammetric manufacturers have developed their own integrated systems, such as Kern's MAPS 200 and 300 systems based on DEC PDP-11 series of computers and Imlac or Tektronix screens which are attached to the firm's well-known PG-2 stereoplotting machines. The vast majority of such integrated systems have been supplied by Intergraph working in collaboration with leading European manufacturers of stereoplotting machines such as Zeiss Oberkochen and Wild. The Zeiss Oberkochen/Intergraph system is especially interesting in that the plotted contours are displayed on a graphics display mounted on the side of the stereoplotting machine. The graphical image of the plotted or measured information is then projected via a semi-reflecting mirror into the optical channel of the

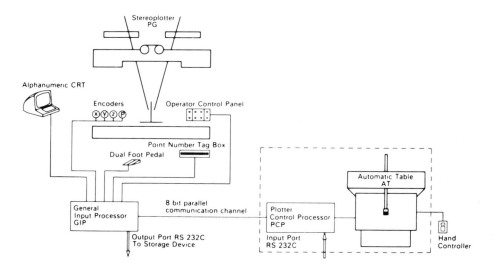

Figure 4.8 The Kern DC 2-B computer-supported plotting system.

stereoplotting machine so that the operator may see exactly what has been measured or plotted and may correct any errors or omissions as required.

4.2.7 Analytical plotters

The latest development in this area is of *analytical stereoplotting machines* in which the optical or mechanical models are replaced by equivalent purely mathematical or numerical solutions which are executed in real time by a suitably programmed high-speed mini-computer (Fig. 4.9). The X, Y and Z terrain coordinates of a point measured by the operator are passed to the control computer which computes the corresponding position of the same point on each photo of the stereopair in real time. Thus the point being measured can be viewed by the operator in 3D (stereo) since the positions of the viewing optics over the stereopair of photographs are also controlled by the computer. The latter also controls the output which may be in the form of a plotted map or purely digital terrain data.

One of the major advantages of using an analytical plotter for the acquisition of terrain elevation data is that it can be programmed to drive to *any desired position* or series of positions corresponding to the required pattern and density of points in the stereomodel. Also techniques such as the progressive sampling of digital terrain models (DTMs) can readily be implemented. As soon as the operator has measured the height of an individual terrain point, the computer automatically moves the measuring mark to the next point in the stereomodel which has to be measured. Thus, with the use of analytical plotters, the rate of acquisition of the terrain elevation data is greatly increased as compared with that of analogue stereoplotting machines. However, the latter continue to be used because of the large capital sums required to purchase analytical plotters at the present time. However, as time goes on and the cost of computers continues to fall, the use of analytical stereoplotters must become more widespread.

The integration with *graphics workstations* which has occurred with analogue (optical and mechanical projection) stereoplotters is also taking place with these analytical plotters. Thus, for example, the Kern DSR-1 and DSR-11 analytical plotters can be supplied with MAPS 200 or 300 workstations (Fig. 4.10) and the Zeiss Oberkochen Planicomp series with Intergraph InterAct dual-screen workstations. In spite of the current enormous expense of such units, there has been sufficient demand for them for Intergraph to announce its entry into the field on its own account with the recent introduction of its InterMap Analytic machine with complete integration of the analytical plotter and the Interpro 32 desktop graphics workstation into a single unit. A similar development and integration has taken place in the new Wild System

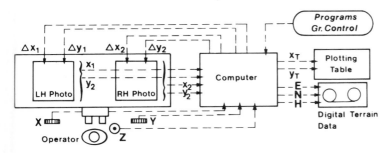

Figure 4.9 Overall concept of analytical stereoplotting machine.

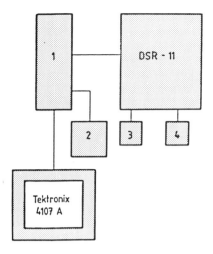

Figure 4.10 Kern DSR – 11. (1) MAPS300;
(2) VT terminal; (3) foot disk; (4) tracker bali.

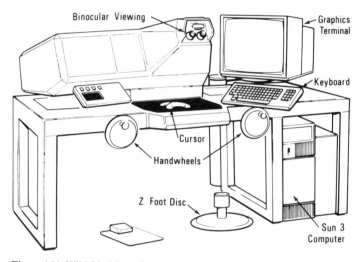

Figure 4.11 Wild S9–AP analytical plotter.

9-AP analytical plotter (Fig. 4.11) which utilizes a Sun graphics workstation as the graphics component of the machine.

4.2.8 Progressive sampling
As mentioned above, the obvious shortcoming of the *regular grid-based approach* to data acquisition is that the distribution of data-points is not related to the characteristics of the terrain itself. If the data point sampling is conducted on the basis of a regular grid, then the density must be high enough to portray accurately the smallest terrain features present in the area being modelled. If this is done, then the density of data collected will be too high in many areas of the model, in which case there will be an embarrassing and unnecessary data redundancy in these areas. In this situation, filtering of the pre-measured data may need to be carried out as a pre-processing activity before the terrain model can be defined.

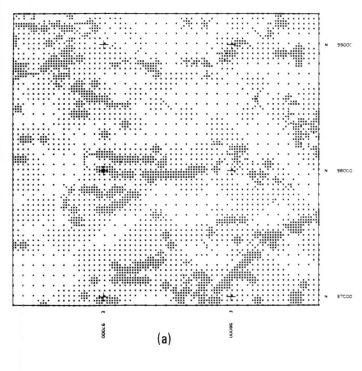

(a)

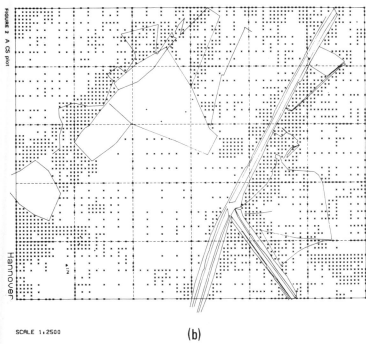

(b)

Figure 4.12 Progressive sampling patterns.

Since this grid-based approach is most easily implemented in computer-controlled photogrammetric machines such as analytical plotters, a solution to the above shortcomings has come from photogrammetrists in the form of *progressive sampling*, and its development, *composite sampling*, originally proposed by Makarovic of the ITC in the Netherlands (Makarovic 1973, 1975). Instead of all the points in a dense grid being measured, the density of the sampling is varied in different parts of the grid, being matched to the local roughness of the terrain surface (Fig. 4.12).

The basis of the method is that one starts with a widely spread (low-resolution) grid which will give a good general coverage of height points over the whole area of the model. Then a progressive increase in the density of sampling (or measurement) takes place on the basis of an analysis of terrain relief and slope using the on-line computer attached to the photogrammetric machine. Thus the basic grid is first densified by halving the size of the grid-cell in certain limited areas based on the results of the preceding terrain analysis. Measurements of the height points at the increased density are carried out under computer control only in these predefined areas. A further analysis is then carried out for each of these areas for which the measured data has been increased or densified. Based on this second analysis, an increased density of points may be prescribed for still smaller areas. Normally, three such runs or iterations are sufficient to acquire the terrain data necessary to defined a satisfactory model.

In this way, the progressive sampling technique attempts to optimize automatically or semi-automatically the relationship between specified accuracy, sampling density and terrain characteristics. Recently the method has been widely implemented in *grid-based terrain modelling packages* devised by photogrammetrists, e.g., HIFI (Ebner *et al.*, 1980, 1984), SCOP (Assmuss, 1976; Stanger, 1976), the package devised in the ITC by Huurneman and Tempfli (1984) and the Graz Terrain Model (GTM) implemented at the Technical University of Graz.

4.3 Photogrammetric data acquisition based on orthophotograph production

Another photogrammetric approach to the acquisition of digital terrain information is that associated with the production of orthophotographs. An orthophotograph is a photographic image of the terrain which has the geometric characteristics of a map, i.e., the positional displacements due to terrain relief and camera tilt and the resultant variations in scale encountered in aerial photographs are eliminated. A composite of two or more orthophotographs made to fit one another is termed an *orthophoto-mosaic*; and if a substantial amount of cartographic enhancement in the form of symbols, names, colours, etc., is added, then the final product is called an *orthophotomap*. While this type of product is little known in the UK, where it has not been adopted by the Ordnance Survey, it has been produced widely in other European countries such as Sweden, West Germany, Belgium, etc., where it forms the basis for national map series.

The process of producing an orthophotograph depends on the systematic scanning and measurement of terrain height over the whole of a stereomodel. Thus the technique produces a complete record of the height variations in the landscape as a byproduct of the orthophotograph production process. First of all, a 3D stereomodel is formed in a stereoplotting machine as described above. It is then systematically scanned in a series of parallel scans (Fig. 4.13) with the operator keeping the

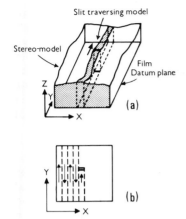

Slit traversing model

Stereo-model

Film
Datum plane

(a)

(b)

Figure 4.13 (a) Orthophotograph production using profiles along which heights are continuously measured. (b) Series of parallel (raster) scans used for orthophotograph production.

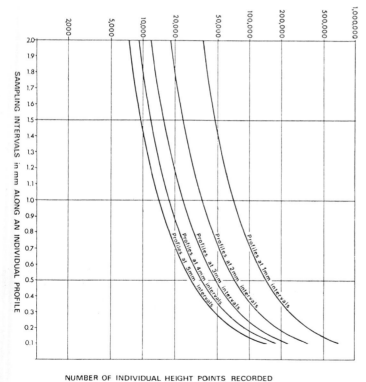

Figure 4.14 Total amount of elevation data generated from production of a single orthophotograph depending on the interval between profiles and the sampling interval along each individual profile.

measuring mark at the correct height throughout each scan. The result is that he is in effect measuring a continuous *profile* of the terrain along a single scan line, the data being recorded digitally during measurement. When the elevation data from each of these profile lines are combined, a complete terrain height model will result besides the orthophotograph which is being exposed simultaneously. As Fig. 4.14 shows, the density of the terrain elevation data collected during these scanning operations can be very high, amounting to tens of thousands of points if narrow intervals are used between profiles and close sampling intervals are selected along each individual profile.

4.3.1 Correlator-based approach to terrain elevation measurements

The stereoplotting machine used during scanning may be of the analogue type (using optical or mechanical models) or of the analytical type (using a purely mathematical model). In both types, the height measurement is carried out by a human operator. However, it is also possible to automate the measuring process, using the so-called *correlator* (Fig. 4.15). A cathode-ray tube (CRT) is used to scan a small area on each of the aerial photographs comprising the stereopair converting the photographic images into a matrix of intensity (or brightness) values. These values are then compared in the correlator to give measurements of parallax which in turn can be converted to height values. The scanning may again be carried out systematically in a raster-scan pattern till the whole stereomodel has been covered, once again resulting in a 3D terrain model of the area covered by the stereopair.

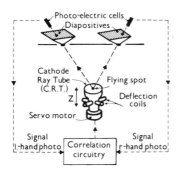

Figure 4.15 Electronic correlator used with optical projection machine.

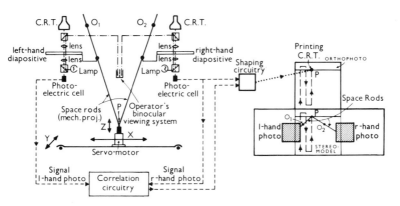

Figure 4.16 Wild-Raytheon B-8 Stereomat automated orthophotograph production system.

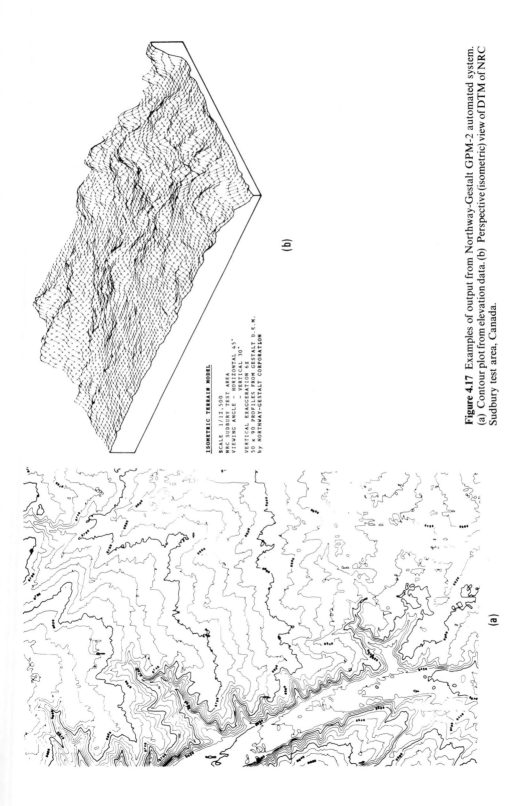

ISOMETRIC TERRAIN MODEL

SCALE 1/12,500
NRC SUDBURY TEST AREA
VIEWING ANGLE – HORIZONTAL 45°
          – VERTICAL 30°
VERTICAL EXAGGERATION 6X
50 x 90 PROFILES FROM GESTALT D.E.M.
by NORTHWAY–GESTALT CORPORATION

(b)

(a)

Figure 4.17 Examples of output from Northway-Gestalt GPM-2 automated system. (a) Contour plot from elevation data. (b) Perspective (isometric) view of DTM of NRC Sudbury test area, Canada.

The most successful implementation of this correlator-based approach was the Wild/Raytheon B8 Stereomat of which a dozen or so examples were built and operated by various national mapping agencies (USA, Australia, France, Mexico, etc.) in the 1970s. This machine was basically a specially modified version of the standard Wild B8 analogue (mechanical projection) stereoplotting machine, on which correlators were mounted together with the orthophotograph printing unit and the necessary electronics (Fig. 4.16). While the B8-Stereomat system was technically and operationally successful, the capital cost involved in the acquisition and the operation of such a system prevented it being adopted on a widespread scale.

Currently the most highly developed implementations of this automated approach to terrain data acquisition are analytically based systems such as the OMI-Bendix AS11B-X (used mainly by military mapping agencies such as the US Defense Mapping Agency) and the Northway-Gestalt Photo Mapper or GPM. The latter system has been used extensively by the main government mapping agencies in North America such as the United States Geological Survey (USGS) and the Canadian Topographic Survey (Brunson and Olsen, 1978; Allam, 1979). With machines such as the GPM-2, the approach is almost entirely digital, with the photographic image converted to a raster of finely-spaced intensity values and the correlation between the respective photographs of the stereopair being carried out digitally in the machine's on-line computer system.

The density of the terrain elevation data acquired with such a device is extremely high—in the case of the GPM-2, some 500 000 to 750 000 elevation points can be obtained from a single stereomodel, which of course gives an excellent representation of the terrain surface in the area covered by the photographs. Such a density makes it relatively easy to generate accurate contours (Fig. 4.17) and perspective views representing the terrain surface of the area. Unfortunately the cost of buying and running such automated systems is extremely high and can normally only be justified where there is a very large programme of mapping to be carried out as is the case with national mapping agencies, or where the overriding requirement is for rapid measurement and cost is a subsidiary factor, as is the case in military mapping agencies, where data has to be acquired on a vast scale for use in aircraft simulators and in aircraft and cruise missile terrain following and avoidance systems.

As with mass cartographic digitizing, a few commercial mapping firms offer an automated photogrammetric terrain data acquisition service based on the use of GPM machines, e.g., Teledyne Geotronics (USA) and Northway (Canada). Although early examples of the GPM machines were installed in bureaux in Europe in the 1970s, the demand for their services was too low for them to be operated economically and the machines were returned to North America.

A very recent development has been the introduction of relatively inexpensive correlator units which can be mounted on the Kern DSR series of analytical plotters. In functional terms, this allows the execution of terrain elevation data acquisition in much the same manner as the purpose-built GPM-2 machines, albeit with some compromises in performance, but at a decidedly lower cost.

4.4 Classification of photogrammetrically-based DTM collection systems

An attempt has been made in Table 4.3 to summarize and classify all the different approaches to acquiring height data by photogrammetric means for terrain

Table 4.3 Comparison of photogrammetrically-based methods of data capture for DTMs

		Type of terrain coverage	Associated with	Sampling	Automated	Break lines	Height measurement by	Speed	Height meas. mode	DTM accuracy	Terrain representation	System cost
I	A	Comprehensive high-density	Orthophoto production	Grid	Yes (X, Y, Z)	No	Correlator (e.g. GPM-2)	Very high	Static	Medium	Very good	Very high
	B	Parallel profiles (automatic)	Orthophoto production	Grid	Yes (X, Y, Z)	No	Correlator (e.g. Planimat + EC5)	High	Dynamic	Medium to low	Fair	Very high
	C	Parallel profiles (semiautomatic)	Orthophoto production	Grid	Yes (X, Y only)	No	Operator (e.g. Topocart/Orthophot)	Medium	Dynamic	Medium to low	Fair	Medium
II	D	Regular grid (automated)	DTM collection	Grid	Yes (X, Y, Z)	No	Correlator (e.g. Kern DSR + VLL)	High	Static	High	Good	High
	E	Regular grid (semi-automated)	DTM collection	Grid + PS	Yes (X, Y only)	Yes	Operator (e.g. analytical plotter)	Medium	Static	High	Good (if progr. sampling)	High
	F	Regular grid (manual)	DTM collection	Grid + PS	No	Yes	Operator (e.g. analogue plotter)	Low	Static	High	Good (if progr. sampling)	Low
III	G	Random, but specific locations	DTM collection	Selective	No	Yes	Operator (analogue or analytic plotting)	Low	Static	Very high	Very good	Low
IV	H	Contouring	Topo (line mapping)	Equal Z-values	No	Yes	Operator (analogue or analytic plotting)	Low	Dynamic	Low-DTM by interpolation	Very good	Low

modelling. This table also indicates in rough terms the speed, accuracy, terrain representation and system cost of all the different types of photogrammetric systems used for data collection.

Class I: Machines and systems associated with orthophotograph production

The Class I machines are those producing terrain elevation data as a byproduct of orthophotograph production. The sampling will invariably be carried out on a grid basis with automated location of successive nodes on the grid under computer control. In view of this, no break lines can be incorporated in the data set. Furthermore, as a consequence of the dynamic mode of operation of most of these devices, the resulting accuracy of height measurement will normally be medium to low.

The main distinguishing feature of the different subclasses within Class I is the means by which the height measurement is achieved. In Classes IA and IB, the measurement is carried out automatically using correlators. Class IA comprises analytically-based machines such as the OMI-Bendix AS-11B-X and Northway-Gestalt GPM-2, while Class IB includes analogue types of stereoplotting machines with purpose-built electronic analogue correlators. The system costs are very high and the speed with which a single stereomodel is measured is also very high in both types of machine. By contrast, the Class IC machines feature height measurements being carried out along profiles using a human operator (i.e., without correlators), the main advantage being the reduced cost of the system. This will result in a reduced speed of operation.

Class II: Machines and systems carrying out systematic DTM measurement

Class II covers machines which are being used specifically to collect terrain model data. As usual with photogrammetrically-based data collection, grid sampling is standard but may be supplemented by spot heights measured on critical points and along breaklines. Most of the systems falling in this category depend on the height measurement being carried out by human operators in the static rather than dynamic mode of operation, though the use of the VLL correlator on the Kern DSR series of machines is a recent development. Since the machine is stationary during the actual measurement of the spot height, the accuracy of height measurement is high. System cost will be high if an analytically-based plotter is used, whereas it will be low if a traditional type of analogue (projection-type) stereoplotting machine equipped with encoders/digitizers is used.

Class III: Machines and systems with selective (i.e. operator-controlled) sampling and measurement

This type of measurement can be based on either analytical or analogue types of stereoplotting machine, with the operator selecting the most appropriate position at which measurements will be carried out. In general, accuracy is high, the resulting terrain representation will be very good and the system cost very low. However, the overall speed of data collection may be low.

Class IV: Contour-based data collection

Since almost all topographic maps produced in the modern (post-World War II) era contain photogrammetrically-measured contours, obviously if the precaution has been taken to digitize the measurements, then the digitized contour-line data is available for digital terrain modelling. The situation is therefore not too different to that where cartographic digitizing of existing contour lines is being carried out. The lower accuracy with which contours are measured initially and the subsequent further deterioration resulting from the generation of a grid-based DTM by interpolation has already been noted.

4.5 Photogrammetrically-based DTM data acquired on a regional or national scale

Using the equipment and methods described above, the elevation data required for terrain modelling will often have been acquired photogrammetrically. In the UK, such operations will almost always have been carried out by commercial air survey companies, since the national surveying and mapping organization, the Ordnance Survey (OS), is not in any way active in the acquisition of digital elevation data needed for terrain modelling at the present time. Unfortunately the OS has just completed its resurvey of much of the upland areas of the country at 1 : 10 000 scale and the opportunity was not taken to digitize the photogrammetric contouring. Neither has the OS implemented a programme of orthophotomapping—topographic maps are still of the traditional type using lines and symbols. However, the situation in many other highly developed countries is very different. In particular, many countries in Western Europe (West Germany, Sweden, Belgium, etc.) and North America (USA and Canada) have large programmes of orthophotomapping at medium scales giving national or regional coverage. A byproduct of these photogrammetric operations is the systematic height profiles from which elevation data can be derived to cover large areas of terrain.

4.5.1 Digital terrain model data in West Germany
Typical of this situation is that existing in certain of the provinces or states in West Germany such as North Rhine Westphalia, Baden-Württemberg, etc., where 1 : 5 000 or 1 : 10 000 scale orthophotographs are produced at regular intervals of several years, either to act as the basis for the construction of orthophotomaps or for the revision of existing topographic maps (Sigle, 1984). The height data from the profiling has often been stored on the scribed glass plates which are an integral part of the Zeiss Oberkochen GZ-1 Orthoprojector systems used for the orthophotograph production. This data can relatively easily be scanned and converted to digital form.

The state of Baden-Württemberg, with an area of 35 751 km^2 is covered by about 75 sheets of the standard West German topographic map series at 1 : 50 000 scale, each sheet covering an area of 22 × 24.5 km. Each of these sheets is covered by a pattern of 5 × 5 (= 25) overlapping orthophotographs at 1 : 10 000 scale. In turn, each orthophotograph has been produced from two stereopairs of wide-angle aerial photographs taken at 1 : 30 000 scale. The total number of orthophotographs involved in the state-wide coverage is about 1 800. Each stereopair had been measured in a high-accuracy stereoplotting machine to give profiles at 80 m interval (8 mm at 1 : 10 000 scale) which had been recorded on glass plates. These profiles have now been digitized to give elevations at 10 m intervals along each profile. Thus 34 000 height points are produced from each storage plate. A comparison of these profile heights with measured data in test areas showed mean height errors of between 1.5 m (flat terrain) to 5 m (rough terrain) with an overall mean error of 2 m.

From this input data, a digital elevation model has been interpolated from the profile height data to give points at a regular grid of 50 m over the whole state using the SCOP DTM program suite developed at the University of Stuttgart and the

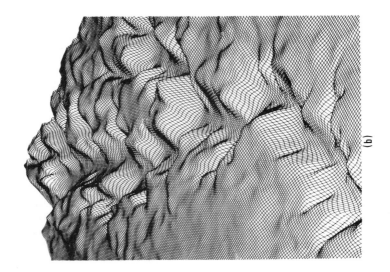

(b)

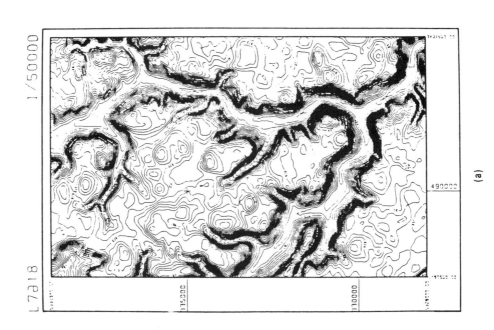

(a)

Figure 4.18 Baden-Württemberg DEM shown (a) as contour map, and (b) as perspective plot.

Technical University of Vienna (Assmus, 1976; Stanger, 1976). About 16 000 000 elevation values at 50 m interval give complete coverage and are available from the state survey organisation of Baden-Württemberg to produce contours and perspective views of the terrain (Fig. 4.18) for use by earth scientists, by photogrammetrists for the production of orthophotographs for map revision and land consolidation projects, for the positioning of transmitters, etc. The project is a quite astonishing and visionary development and an example which others will want to emulate.

Very similar procedures have been put into operation in Sweden for much the same purposes, and indeed, a nationwide DTM based on orthophotograph profile data derived from the 1 : 10 000 and 1 : 20 000 scale Economic Map series covering the whole country was scheduled for completion by 1988. This gives terrain elevation values at a grid interval of 50 m over the whole of Sweden.

4.5.2 Digital terrain model data in North America
In Canada (Allam, 1978) and the United States (El Assal, 1978), much of the data acquisition for terrain modelling and orthophotograph production by the respective national map agencies (Topographic Surveying Division, EMR in Canada, and the USGS) is being carried out using the Gestalt GPM-2 machines which basically are highly-automated correlator-equipped analytical stereoplotters optimized for these purposes. El Assal mentions that the productivity of the two GPM-2 machines operated by the USGS is 5 000–6 000 stereomodels annually, each model comprising 750 000 individual elevation points. Prior to this, similar terrain data was also produced by the Defense Mapping Agency (DMA) utilizing the UNAMACE machines, again using automatic stereocorrelation, but in a series of parallel profiles rather than in patches as in the GPM-2 (Noma and Spencer, 1978).

4.6 Conclusions

Photogrammetric methods offer the greatest flexibility in data collection for terrain modelling in that the density of sampling, the accuracy of the elevation data, etc., can all be varied according to the needs of the user. This can be done by utilizing aerial photography taken at different scales and at different flying heights and by employing various alternative types of photogrammetric instrumentation of varying capability in terms of accuracy, degree of automation, etc. By contrast, field survey methods are constrained to large-scale modelling of limited areal extent and are less flexible in terms of the accuracy, data sampling patterns, etc., that can be implemented, though these deficiencies will be less apparent to a civil engineer or surveyor often interested in a single specific site covering a relatively limited area.

On the wider (regional or national) scale, again photogrammetric methods of data acquisition offer much to surveying and mapping agencies, especially in terms of acquiring dense terrain elevation data of a reasonable metric quality by automated means covering huge areas of terrain. These methods have been implemented in many countries, e.g., in Western Europe in conjunction with statewide programmes of orthophotomapping, and in North America where highly automated systems such as the GPM-2 have been introduced into national mapping agencies and collect elevation data on a massive scale. The alternative method of acquiring digital elevation data on a national scale through the digitization of the contours on existing topographic map series and subsequent interpolation will almost inevitably result in elevation data of a lower metric quality. Nevertheless the latter method appears to be the only practical method of creating a national digital elevation database for those countries such as the UK and France which have not implemented ortho-photographic-based topographic map series at medium scales and have not captured the contour data in digital form during their programs of mapping at medium scales. In this case, photogrammetric methods will be employed in the area between large-scale ground surveys of limited extent and the national databases of lower sampling density and moderate accuracy produced from the contours on existing topographic maps.

References and further reading

Ackermann, F. (1978) Experimental investigation into the accuracy of contouring from DTM. *Photogrammetric Engineering and Remote Sensing* **44**(12), 1537–1548. Also in *Proceedings of the Digital Terrain Models (DTM) Symposium*, 165–192.

Ackermann, F. (1980) The accuracy of digital height models. *Proceedings of the 37th Photogrammetric Week, University of Stuttgart*, 133–144.

Allam, M.M. (1978) DTM's application in topographic mapping. *Proceedings of the Digital Terrain Models (DTM) Symposium*, 1–15. Also in *Photogrammetric Engineering and Remote Sensing*, **44** (12), 1513–1520.

Assmus, E. (1976) Extension of Stuttgart contour program to treating terrain break lines—theory and results. *Presented Paper, Commission III, XIII Congress of the I.S.P.*, Helsinki, 13 pp.

Blais, J.A.R., Chapman, M.A. and Kok, A.L. (1985) Digital terrain modelling and applications. *Technical Papers, ACSM-ASP Fall Convention*, 646–651.

Brunson, E.B. and Olsen, R.W. (1978) Digital terrain model collection systems. *Proceedings of the Digital Terrain Models (DTM) Symposium*, 72–99.

Cogan, L. and Hunter, D. (1984) Kern DSR1/DSR11 DTM collection and the Kern Correlator. *Presented Paper, XVth ISPRS Congress, Rio de Janeiro*, 11 pp.

Ebner, H., Hoffman–Wellenhof, B., Reiss, P. and Steidler, F. (1980) HIFI—a minicomputer program package for height interpolation by finite elements. *Presented Paper, Commission, IV, XIV Congress of the I.S.P., Hamburg*, 14 pp.

Ebner, H. and Reinhardt, W. (1984) Progressive sampling and DEM interpolation by finite elements. *Bildmessung und Luftbildwesen*, **52** (3), 177–182.

Ebner, H. and Reiss, P. (1984) Experience with height interpolation by finite elements, *Photogrammetric Engineering and Remote Sensing*, **50** (2) 177–182.

El Assal, A.A. (1978) USGS Digital Cartographic File Management System. *Proceedings of the Digital Terrain Models (DTM) Symposium*, 493–495.

Huurneman, G. and Tempfli, K., (1984) Composite sampling using a minicomputer supported Analogue instrument. *ITC Journal* 1984 (2), 129–133.

Huegli, P., Steidler, F. and Zumofen, G., 1984. CIP—a program package for interpolation and plotting of digital height models. *Proceedings ACSM-ASP Convention, Washington*, 10 pp. Also in *International Archives of Photogrammetry*, 25 (A36), 1040–1049.

Köstli, A. and Wild, E. (1984) A digital elevation model featuring varying grid size. *International Archives of Photogrammetry*, **25**(A3b), 1130–1138.

Makarovic, B. (1973) Progressive sampling for digital terrain models. *ITC Journal*, 1973 (3), 397–416.

Makarovic, B. (1975) Amended strategy for progressive sampling. *ITC Journal*, 1975(1), 117–128.

Makarovic, B. (1977) Composite sampling for DTMs. *ITC Journal*, 1977(3), 406–433.

Noma, A.A. and Spencer, N.S. (1978) Development of a DMTAC digital terrain data base system. *Proceedings of the Digital Terrain Models (DTM) Symposium*, 493–505.

Petrie, G. (1981) Hardware aspects of digital mapping. *Photogrammetric Engineering and Remote Sensing*, **47**(3), 307–320.

Sigle, M. (1984) A digital elevation model for the State of Baden–Württemberg. *Contributions to the XVth ISPRS Congress, Rio de Janeiro. Schriftenreihe*, Institute of Photogrammetry, University of Stuttgart **10**, 105–115. Also in *International Archives of Photogrammetry*, **25**(A3b), 1016–1026.

Stanger, W. (1976) The Stuttgart contour program SCOP—further development and review of its application. *Presented Paper, Commission III, XIII Congress of the ISP*, Helsinki, 13 pp.

Tempfli, H. (1986) Composite/Progressive sampling—a program package for computer supported collection of DTM data. *Presented Paper, ACSM-ASP Convention, Washington*, 9 pp.

Young, W.H. and Isbell, D.M. (1978) Production mapping with orthophoto digital terrain models. *Photogrammetric Engineering and Remote Sensing*, **44**(12), 1521–1536. Also in *Proceedings of the Digital Terrain Models (DTM) Symposium*, 100–124.

5 The impact of analytical photogrammetric instrumentation on DTM data acquisition and processing

G. PETRIE

5.1 Introduction

The previous chapter dealt with the more general aspects of the photogrammetric methods used to acquire data for DTM work. This chapter concentrates on the developments in computer-based analytical photogrammetric instrumentation which have taken place over the last few years and the possibilities which these developments offer with respect to the acquisition of data for terrain modelling. Although the analytical plotter concept was first introduced by Helava in 1957–58 and realized soon after in the early 1960s in the form of the OMI/Bendix AP-1 and AP-2 instruments, the users were almost exclusively in the area of military mapping and intelligence gathering. This resulted from the sheer expense of the equipment— especially the provision of purpose-built, real-time computers and electronic control components—which could, however, be justified on the basis of defence require-ments.

In 1976, the situation changed with the simultaneous introduction of seven or eight analytical plotters at the ISP Congress at Helsinki (Petrie, 1977). This quite sudden and dramatic change resulted largely from the development of small, fast, reliable and powerful minicomputers such as the DEC PDP-11, Data General Nova and Hewlett Packard HP-1000, which were widely available and supported and which could be used as the control computers instead of the specially built computers employed previously. This allowed analytical plotters to be built at a price which allowed them to compete with the larger, more accurate types of analogue stereoplotter based on the use of optical or mechanical models.

Since 1976, the situation has changed. First of all, manufacturers gradually dropped the production of the larger and more expensive analogue instruments. However, since 1984, with the introduction of powerful microcomputers, especially the IBM-PC series and its numerous clones, the production of even the smaller analogue instruments has largely been terminated until now, in 1989, only three or four analogue instruments are available in the marketplace. This period (1984–89) has seen the introduction of many comparatively low-cost analytical instruments, often of an innovative type, to complement the larger, more expensive analytical instruments which have also developed considerably in terms of capability over this same period.

This chapter reviews these recent developments in the particular context of digital terrain modelling.

5.2 Analytical photogrammetric instrumentation

It must be noted that certain types of analytical photogrammetric instrument are not suitable for the collection of DTM data since they cannot measure height. These comprise the *monoscopic instruments*, e.g., monocomparators and digital mono-plotters, in which only a single photographic image can be measured at one time, so there is no stereomodel available for the measurement of the heights and contours required for terrain modelling.

   Suitable analytical instruments, those in which a stereomodel can be formed and viewed, can be classified into three main types (Helava, 1980; Konecny, 1980):

  (i) *image space plotters* in which x, y image (i.e., photo) coordinates are measured and form the primary input to the photogrammetric solution, but a continuously oriented, parallax-free stereomodel does not result and heights can only be measured on a point-to-point basis;

 (ii) *analytical plotters* in which x, y image coordinates are again the basis of the photogrammetric solution but, in this case, the operator can carry out the measurements of heights and contours for terrain modelling on the basis of a continuously oriented stereomodel; and

(iii) *analytical plotters* in which the object (X, Y, Z) or terrain (E, N, H) coordinates are the primary input to the photogrammetric solution which again results in a parallax-free stereomodel on which measurements of terrain elevation can be carried out either manually and visually by a photogrammetrist or automatically using an image correlator.

5.2.1 Image space plotter

In the *image space plotter* (Fig. 5.1), the handwheels drive the x and y movements of the left photo under the control of the operator. The footwheel is actuated by the operator to impart to the right-hand photo the px (x-parallax) movement required to carry out height (Z) measurement. All three coordinate values (x, y, px) are encoded digitally and sent continuously to the computer which then computes the E, N, H terrain coordinate values. However no correction values are computed nor are any feedback movements imparted to the photographs to ensure an oriented parallax-free

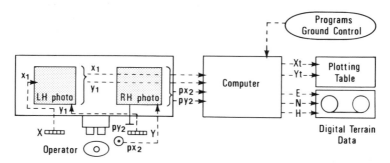

Figure 5.1 Image space plotter.

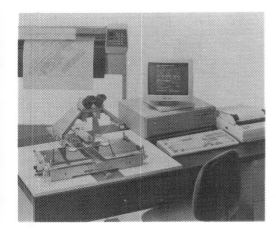

Figure 5.2 Zeiss Oberkochen Stereocord G3.

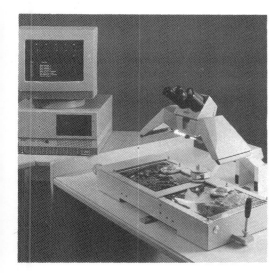

Figure 5.3 Galileo Stereobit.

stereomodel. Therefore the removal of y-parallax (the want of correspondence between the two images) must be carried out manually by the operator as a preliminary operation at every point before the elevation value can be measured. Thus the measurement of the heights required for DTM work can only be carried out on a point-by-point basis. Instruments which fall into this category include the Zeiss Oberkochen Stereocord G3 (Fig. 5.2) and the Galileo Stereobit (Fig. 5.3).

From the specific point of view of the acquisition of DTM elevation data, this is the least capable type of analytical instrument. Since there are no computer-controlled drives or feedback devices, obviously it is not possible to implement systematic, grid-based sampling or methods such as progressive sampling where the positions of the elevation points in the terrain system are predefined and must be driven to. Furthermore, since there is no oriented parallax-free stereomodel, it is impossible to carry out the direct measurement of contours. The acquisition of elevation data for

DTM work is therefore confined to the measurement of individual points at locations chosen by the operator, i.e., effectively only selective sampling can be implemented. Also it must be recognized that the measuring process will be quite slow—again the consequence of the lack of an oriented stereomodel and the need to eliminate y-parallax at every point prior to making the elevation measurement. Contours can of course be derived from the network of individually measured height points by interpolation in a later off-line operation, but the quality, i.e., the accuracy, of such contours will be lower than that of directly measured contours.

5.2.2 Analytical plotter with image coordinates primary

The *analytical plotter* solution based on the use of the *image coordinates* as the primary input is shown in Fig. 5.4. As with the image space plotter, the operator controls the positions of the points being measured by physically moving the left photograph using two handwheels and uses the footwheel to carry out the height (Z) measurements. Again the x_1, y_1 and px_2 values are sent to the controlling computer and the E, N, H terrain values are computed. However the computer is also programmed to calculate in real time the small correction values Δx_2 and Δy_2 which are imparted continuously to the right photograph using small motors. The result is

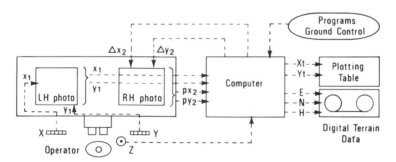

Figure 5.4 Analytical plotter with image coordinates primary.

Figure 5.5 Galileo Digicart 20.

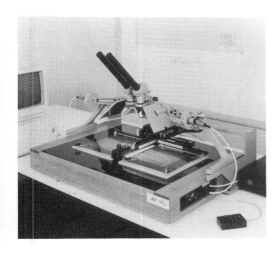

Figure 5.6 Carto Instruments AP-190.

the creation of an oriented, parallax-free stereomodel. Instruments which implement this particular solution include the Galileo Digital Stereocartograph (Inghillieri, 1980) and Digicart 20 (Fig. 5.5), the Autometrics APPS IV (Greve, 1980), the Topcon PA-1000 (Goudswaard, 1988) and the AP-190 produced by Carto Instruments A/S (Carson, 1987) and based on the scanning mirror stereoscope of Cartographic Engineering (Fig. 5.6).

Again, viewing this class of analytical plotter from the point of view of its applicability to the acquisition of DTM elevation data, the situation is somewhat improved over that already described above for the image space plotter. Obviously the fact that these instruments have a parallax-free stereomodel means that the height measurements of individual points can be carried out by the operator in a much more efficient and comfortable manner and that directly measured contouring can also be implemented. However, it is still not possible to implement systematic sampling methods based on the use of the terrain coordinate system or its derivative, progressive sampling, since it is not possible to drive to the specific predefined or predetermined points where heights have to be measured.

5.2.3 Analytical plotters with object coordinates primary

The third approach is that of an *analytical plotter* in which *object coordinates* are the primary input to the photogrammetric solution (Fig. 5.7). It will be seen that the

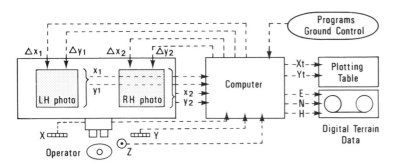

Figure 5.7 Analytical plotter with object coordinates primary.

movements of the X and Y handwheels and the Z footwheel are not applied to the photo carriers, but instead are encoded and pass directly to the control computer. The values $\Delta x_1, \Delta y_1, \Delta x_2$ and Δy_2 are generated and imparted continuously to the left and right photographs to ensure both the correct positioning of the measuring mark and the generation of a stereomodel free from y-parallax. From the computational/algorithmic point of view, the mathematical solution which is implemented using the standard collinearity equations of analytical photogrammetry, is the inverse of that used in the previous two types of instrument in which the image coordinates are the primary input to the photogrammetric solution (Helava, 1980).

The classical type of analytical plotter produced by most of the mainstream photogrammetric instrument manufacturers employs this solution. Thus the Wild Aviolyt AC-1, BC-1 and BC-2 (Fig. 5.8) instruments; the Zeiss Oberkochen Planicomp C-100 and P series of instruments; the Kern DSR series (Fig. 5.9); the Matra Traster analytical plotters; and the OMI AP series all implement this approach. In

Figure 5.8 Wild Aviolyt BC-2.

Figure 5.9 Kern DSR-11.

Kern PG2

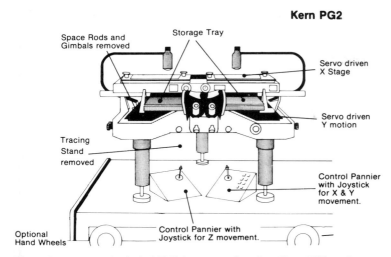

Figure 5.10 Quasco Analytical (QA) instrument based on Kern PG2 analogue stereoplotter.

addition, some smaller manufacturers, e.g., Adam Technology, have also adopted this solution.

It is also worth noting that a number of conversions of analogue stereoplotters into analytical plotters using the object coordinates primary approach have also taken place. These include the Zeiss Jena Topocarts converted by Helava Associates in the USA and the Kern PG2 and Wild B8 instruments converted by the Australian companies Quasco (Fig. 5.10) and Adam Technology. In each of these, the space rods and plotting/tracing stand of the mechanical projection solution have been discarded. Inputs to the PC-based microcomputer are generated either by handwheels and footwheel or by joystick movements. Again the outputs (x and y image coordinates) control the appropriate movements of each plate and ensure correct positioning of the measuring mark within a parallax-free, oriented stereomodel.

Analysing this class of instrumentation in terms of its potential for the acquisition of DTM elevation data, it will be apparent that it offers by far the greatest capability. Since the basic input to the computer can be terrain (E, N, H) coordinates, it offers the easy implementation of any sampling pattern based on these coordinates, since the measuring mark can be driven to a predetermined position. Thus systematic grid-based sampling and progressive sampling can readily be implemented and there is also a straightforward implementation of measured contouring. Furthermore, as will be discussed in more detail later, the object coordinate primary solution is also the basis for more advanced analytical instruments which carry out automatic height measurement using correlators. It will be obvious that the measurement of random but specifically located points, which is the method normally used with the two preceding types of analytical instrument based on image coordinates primary, is also possible in the analytical plotters with the object coordinates primary solution.

5.3 Recent developments in analytical photogrammetric instrumentation

Recent developments in analytical plotter design include the following:

(i) the complete integration of graphics workstations into analytical plotter design;

(ii) the introduction of graphics superimposition and stereosuperimposition—
developments which are closely associated with (i);

(iii) the use of enlarged photostages to handle photographs of up to 25 × 50 cm
(9 × 18 inch) format;

(iv) the addition of correlators for the automatic measurement of heights.

All of these developments impact on the use of analytical plotters for the acquisition of
DTM data from stereoscopic imagery to some extent, but items (ii) and (iv) are of
particular importance.

5.3.1 Superimposition and stereosuperimposition

Graphics workstations were first introduced as attachments to analytical plotters as
in the case of the MAPS 200 or 300 systems which could be attached to the Kern DSR
analytical plotters (Newby and Walker, 1986), and the Intergraph InterAct units
which could be attached to Zeiss Oberkochen Planicomp instruments. One of the
consequences of this development is that it is also possible to have an auxiliary CRT
screen (monitor) which displays the plotted detail in graphical form and is mounted
within the analytical plotter itself. This data will have been transformed from its
terrain coordinate (E, N, H) form back into the corresponding image coordinate (x, y)
values which can then be injected into the optical channel of the analytical plotter via
a semi-reflecting mirror. In this way, a 2D monoscopic image of the plotted detail is
superimposed on the 3D image of the stereomodel and the operator can see what has
been measured and plotted without having to look away from the viewing oculars. An
example of this procedure applied to digital terrain model data using the Zeiss
Oberkochen Planicomp equipped with the Videomap superimposition system is
described by Reinhardt (1986).

While this is a generally useful facility, a further development—that of *stereosuperimposition* (Beerenwinkel *et al.*, 1986)—is of even more direct interest to those
concerned with the measurement of heights and contours. With this approach, the
graphical images of the measured or plotted detail are displayed on two CRT screens
and injected into each of the two optical channels of the viewing system of the
analytical plotter (Fig. 5.11), each being transformed and displaced individually to fit

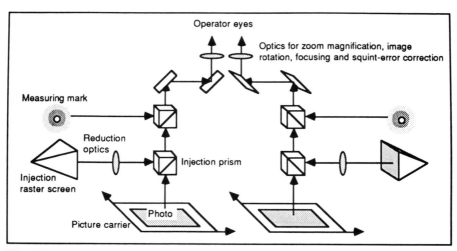

Figure 5.11 Principle of stereosuperimposition as applied to an analytical plotter.

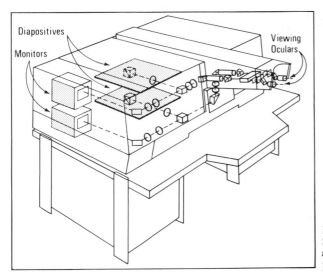

Figure 5.12 Kern Raster Image Superimposition System (KRISS) added to a Kern DSR analytical plotter.

exactly the detail present in the left and right photographs respectively. Thus the orthographic detail of the measured data in the form of terrain coordinates (E, N, H) is transformed into the perspective geometry, i.e., the image coordinates $(x\,y)$, of each of the photographs of the stereopair. This means that the measured or plotted detail can be viewed by the photogrammetrist as a stereo, 3D image against the background of the stereomodel itself. The arrangement used in the Kern Raster Image Superimposition System (KRISS) which can be attached to the firm's DSR series of analytical plotters is shown in Fig. 5.12.

The implications of this recent development for DTM elevation data acquisition could be considerable. In the first place, the accuracy of the individual height points and measured contours can be checked visually to see if they fit the terrain surface exactly as viewed in the stereomodel. If any error is detected, corrective action can be taken. If it is felt that the measured data does not adequately represent the terrain surface morphology as seen in the stereomodel, supplementary measurements can be undertaken to ensure a more effective terrain representation. It can be seen that stereosuperimposition adds considerable capability to an analytical plotter in terms of its potential to monitor height and contour accuracy and completeness, but only at a very considerable additional cost.

5.3.2 The use of large-format photographs
The development of analytical plotters equipped with enlarged photo stages is one which is intimately bound up with the use of large-format cameras (LFC), typically producing photographs 25×50 cm (9×18 inch) in size. This development originally took place in the early 1970s, principally for military mapping and intelligence-gathering activities. If the longer side of the format is oriented in the direction of flight, a much better base:height ratio can be achieved from high-flying military re-connaissance aircraft or satellites while still giving a reasonably high-resolution image. Thus from an altitude of 70 000 ft (21 km), the use of an $f = 30$ cm (12 inch) lens with a large-format camera gives a photographic scale of 1:70 000, while still preserving a base:height ratio of 0.6 to ensure a useful height measuring capability. A

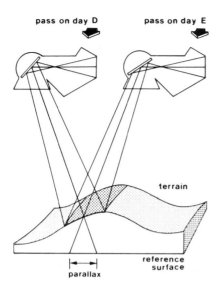

pass on day D   **pass on day E**

terrain

reference
surface

|←—→|
parallax

Figure 5.13 SPOT side-overlapping images forming a stereomodel of the terrain.

few American civilian mapping agencies such as the National Ocean Survey (NOS)
have had access to large-format aerial photography and have suitably modified
analytical plotters, e.g., NOSAP (Fritz, 1977; Slama, 1980), which can handle it, but
the majority of civilian users have gained only limited experience of large-format
photography through NASA's LFC experiment based on the use of the Space Shuttle
as the camera platform. From an altitude H of 250 km, this resulted in a photo scale of
1:850 000 and a spot height accuracy of $\pm$ 15 m, which is of course only of interest for
reconnaissance engineering surveys and small-scale topographic mapping at scales
between 1:100 000 and 1:250 000.

It is worth noting that the use of analytical plotters has also allowed the acquisition
of elevation data from stereoscanner imagery taken from space, especially that
acquired by the French SPOT satellite using side-overlapping images taken on
different runs (Fig. 5.13). In general, these have resulted in spot height data of a
quality and accuracy of the same order as that of the LFC camera.

5.3.3 Analytical plotters with correlators

Although certain types of analogue stereoplotters such as the Wild B8 Stereomat, the
Zeiss Oberkochen Planimat with the Itek EC-5 system and the Zeiss Jena Topocart
(as the Topomat) have all been equipped with correlators and have been used during
the 1970s for the measurement of terrain height profiles in conjunction with the
production of orthophotographs, the current development of correlator-equipped
devices is largely concentrated on analytically-based photogrammetric instruments.
From the algorithmic or computational point of view, the approach followed is that
of the object coordinates primary solution outlined above for manually-operated
analytical instruments. Thus the inputs are the terrain (E, N, H) coordinates while the
outputs are the image coordinates (x_1, y_1, x_2, y_2) of each photograph.

The current types of analytical plotter equipped with correlators can be different-
iated into two main types:

 (i) these employing *area correlation techniques*; and
(ii) those which are based on correlation along *epipolar lines*.

5.3.3.1 *Instruments using area correlation*

Various approaches have been taken in this area. In the early analytical instruments such as the Bunker Ramo UNAMACE instruments of the 1960s (Bertram, 1963, 1965), the correlation was carried out along a series of parallel (raster) scans in the manner typical of all instruments concerned with orthophotograph production at that time, whether analogue or analytically-based. A series of parallel elevation profiles was measured in a dynamic mode with the correlator replacing the human operator. The digital terrain model data was in profile form and virtually the same as that provided by analogue stereoplotting instruments.

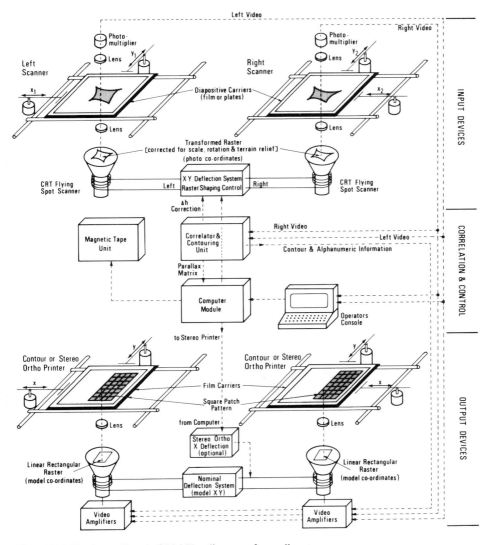

Figure 5.14 Northway-Gestalt GPM II—diagram of overall system.

In the 1970s, the *Northway-Gestalt GPM II and IV instruments* designed by Hobrough carried out the area correlation on a patch-by-patch basis (see chapter 4). As shown in the GPM II system (Fig. 5.14), the input consists of two scanner/transport units equipped with cross-slide movements on which the stereopair of photographs is mounted. At any one time, small corresponding parts (9 × 8 mm) of these photographs are being scanned simultaneously using a pair of CRT-based Flying Spot Scanners, the output being converted to video signals by photomultipliers. The operator's console is equipped with an alphanumeric terminal which communicates with the controlling Data General minicomputer and a monitor displays the video image of the photography both during set-up to measure parallaxes and control points and during the correlation process itself. The output side comprises two scanner/transporter units, again based on the use of cross-slides, on which the final orthophotograph and the contour sheet may be written separately on photographic film.

The actual heighting and contouring operations are carried out in the *GPM II correlator* which measures the x-parallaxes between the two photographs and converts them into the corresponding terrain elevation values. These parallax values are calculated on a grid of points separated by 182 μm at photo scale resulting in 2400 height values being generated for each 9 × 8 mm patch of the stereomodel. The correlator also contains the contouring circuitry which converts the grid of spot heights into contours in real time so that they can be output on the contour printer. The matrix of spot heights forming the DTM is stored in the computer memory and then transferred and recorded on to magnetic tape. As mentioned previously, the process results in a dense elevation matrix of between 500 000 and 600 000 points per stereomodel, the complete scanning/processing/output operation averaging just over one hour for a single model (Bethel *et al.*, 1978).

Undoubtedly the capabilities of the dedicated GPM II and IV analytical instruments are such that they are ideal for the generation of dense DTMs over large areas of terrain. However, they are also extremely expensive to purchase and maintain and relatively few have been built and operated. Thus the present development of much less expensive and more compact correlator units which can be bolted on to existing standard analytical plotters is an important advance in DTM data acquisition technology. An example is the correlator for the Kern DSR analytical plotter (Cogan and Hunter, 1984; Bethel, 1985) based on the use of small compact cameras using CCD areal arrays to transform the photographic image into digital values and a digital frame store or memory to store the images (Fig. 5.15). The algorithm used in the image matching or correlation process is the *vertical line locus (VLL) method* in which a number of small windows are defined in the model space at different Z-values along the vertical linear locus through the point located at a specific planimetric position. Next the corresponding image points of the corners of each small window are defined for both photographs and the digital values of the image densities (grey levels) are obtained using the CCD cameras. These are then matched for each model window using computational methods and a best estimate made of the terrain elevation value for the specific point location (Benard *et al.*, 1986). The DSR analytical plotter is then driven under computer control to the next point in the predefined grid required for the DTM, each spot height being determined automatically using the correlator. Alternatively the control computer of the DSR can drive the instrument to a series of points situated at predefined intervals along a specific profile and its cross-sections within the stereomodel, the height again being measured automatically using the correlator.

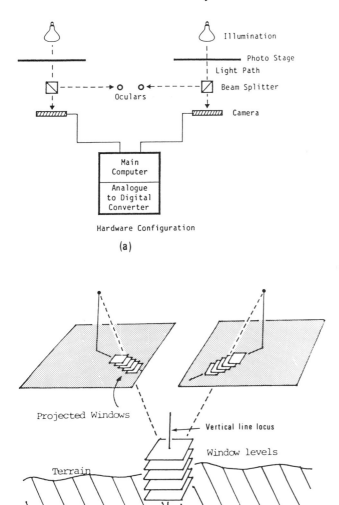

Figure 5.15 Kern VLL Correlator: (a) Hardware configuration, and (b) correlator concept.

5.3.3.2 *Instruments using correlation along epipolar lines*

The basic concept of this alternative approach to the automatic measurement of the terrain elevations for DTMs was introduced simultaneously but independently by Helava and Chapelle (1972) and by Masry (1972). The basic idea is to first establish the positions and directions of the corresponding sets of epipolar lines on each photograph of the stereopair being measured (Fig. 5.16) using the dedicated control computer of the analytical plotter. The scanning of the image densities (grey levels) then takes place along each pair of corresponding epipolar lines using, for example, a CCD-based linear array or areal array camera mounted below each photograph. Correlation of the corresponding images for height measurement is greatly simplified

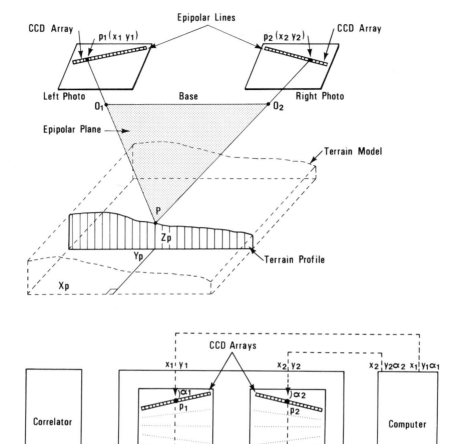

Figure 5.16 Automated analytical plotter based on epipolar line correlation. Top: perspective view showing geometric concept, and (bottom) plan view showing hardware concept.

since the search for and matching of common points merely has to be carried out along specific predetermined lines instead of corresponding areas as in the area correlation method. This technique was first implemented in the OMI/Bendix AS-11-BX analytical instrument developed for the US Defense Mapping Agency in the mid-1970s and then in the Raster device designed and built by Hobrough and attached to an OMI AP/C-3 analytical plotter at the University of Hanover in 1978.

5.3.3.3 *Quality control of DTM data acquired by automatic correlation*
Obviously the possibility of errors in the DTM elevation data acquired by automatic

correlation methods must be considered since there is a minimum of man–machine interaction and unexpected errors or omissions may occur, e.g., when the stereocorrespondence algorithm produces mismatches. Since much of the use of correlator-equipped analytical plotters has been by military mapping agencies in the United States, it is interesting to note the intensive effort made by the US Army Engineer Topographic Laboratories (USETL) to superimpose DTM elevation data on the stereomodels from which they have been derived (Nagel and Camp, 1985). This forms part of the so-called Computer Assisted Photographic Interpretation Facility (CAPIR) at ETL and comprises an Elevation Data Editing Terminal (EDET) attached to a Data General Eclipse 5/250 minicomputer and an Autometric APPS IV analytical plotter. Essentially it is an off-line method rather than an on-line operation carried on in real time in the correlator-equipped analytical plotter.

As with the Kern and Wild stereosuperimposition systems (see 5.3.1), the DTM height matrix expressed in terrain (E, N, H) coordinates is transformed into photo (x, y) coordinates and appears superimposed as a grid-like pattern of floating dots on the stereomodel. However, instead of superimposing the DTM data displayed on CRTs on the actual hard-copy film diapositives, the photographic images are scanned using a Flying Spot Scanner (FSS) and the dots of the DTM elevation data matrix are superimposed on these. The output from the two overlapping images and their superimposed data patterns are displayed on a single CRT display where they can be viewed in stereo; up to 4000 data points can be seen in 3D in this way.

This operation allows the operator or editor to evaluate accurately the quality of the DTM data and to carry out various types of editing, viz.:

(i) *point edit*, which allows the correction of a specific elevation value;
(ii) *segment edit*, which permits the correction of an entire segment by a specific amount;
(iii) *sequential edit*, which permits the editor to step through the whole elevation data matrix automatically; and
(iv) *area edit*, which gives the capability of allowing a specific area such as a forest to be defined by the operator as a polygon so that it can be altered as a whole by a specific amount.

Obviously this gives the capability of carrying out a check on the accuracy of the DTM elevation data which is a real requirement when it has been acquired using automatic correlation techniques.

5.3.3.4 *Possible future developments*

From the preceding parts of this section, it will be obvious that, at the present time, the combination of an analytical plotter and a digitally-based correlator offers the only practical method of achieving very dense, high-quality terrain elevations over large areas of the Earth's surface. For the future, it is expected by many photogrammetrists that *all-digital photogrammetric systems* will be developed and used for the acquisition of DTM information based on digital image data of the terrain instead of the present hard-copy film transparencies produced from the negative aerial photographs. This will eliminate the requirement for the high-quality mechanical and optical components needed for the measurement and viewing of stereomodels in the present types of analytical photogrammetric instrumentation, the whole process of height determination being carried out using digital image processing techniques.

5.4 The implementation of alternative DTM sampling/measuring strategies using analytical photogrammetric instruments

The various alternative sampling strategies which can be used for the photo-grammetric measurements of height data for DTM have already been set out in chapter 4. These include the acquisition of data by means of:

   (i) *systematic* (usually grid-based) *sampling* which may be implemented as a square or triangular grid or in the form of a series of parallel profiles;

  (ii) *progressive sampling* which is a development of systematic sampling comprising a basic widely-spaced grid of height points which are supplemented by additional measurements made in certain limited areas of the stereomodel after terrain analysis of the initial data;

 (iii) *selective sampling* in which the measured elevation points are randomly located in terms of absolute position, but are in fact located at specific positions selected by the photogrammetrist;

 (iv) *composite sampling* which combines both grid (and progressive) sampling supplemented by height measurements made at significant points and along breaklines; thus it is a combination of the three previous strategies—systematic, progressive and selective sampling; and

  (v) *measured contours.*

5.4.1 Systematic sampling

With an *analogue stereoplotting instrument*, the measurement of the grid of spot heights can only be carried out on the basis of the X, Y model coordinate system of the instrument itself. This is the case in almost all analogue instruments since the positioning is controlled manually by the operator and normally he has no direct reference to the terrain (E, N) coordinate system except by using an existing map laid out on the instrument's plotting table or coordinatograph. Reference to the terrain coordinate system can then be made using a grid overlaid on the map and positioning can be achieved by use of a profiloscope or a small TV camera mounted on the plotting table and a monitor mounted on the instrument in front of the operator (Fig. 5.17). Alternatively, if an on-line computer is available, the terrain (E, N) coordinates can be generated as numerical values on the coordinate display or terminal, but once again the actual positioning has to be carried out by the operator. Since both of these procedures are somewhat demanding or even exhausting, they are seldom implemented, so that the model (X, Y) coordinate system is almost invariably used as the reference system. It will be noted that the consequence of this is that the data generated from each stereomodel will have a slightly different orientation to that of its neighbours (Fig. 5.18). If the DTM data is required in the form of a systematic raster pattern within the terrain coordinate system, then it will have to undergo an additional transformation and interpolation process so that elevation values are generated for each required position. This will of course reduce the accuracy of the final height values.

   With *analytical photogrammetric instrumentation*, the situation with regard to both the image space plotter and the analytical plotter based on image coordinates primary is only a little different to that described for analogue instruments. The terrain coordinates can be generated continuously by the on-line computer which is an integral part of the instrument but since the instrument's measuring/viewing systems are image coordinate-based, no method exists of driving them to the required terrain

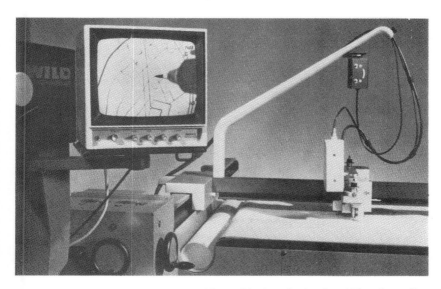

Figure 5.17 TV camera and monitor used for positioning of points in grid-based sampling.

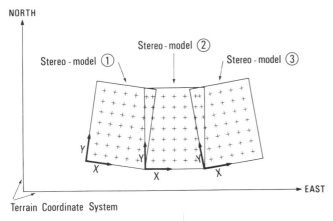

Figure 5.18 Different orientations of successive stereomodels within the terrain coordinate system.

position. By contrast, those analytical plotters based on the object coordinates primary solution can use terrain coordinate positions as the input to the computer which can therefore drive the instrument to the specific positions needed to implement any conceivable systematic sampling strategy. Not only can terrain-oriented grids be used but it is also possible to conduct sampling along specific profile lines and cross-section lines as is required for input to certain road design program packages. Here, undoubtedly the use of an analytical plotter based on object coordinates primary is a more satisfactory solution since the location of the starting and end points, the direction (azimuth) of scan and the sampling interval along each profile or cross-section can be implemented readily on such an analytical plotter.

From the previous discussion (section 5.3.3) it will be apparent that the dense matrix of terrain elevations generated by analytical plotters equipped with *area-based correlators* will constitute a close-spaced regular grid of heights for DTM purposes. Indeed the systematic sampling strategy is the only obvious one to apply with such automated instruments. However, based on the geometric characteristics of *epipolar line scanning*, the grid of heights generated from the application of this type of scanning will be slightly irregular in shape since points that lie at different elevations along the same epipolar plane do not have their orthogonal projection lying along the same straight line (Jaksic, 1976). So if a regular grid of terrain elevation data is a specific requirement, e.g., for input to a DTM-based civil engineering design package, it must be generated by interpolation from the measured data.

5.4.2 Progressive sampling

Most of the remarks made above (section 5.4.1) apply equally to progressive sampling which attempts to optimize (on the basis of the terrain analysis) the sampling density in different parts of the stereomodel in relation to the characteristics of the specific piece of the terrain surface being modelled. While it is possible to implement this procedure on *analogue stereoplotting instruments* by adding stepping motors and drives to the lead screws of the instrument and controlling the positioning of the measuring mark using an on-line computer, the difficulties and expense associated with such modifications are such that they have only been implemented in a very few isolated cases. For example, Roberts and Cogan (1978) installed motors and drives on a Kern PG2 and used a Wang desktop computer to control the positioning in the stereomodel, and Huurneman and Tempfli (1984) made similar additions to a Zeiss Oberkochen Planimat attached to a PDP-11 minicomputer specifically for the implementation of progressive sampling. It is also worth noting that, at the 1976 ISP Congress held in Helsinki, Wild showed the AMU version of their Aviomap analogue stereoplotter which featured the setting of X, Y coordinates under computer control.

However, in total, very few of these instruments have been built, and indeed it would be fair to say that the concept of progressive sampling for DTM data acquisition had to await the widespread adoption of the analytical plotter with its built-in computer and integral drives and feedback controls before it could be adopted to any real extent. Now many of the photogrammetrically oriented DTM packages, e.g., HIFI with PROSA (PROgressive SAmpling) (Reinhardt, 1984), and the Kern DTM package (Tempfli, 1986), are incorporating progressive sampling as an optional module wherever they are intended for use in an analytical plotter. From the previous discussion, it will be apparent that both the image space plotter and those analytical plotters based on image coordinates primary cannot have this strategy implemented in any really practical way and, once again, the method of progressive sampling, as with all computer-controlled, terrain-oriented strategies, is only really practicable with those analytical plotters which are based on the object coordinates primary solution. However it should also be noted that the introduction of progressive sampling will of course add to the computational load on the controlling computer of the analytical plotter. This results from the fact that it must analyse the existing measured height data and provide the terrain coordinate information required to drive the instrument to each new position required for the DTM, besides carrying out the standard real-time computations needed to allow the analytical plotter to function at all.

If progressive sampling is implemented, the situation regarding both the data structure and the representation of the DTM becomes more complicated arising from

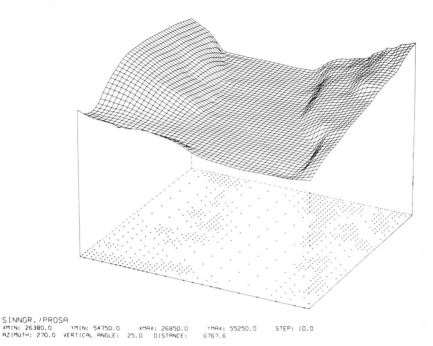

SINNGR./PROSA
XMIN: 26380.0   YMIN: 54750.0   XMAX: 26850.0   YMAX: 55250.0   STEP: 10.0
AZIMUTH: 270.0 VERTICAL ANGLE: 25.0   DISTANCE: 6767.6

Figure 5.19 Data acquisition using PROSA program with interpolated DTM. Shown in a perspective view using HIFI (Ebner and Reinhardt, 1984).

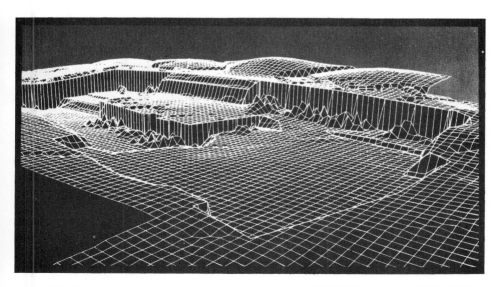

Figure 5.20 Perspective view of open-cast mine. Produced by PERSPECT module of SCOP DTM program package.

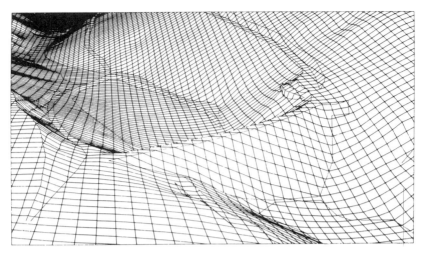

Figure 5.21 Perspective view of terrain generated by SCOP DTM package operating within the Siemens SICAD system.

the variable size of the grid in different parts of the model. It is noticeable that, in the published examples of perspective views of terrain surfaces produced by HIFI, and based on progressively sampled data from analytical plotters, e.g., Fig. 5.19, the fishnet pattern imposed on the surface is completely regular and has a high-density grid. This suggests that the height values of the non-measured points have been derived by interpolation. The well-known SCOP program has also been modified to accommodate progressively sampled data, though here the variable density of the gridded data is still preserved in the perspective plots (Figs. 5.20 and 21).

5.4.3 Selective sampling

In this alternative approach, heights are measured selectively at significant points only, i.e., on top of hills, in hollows, on saddles and along breaks of slope, ridge lines and streams. Thus all the points to be measured are identified by the photogrammetrist on the basis of his inspection and interpretation of the terrain features seen in the stereomodel. Therefore all the measured points will be randomly located, i.e., an irregular network of points, strings and polygons will result from the adoption of this procedure. The method is very well suited to implementation on analogue stereo-plotting instruments, especially if the instrument has a freehand movement to allow rapid scanning of the stereomodel without loss of coordinate count so that the operator can make the required decisions about the location and density of the measured points.

Since the whole operation is non-automated and depends entirely on the interpretative skills of the photogrammetrist, it makes no use of the built-in-computer or feedback systems of an analytical plotter. Thus it is difficult to see any major impact or improvement resulting from the adoption of analytically-based photogrammetric instrumentation in place of its analogue equivalent other than the fact that the result may be slightly better in terms of heighting accuracy.

5.4.4 Composite sampling

Since this strategy combines both grid and progressive sampling, supplemented by spot height measurements made at significant points and along breaklines, essentially it is a combination of systematic, progressive and selective sampling. Normally the selective sampling is carried out first, the heights of significant high and low points, and the elevations along breaklines and streams and around lakes and dense forests being measured by the operator as an initial operation. This needs to be done in a very systematic manner, to ensure both that significant points are not missed and that the operation does not take an unreasonable amount of time to execute, otherwise it would be uneconomic to implement. This initial coverage is then followed by the grid-based sampling which obviously is greatly assisted by the use of an analytical plotter with object coordinates primary especially if progressive sampling is to be employed in the data-gathering process.

As noted in section 5.4.3, this does add to the computational load, but the advantages are such that both grid-based and triangular-based DTM packages have been modified to handle elevation data collected by composite sampling. In the case of HIFI, which is a grid-based DTM program, supplied with the Zeiss Oberkochen Planicomp series of analytical plotters, the inclusion of significant points, breaklines, etc., in the measured data set, are coped with by using a specific version of the program based on the use of smaller patches modelled by bilinear finite elements instead of the much larger patches modelled by bicubic elements which are used if purely gridded data is collected by systematic sampling (Ebner and Reinhardt, 1984). In the case of *triangular-based packages* such as CIP, which can be purchased and used in conjunction with Wild BC and S9-AP series of analytical plotters, there is little or no modification needed to accommodate composite sampled data. It is interesting to note that in the latest version of HIFI, called HIFI-88, the data from all the different sampling strategies can be combined. Thus there is a basic gridded structure of cells related to systematic and progressive sampling. But within each cell, the points lying along breaklines, structure lines, etc., are connected up using an interior triangular structure (Fig. 5.22).

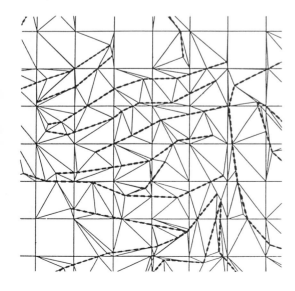

Figure 5.22 HIFI 88 network generation.

5.4.5 Measured contours

The use of measured contours for the data collection does give a systematic coverage of the whole stereomodel with strings of height data collected along lines of constant elevation, either at a constant time interval; at a predefined distance interval (i.e., with a constant vector); or alternatively at operator selected positions. These can of course be supplemented by individual spot heights measured along significant features such as terrain breaklines or streams or on hill tops, saddles and hollows.

It must be said that, from the point of view of accuracy of the subsequent DTM, the data derived from direct photogrammetric contouring with a dynamic method of measurement will be significantly lower than that achievable in other methods of photogrammetric heighting where the measurement of individual height points is carried out in a static mode with the instrument stationary. Obviously the use of an analytical plotter offers no significant advantage over an analogue instrument since there is no automated or pre-positioning component to the contouring procedure. Accuracy may, however, be better due to the intrinsic higher accuracy of the analytical plotter rather than any other factor.

5.5 Conclusion

As demonstrated the impact of the new types of analytical plotter on the collection of DTM data has been very considerable. Since photogrammetric methods are the only practical means of acquiring high-quality elevation data over large areas of terrain, the development of both operator-controlled and fully automated analytical stereophotogrammetric instrumentation is already extremely significant. Its importance can only increase as analytical instruments become more widespread and gradually displace the traditional type of analogue stereoplotter.

In the UK, quite a large number of commercial air survey firms, which are the main providers of high-quality DTM data over large terrain areas to the civil engineering industry, have already invested in analytical plotters. Thus Wild BC-2 instruments have been installed at BKS Surveys, Mason Land Surveys, Survey International, etc; Galileo Digicart analytical plotters are used by Engineering Surveys, Plowman, Craven Associates, AMC, etc; and a single Zeiss Oberkochen Planicomp instrument has been acquired by Clyde Surveys. Their various capabilities have already been discussed above but, taken both individually and collectively, they offer a significant additional capability to the acquisition of photogrammetrically-based DTM data in the UK, which hopefully will be exploited by engineering consultants and contractors.

It should be noted that recently the main UK government mapping agencies have also acquired analytical photogrammetric instrumentation, e.g. the Ordnance Survey (OS) has purchased Kern DSR instruments and the Mapping and Charting Establishment (MCE) of the Ministry of Defence has installed a number of Intergraph InterMap Analytic instruments. However these are less likely to have an impact on DTM data acquisition since, at the present time, these agencies (principally MCE) are generating lower-resolution, lower-accuracy DTM data for the UK from the contours present on existing small-scale topographic maps. The equipment and methods which are used to acquire DTM elevation data from existing maps will be discussed in chapter 7.

References and further reading

Beerenwinkel, R., Bonjour, J.-D., Hersch, R.D. and Kölbl, O. (1986) Real time stereo image injection for photogrammetric plotting. *International Archives of Photogrammetry*, **25**(4) 99–109.

Benard, M., Boutaleb, A.K., Kölbl, O., Penis, C. (1986) Automatic photogrammetry: implementation and comparison of classical correlation methods and dynamic programming based techniques. *International Archives of Photogrammetry*, **26**(3/3), 131–140.

Bertram, S. (1963) The automatic map compilation system. *Photogrammetric Engineering*, **29**(4), 675.

Bertram, S. (1965) The universal automatic map compilation equipment. *Photogrammetric Engineering*, **31**(2), 244–260.

Bethel, J. (1986) The DSR 11 image correlator. *Technical Papers, ACSM-ASPRS Annual Convention*, **4**, 44–49.

Bethel, J., Crawley, B., Shepphird, G. and Hussin, M. (1978) The automated generation and processing of digital terrain data for engineering planning. *Proceedings of the Digital Terrain Models (DTM) Symposium*, 469–490.

Blais, J.A.R., Chapman, M.A., and Kok, A.L. (1985) Digital terrain modelling and applications. *Technical Papers, ACSM-ASP Fall Convention*, 646–651.

Carson, W.W. (1987) Development of an inexpensive analytical plotter. *Photogrammetric Record*, **12**(69), 303–306.

Cogan, L. and Hunter D. (1984) Kern DSR 1/DSR 11 DTM collection and the Kern correlator. *Presented Paper, XVth ISPRS Congress, Rio de Janeiro*, 11 pp.

Ebner, H. and Reinhardt, W. (1984) Progressive sampling and DEM interpolation by finite elements. *Bildmessung und Luftbildwesen*, **52**(3), 177–182.

Edwards, D.L. (1983) Terrain analysis generation through computer-assisted photo-interpretation. *Technical Papers, 49th Annual Meeting, ASPRS*, 152–159.

Fritz, L.W. (1978) The NOSAP—a unique analytical stereoplotter. *Presented Paper, ISP Commission II Symposium*.

Goudswaard, F. (1988) A new concept for quantitative interpretation of stereo photographs by non-photogrammetric specialists. *Presented Paper, XVIth ISPRS Congress, Kyoto*, 7 pp.

Greve, C.W. (1980) The analytical photogrammetric processing system—IV (APPS-IV) *Proceedings A.S.P. Analytical Plotter Symposium and Workshop*, 79–85.

Helava, U.V. (1957/8) New principle for photogrammetric plotters. *Photogrammetria*, **14**(2), 89–96.

Helava, U.V. (1959/60) Mathematical methods in the design of photogrammetric plotters. *Photogrammetria* **16**(2), 41–56.

Helava, U.V. and Chapelle, W.E. (1972) Epipolar scan correlation. *Bendix Technical Journal*, **5**(1), 19–23.

Helava, U.V. (1980) The concepts of the analytical plotter. *Proceedings ASP Analytical Plotter Symposium and Workshop*, 12–29.

Huurneman, G. and Tempfli, K. (1984) Composite sampling using a minicomputer supporting analogue instrument. *ITC Journal*. 1984(2), 129–133. (Also published in *International Archives of Photogrammetry and Remote Sensing*, **25**(A2), 253–261).

Inghilleri, G. (1980) Theory of the DS analytical systems. *Proceedings ASP Analytical Plotter Symposium and Workshop*, 101–111.

Jaksic, Z. (1976) The significance of analytical instruments for the development of methods and techniques in photogrammetric data processing. *International Archives of Photogrammetry*, **21**(3), Commission II, Paper 04, 23 pp.

Konecny, G. (1980) How the analytical plotter works and differs from an analog plotter. *Proceedings ASP Analytical Plotter Symposium and Workshop*, 30–75.

Masry, S.E. (1972) The analytical plotter as a stereo-microdensitometer. *Presented Paper, Commission II, XII I.S.P. Congress, Ottawa*. 17 pp. Also published as 'Digital Correlation Principles' in *Photogrammetric Engineering* (1974), **40**(3), 303–308.

Nagel, R.W. and Camp, D.M. (1985) Superpositioning of digital elevation data with analog imagery for data editing. *Technical Papers, ACSM-ASPRS Fall Convention*, 596–602.

Newby, P.R.T. and Walker, A.S. (1986) The use of photogrammetry for direct digital data capture at Ordnance Survey. *International Archives of Photogrammetry and Remote Sensing*, **26**(4), 228–238.

Petrie, G. (1977) Commission II: instrumentation for data reduction. *Photogrammetric Record*, **9**(49), 7–13.

Reinhardt, W. (1984) A program for progressive sampling for the Zeiss Planicomp. *Proceedings of the 39th Photogrammetric Week*, Special Publication No. 9, Institute for Photogrammetry, University of Stuttgart, 83–90.

Reinhardt, W. (1986) Optical superimposition of stereomodel and graphic information as a tool for DEM quality control. *International Archives of Photogrammetry*, **26**(4), 207–215.

Roberts, T. and Cogan, L. (1978) Practical results with the PG2-H Servo Drive Unit. *Presented Paper, 44th Annual Meeting, ASP.*

Steidler, F., Dupont, C., Funke, G., Nuattoux, C. and Wyatt, A. (1986) Digital terrain models and their applications in a database system. *International Archives of Photogrammetry*, **26**(4)

Slama, C.C. (1980) Special application of analytical stereoplotters. *Proceedings, ASP Analytical Plotters Symposium and Workshop*, 355–361.

Tempfli, H. (1986) Composite progressive sampling—a program package for computer supported collection of DTM data. *Presented Paper, ACSM*–ASP Convention, Washington, 9 pp.

6 Creation of digital terrain models using analytical photogrammetry and their use in civil engineering

HELEN WEBB

6.1 Introduction

A digital terrain model (DTM) can be described as a three-dimensional represent-ation of a terrain surface consisting of X, Y, Z coordinates stored in digital form. However the idea of surface modelling can in fact be applied to other disciplines using two or three variables, as for example in census data analysis or other statistical applications. The digital terrain models considered here are representations of both existing and proposed (design) ground surfaces. With the increasing use of computers in engineering and the development of fast three-dimensional computer graphics, the DTM is becoming a powerful tool for both the design and the construction phases of civil engineering projects.

The raw data of the DTM, in this case, sets of three-dimensional ground coordinates, can be captured by a number of different methods, depending on the site itself and the resource availability (Fig. 6.1).

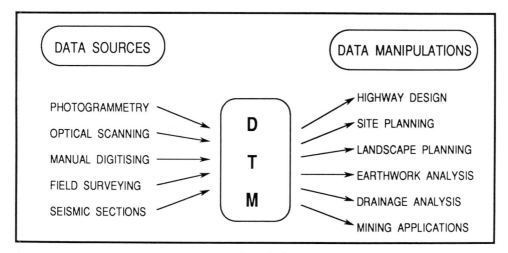

Figure 6.1 Digital terrain model data sources and manipulations.

From this, it can be seen that analytical photogrammetry is just one of the data capture techniques available for the creation of DTMs. Furthermore, the photogrammetrically derived data can be combined later with field-surveyed control points and profiles, and also with contour data which has been manually or automatically digitized or optically scanned from existing maps, to represent an existing terrain surface.

6.2 Analytical photogrammetry as a data capture technique

The capture and recording of three-dimensional ground coordinates in digital form by photogrammetry involves the use of one or other of three different instrument systems:

(i) a *monocomparator* or *stereocomparator* with digital output of image (x, y) coordinates and the subsequent computation of the corresponding model and ground $(X, Y, Z$ and $E, N, H)$ coordinates;

(ii) an *analogue stereoplotter* fitted with encoders and digital storage capacity which generate and record model (X, Y, Z) coordinates which can be transformed easily into the corresponding terrain (E, N, H) coordinates; and

(iii) an *analytical stereoplotter* such as the Intergraph InterMap Analytic in which the terrain (E, N, H) coordinates can be the primary input both to the analytical photogrammetric solution and to the DTM package itself.

In practice, only the second and third approaches are used to capture digital coordinate data for digital terrain models. While the analogue type of stereoplotter has been used extensively for DTM data collection in the past, analytical plotters are being used increasingly for the purpose. This results from the availability of a high-speed computer carrying out computational processes in real time which is an integral component of the analytical plotter. This allows the implementation of many procedures such as back-driving to a grid of predetermined positions, progressive sampling, etc., which are very difficult or impossible to execute in an analogue plotter.

The InterMap Analytic instrument (Fig. 6.2) is an analytical plotter which is based on the use of object coordinates (X, Y, Z) as the primary input to the analytical photogrammetric solution. These are provided by an ergonomically designed two-handed cursor (Fig. 6.3) operating over a digitizing tablet to generate the X, Y

Figure 6.2 InterMap Analytic Instrument— overall view.

(a)

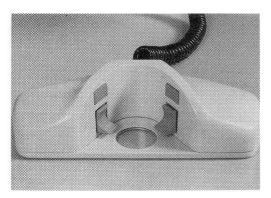

(b)

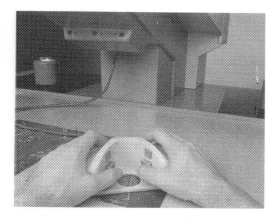

(c)

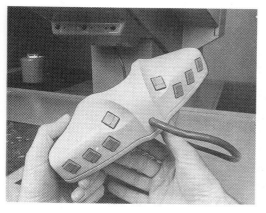

Figure 6.3 (a), (b) Two-handed cursor, (c) Special function buttons on the back of the cursor.

coordinates, together with a thumb control to provide the Z coordinate movement. Alternatively, X, Y handwheels and a Z footdisk can be provided with the instrument for measurement purposes. The measured X, Y, Z coordinates are transferred directly to the main processor where the corresponding image coordinates (x_P, y_P) are computed for each plate in real time using the well-known collinearity equations of analytical photogrammetry:

$$x_P = -f\frac{M_{11}(X_A - X_L) + M_{12}(Y_A - Y_L) + M_{13}(Z_A - Z_L)}{M_{31}(X_A - X_L) + M_{32}(Y_A - Y_L) + M_{33}(Z_A - Z_L)}$$

$$y_P = -f\frac{M_{21}(X_A - X_L) + M_{22}(Y_A - Y_L) + M_{23}(Z_A - Z_L)}{M_{31}(X_A - X_L) + M_{32}(Y_A - Y_L) + M_{33}(Z_A - Z_L)}$$

where x_P, y_P are photo-coordinates of the image point

X_A, Y_A, Z_A are ground coordinates of the corresponding point on the terrain surface

X_L, Y_L, Z_L are coordinates of the camera in the ground system

f is the camera principal distance

M_{IJ} are the elements of rotation matrix

The processor uses these computed x_P, y_P image coordinates to control the

Figure 6.4 InterMap Analytic instrument showing the operator's position with binocular viewing, digitizing tablet and cursor.

Figure 6.5 Stereoviewer of InterMap Analytic instrument seen from the rear.

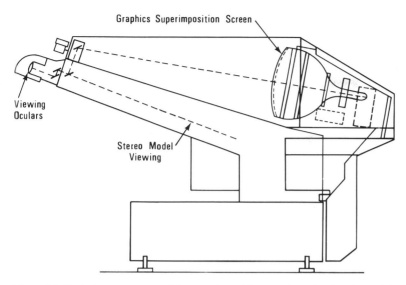

Graphics Superimposition Screen

Viewing
Oculars

Stereo Model
Viewing

Figure 6.6 Optical arrangement of the superimposition monitor, showing image injection.

positioning of the photographs so that the operator is presented with a continuously oriented, parallax-free stereomodel when viewing through the instrument's oculars.

The InterMap Analytic instrument (Fig. 6.4) includes the stereoviewer, the analytical processor and the superimposition monitor. The *stereoviewer* (Fig. 6.5), developed and built to Intergraph's specifications by Carl Zeiss, West Germany, includes the optics, main chassis, and eight relative and six absolute positioning systems. These positioning systems control the stage movement and illumination, rotation and optical zoom.

The InterMap Analytic's built-in electronics provide capabilities for analytical stereoplotting and computer graphics integration. The graphics image from the 15-inch colour monitor is 'injected' into the stereoplotter to create colour superimposition on the stereomodel (Fig. 6.6). The InterMap Analytic also supports detailed photogrammetric work by maintaining image quality at all levels of zoom magnification.

6.2.1 Processing capability

The InterMap Analytic can be connected to a host processor such as one of the Intergraph 200/250 series machines or a DEC VAX or MicroVAX computer. However, the instrument also has its own internal processing power which allows it to carry out most calculations locally. Thus two powerful processors allow the InterMap Analytic to conserve host processor resources, making the Analytic a highly efficient network node.

The *math processor*, which is a National Semiconductor 32032, relieves the host processor of the real-time functions of the stereoplotter, performing the calculations for software-controlled tasks such as calibration, orientation and continuous zoom. This 10-MHZ processor performs approximately one million instructions per second (MIPS) and includes a 64-bit floating point unit to speed up arithmetic calculations and increase computational accuracy.

The *control processor*, based on the Intel 80186, supports input/output processing

Figure 6.7 External graphics monitor.

and most system diagnostics. This 8-MHZ processor controls internal communications, provides the interface to the servo electronics and monitors the status of all buttons and switches. The control processor includes two direct memory access (DMA) channel controllers and performs 0.5 million instructions per second.

The InterMap Analytic also includes other processors for the control and support of graphics displays, supporting local operations which include dynamic pan, continuous zoom and the real-time rotation of three-dimensional elements. An additional *communications processor* offloads the host for local and remote communications.

6.2.2 Operator interface
The InterMap Analytic's ergonomically designed operator interface features colour superimposition, on-screen menus and an integrated software environment. Colour superimposition provides the capability for simultaneous through-the-optics viewing of stereomodels, superimposed line work and on-screen menus which are generated on the special *superimposition monitor* (Fig. 6.6).

The *external graphics monitor* (Fig. 6.7) is a 19-inch colour workstation displaying simultaneously the same image as the superimposition monitor. This external workstation can also be used for reviewing, editing and manipulating digital models outside the superimposition environment. It features a single-screen colour display, an alphanumeric keyboard, three internal processors and more than 1.25 megabytes of memory.

To record a particular set of ground coordinates corresponding to a chosen image point, a data record button must be pressed. The information is then stored in digital form, as ground coordinates, in a simple ASCII data file and can be used in the development of a DTM or for some other data manipulation such as on-line or off-line plotting.

6.3 Photogrammetry in digital terrain modelling

Since the DTM process used by Intergraph does not attempt to fit a mathematical equation to the measured data in order to produce the terrain model, the surface is represented by individual measured elevation points. The terrain coordinates of these points can be collected by any one of three general methods, according to the algorithm used. These can be:

(i) points located on a *regularly spaced grid*;
(ii) *non-regular grid or random points* (which are however located in specific locations by the operator); and
(iii) *three-dimensional line strings* which may be measured along breaks of slope, rivers, road edges, etc.

The Intergraph DTM system will accept all three types of coordinate data, either from an ASCII coordinate listing or a three-dimensional graphics file or a tagged two-dimensional graphics file. As with many other packages, the point coordinate data is first translated into a triangular data file, i.e., a triangulated irregular network (TIN) is generated. From this, an interpolated grid surface may be derived. Data input to the Intergraph DTM modelling process can be from a number of different sources, including third-party data collection methods. However, photogrammetry is probably the most suitable method for intensive data collection over large areas of terrain.

After the initial acquisition and processing of the photographs, the coordinate measuring and recording process in photogrammetry is much quicker, and can be more intensive, than in traditional field survey. This is an advantage, particularly at the planning and design stage of a civil engineering project. For example, a number of road alignments can be designed and analysed on the same terrain model before a final planning decision is made. Photogrammetry also requires less time on site, a major consideration for overseas projects when the design process is carried out off-site.

It is important to note that the surface described by photogrammetrically measured point coordinates is the existing ground surface, as opposed to a design or extrapolated surface.

6.4 Collection of coordinates for a DTM with the InterMap Analytic

The Intergraph InterMap Analytic (IMA) DTM data collection software is part of the IMA Applications package and is purely a data capture process recording regular, random and line string coordinates from the stereomodel produced in the plotter. However, it has been specifically written to interface directly with the other DTM packages produced by Intergraph. For this reason, the point coordinate data is stored in a standard Intergraph (IGDS) three-dimensional graphics file, with each different data type being stored on a different level within the file. One file is created for each stereomodel but may be referenced by adjacent models to ensure complete coverage of an area. After data capture is complete, this graphics file can be used directly by the Intergraph DTM software for further manipulation, editing and modelling work. In addition, an ASCII file of three-dimensional coordinates may be generated for use in orthophoto production.

Four different feature types are used by the InterMap Analytic;

(i) *ridge/stream breaklines*, which are recorded as 3D space line strings;
(ii) *spot elevations*, which are treated as being individual random points;
(iii) *obscure areas/lake lines*, which are also recorded as 3D space line strings; and
(iv) *profile points*, which are regarded as being either a grid or a random feature.

(i) Ridges and stream breaklines. These features are stored in the IGDS graphics file as three-dimensional line strings defining specific breaks in the terrain (Fig. 6.8a). Vertical faults can also be defined where different Z values are defined on either side of a linear feature. Although the fault does not have to be truly vertical, this routine will allow an X, Y location to possess two different Z values.

(ii) Spot elevations. These comprise a single point coordinate set, the location of which has been decided completely by the operator, adding to the non-random pattern of coordinate data.

(iii) Obscure areas. An example of this is a polygon defined around an area such as a hill-shadow or a wooded area, where the ground elevation either cannot be measured or is not easily defined. The area within the polygon can be left as an area requiring additional information to be obtained, for example from other sources such as field survey. This feature is particularly useful for defining lake surfaces, where the obscure area is drawn around the shoreline, or as an area of null information shown as a hole in the interpolated grid, or as a flattened surface (Fig. 6.8b).

(iv) Profiling. This is the main form of DTM data capture in the InterMap Analytic (IMA). The three-dimensional coordinates of individual points are collected in a semi-automatic mode, with the plotter for the most part selecting the appropriate X, Y location of the points and the operator setting the Z value and making the final decisions. The stereoplotter will either drive to each individual point or it will allow the operator to move along a fixed direction line or profile, either selecting points as required, or operating in a stream digitizing mode measuring and recording points during the whole time the data button is depressed.

There are a number of different ways in which the plotter selects the points or profile lines to be observed.

(i) *Pre-selected locations.* A formatted ASCII file of three-dimensional coordinates can be used to drive the plotter to pre-selected locations in the stereomodel. This may be applied either to a grid or to a random pattern of points, but it is particularly useful for monitoring or for time-sequence projects where the same X, Y points may have to be visited for each epoch.

(ii) *Single profile.* The operator can digitize a line-string to indicate a line of interest

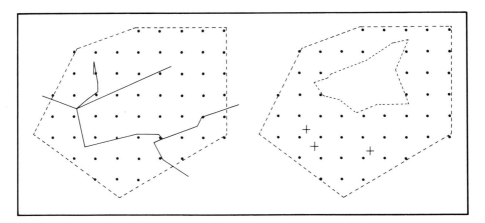

Figure 6.8 (a) Ridge/stream breaklines. (b) Spot elevations and obscure areas.

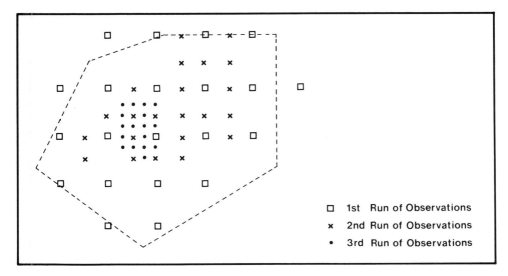

Figure 6.9 Progressive sampling inside a boundary area.

requiring profile points, e.g., a road alignment. The plotter will then drive along the track selecting the vertex points to be observed. This profile line may be copied to produce a line parallel to the original which can also be observed.

(iii) *Internal area.* A number of collection parameters can be set which allow the plotter to drive automatically to points within a prescribed area (the boundary). This method of data collection will produce a near-regular grid of points and is the most generally used method of photogrammetric DTM data collection with the InterMap Analytic. These user defined parameters define the distance between points along a profile line, between profile lines and the angle or azimuth of the profiles. Individual locations may be 'skipped' or moved slightly to take account of inappropriate terrain. These parameters may also be changed part way through the measuring/data collection process.

(iv) *Progressive sampling.* A progressive sampling routine has recently been incorporated into the basic DTM data collection package for the InterMap Analytic. According to the parameters set by the operator, the system will automatically densify the grid sampling in a local area (Fig. 6.9) to match the roughness of the terrain. The main parameter used is the difference in Z between an observed point and its nearest neighbours. The user can decide whether to use one, two or three sampling runs within a particular boundary area (see Fig. 6.9).

6.5 Creation of the DTM surface—Intergraph DTM nucleus

The different features used for the photogrammetrically derived data points are interpreted differently within the main DTM generation process. The X, Y, Z coordinates are first converted into a point data file which may be viewed interactively at the workstation. Also at this point, a number of design files from adjacent stereomodels can be amalgamated and/or a particular area extracted from the main file. (Since the original data is stored in a graphics file, this may also be viewed and manipulated interactively.)

The point profile is then processed to produce a triangulated data file, or this can be generated directly from the IMA file. Again this data can be displayed and edited and contour line strings generated from it. Recently incorporated in the IMA is the ability to generate a triangular network or to interpolate contours directly from this network according to a defined contour interval. These can then be displayed graphically in the viewing optics directly over the stereomodel using the superimposition facility of the instrument. This can be used to check the density of point information quickly before the data is passed into the DTM program.

Finally, an interpolated grid surface can be generated from the triangular data, and the DTM is complete. The model is now defined as a gridded surface and can be used by the basic DTM package to carry out a number of computational procedures. In addition a number of programs may be run to manipulate the DTM and how it is displayed. Some of these computational and display procedures are as follows:

(i) the calculation of the earth volumes contained between existing model surface and another surface, such as a datum plane or a design surface;

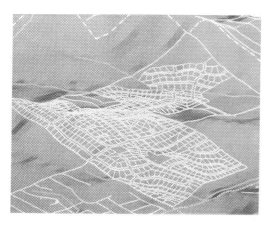

Figure 6.10 Shaded relief with superimposed road and plot layout shown on a perspective view.

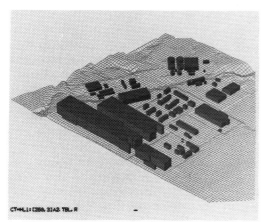

Figure 6.11 Perspective view of buildings overlaid on gridded DTM data.

(ii) the creation and display of an arbitrary cross-section through this surface;

(iii) the generation of contours at user-specified intervals;

(iv) the creation of a shaded and coloured display using hidden line removal techniques to give an accurate visual impression (Fig. 6.10);

(v) the overlaying of interpolated contours and other three-dimensional map feature data such as buildings, field lines, etc. (Fig. 6.11); and

(vi) the whole model may be rotated and viewed in any one of eight different directions from different viewpoints and using different scales.

6.6 Digital Terrain Models in civil engineering

Digital Terrain Models are useful tools in both the design and the execution or construction of an engineering project. The surface model derived from photogrammetric data can be used in both phases. Examples of possible manipulations include:

(i) horizontal, vertical or 3D roadway alignment design;

(ii) the creation of design surfaces overlaying the original;

(iii) the calculation of optimum earthwork cut-and-fill volumes;

(iv) storm drainage and watershed analysis;

(v) dam and reservoir siting (including monitoring once built); and

(vi) landscape planning and visual impact analysis.

These various operations are supported by a number of Intergraph application software packages which are designed specifically for use by civil engineers. The capabilities of some of these are outlined below.

6.6.1 Intergraph DTM-W (earthworks)

The Intergraph DTM-W package is designed specifically for use by the civil engineer in preparing site plans and corridor designs. Four different types of surfaces are used: the original surface as input; and the design surface, the cut-and-fill surface and the final surface as output. From these created surfaces, it is possible to generate two- or three-dimensional contour maps, three-dimensional grid maps and cross-sections of the site displayed at different angles of rotation. The design surface may be created and edited interactively for the rapid calculation of cut-and-fill volumes.

6.6.2 Intergraph DTM-G (graphics package)

Where large areas have been modelled using a digital grid file, the use of DTM-G will speed up the processing time required for a number of interactive editing and display options, such as contour line interpolation, shaded relief, colour-coded elevations, profiles, stereo (anaglyphic) display and the flashing or blinking of highlighted features. DTM-G uses vector data stored in the standard IGDS graphics file and converts this to raster grid data which is faster to display.

6.6.3 Engineering site package

Intergraph's civil engineering product line now includes a comprehensive system for single-facility site design, the *Engineering Site Package* (ESP). ESP supports complete three-dimensional, multidisciplinary site modelling. The design process supported by the ESP environment allows the site designer to readily incorporate three-dimensional graphic information from architectural, structural, plant design and

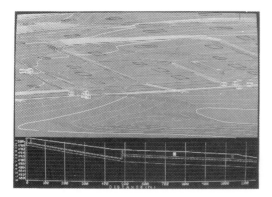

Figure 6.12 Perspective view and profile generated by the storm drainage design package.

other facility designs. The site engineer or designer may choose from ESP's complete offering of design and modelling functions, including:

- survey plotting;
- coordinate computation and traverse adjustments;
- property line and easement layout;
- geometric design;
- terrain modelling and display;
- contouring and sectioning;
- slope alignment and flow vector display;
- surface modelling;
- volume calculations;
- balancing cut-and-fill;
- storm drainage design.

6.6.4 Storm drainage design

Intergraph's *storm drainage design* (SDD) package (Fig. 6.12) enables civil engineers to automate the design and revision of a storm drainage system. The design can be sized automatically, or the engineer may override the design as required by special situations. Drafting is handled automatically as the storm drainage system is laid out. A full range of reports can be generated from the storm drainage design model. The storm drainage design package can be used independently or with other Intergraph software products (such as ESP, DTM and ICS) to build a customized design system which meets the specific needs of civil engineers.

6.7 Conclusion

The Intergraph InterMap Analytic stereo-workstation is an advanced interactive photogrammetric device which is well suited to the collection of accurate DTM elevation data. It is supported by a wide range of Intergraph DTM software available for data capture, processing and display and by a number of application packages produced specifically for use by civil engineers engaged in design work.

7 Terrain data acquisition and modelling from existing maps

G. PETRIE

7.1 Introduction

The acquisition of the basic terrain data required for terrain modelling using computer-based methods is often a major task which has to be undertaken before any progress can be made in the planning and design stage of a particular project. Unfortunately most countries are still very far away from having a *national topographic database* from which the terrain elevation data required for a particular project at a specific geographical location can be obtained in digital form. The reasons for this situation are not hard to find. The national mapping agencies in all countries have a vast archive of topographic maps which are the graphical records of between 50 and 200 years (depending on the specific country) of intensive and extensive surveying and mapping. The sheer size of such a graphical archive often goes unappreciated. Taking the Ordnance Survey (OS) as an example, the numbers of map sheets which are required to cover Great Britain at its basic mapping scales are as shown in Table 7.1.

Since a single map sheet may contain a very large amount of information in the form of lines, graphic symbols and area tints, the task of converting this into a digital form structured in a manner suitable for numerous users with differing requirements is a considerable one. When this conversion has to be carried out for thousands or tens of thousands of sheets to give national coverage, it becomes a formidable undertaking.

A further point which has to be made about this task is that the vast majority of the OS map sheets digitized to date are at the 1:1 250 and 1:2 500 scales. Since these comprise planimetric detail only, virtually no height or contour information is available from this source. Moreover, since the first OS map series which has *systematic contouring* is the 1:10 000 scale series, and the digitizing of this series has barely commenced, it is obvious that the acquisition or generation of suitable digital terrain data does indeed become a matter of very considerable importance to the civil

Table 7.1 Ordnance Survey basic mapping scales and sheet numbers

Map scale	Area of single sheet	No. of sheets	Coverage of Gt. Britain (%)
1:1 250	0.5 × 0.5 km	53 536	5 (urban areas)
1:2 500	1 × 1 km	164 555	60 (rural areas)
1:10 000	5 × 5 km	3 859	35 (highland areas)

engineer, landscape architect, planner or simulator system supplier who wishes to use such data. Either he must acquire the capability and carry out the task within his own organization, or he must employ the services of a specialist agency capable of carrying out the digitizing operations to the required specifications.

Since the acquisition of digital elevation data by field survey methods is only feasible and economic over relatively small areas of terrain, essentially there are two main methods of obtaining the required information about the terrain or physical landscape if a large areal coverage is required:

(a) by employing photogrammetric methods of measuring *stereomodels* of the terrain in a systematic fashion to give the required 3D data; or
(b) by digitizing existing *topographic maps*.

7.2 Digitizing of contour lines on existing topographic maps

This chapter is concerned with the technology and procedures used to acquire terrain elevation data using method (b). Since existing topographic maps contain very few spot heights or elevations, one is dealing with the *measurement of contour lines* so that they are represented by suitably structured strings of digital coordinate data, and with the subsequent processing or derivation of the spot height or elevation data from the digitized contour lines. It must be recognized from the outset that such a procedure will never produce the same metric accuracy as the direct measurement of spot heights carried out by field survey or photogrammetric methods.

In the first place, even if they have been measured photogrammetrically, the contour lines shown on a topographic map will be considerably less accurate than the measured spot heights produced by field survey or photogrammetric procedures. Typically the *accuracy* of such contours is only one-third of that of directly measured spot heights obtained from the same aerial photography. This is because the contour measurements in a stereoplotting machine are conducted in a dynamic mode, whereas the spot heights are measured in a static position, giving the operator time to achieve greater accuracy, if necessary, through repeated measurement at a single location. Furthermore, it follows that the individual spot heights derived by interpolation from such contours for the purpose of forming a digital terrain model will be still less accurate in terms of metric accuracy. Such considerations would apply for example to the *1:10 000 scale* OS topographic sheets which are the first OS series to give nation-wide coverage and provide photogrammetrically measured contours.

With smaller-scale topographic maps, e.g., the *1:63 360 scale* and first edition *1:50 000 scale* OS series, the contour data has been acquired through the field survey operations of the late 19th century, by which only certain specific contours were measured on the ground, the rest being interpolated. Thus, in mountainous areas in particular, their metric quality was poor. The generation of digital terrain model data in the form of spot heights or elevations from such contours has obvious limitations.

Turning to still smaller scales, the contours shown on such topographic sheets (or aeronautical charts), e.g., at 1:250 000 scale will almost certainly have still lower metric quality since they will have been subjected to the generalization inherent in such scale reduction. Thus, although the distribution and appearance of terrain elevation data derived from contours may be the same as that derived by direct measurement of spot heights, its quality in terms of metric accuracy is very much poorer. However, as long as users are aware of this limitation, this lower-quality data

Table 7.2

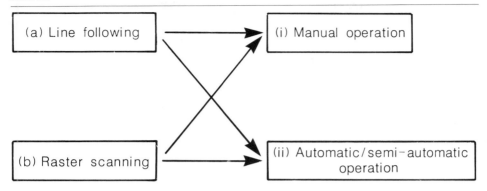

is quite acceptable—many users such as regional planners, landscape architects, geologists and geophysicists, aircraft simulator operators and civil engineers engaged in preliminary project or reconnaissance studies find it quite adequate for their purposes.

7.2.1 Contour digitizing methods

There are two alternative modes for the measurement and digitizing of contour lines. These are (a) line following and (b) raster scanning.

In principle at least, the measurement may be executed either (i) by manual means using a human operator; or (ii) by automatic or semi-automatic means using a suitably designed machine.

Combining these different modes leads to the following four possible solutions for the measurement of contour lines (Table 7.2).

The characteristics of these four possibilities are summarized in Table 7.3.

It will be seen from Table 7.3 that only three of the four possible methods of digitizing the contour lines on existing topographic maps can be implemented in practice. These are:

(1) manual line-following methods;
(2) automatic or semi-automatic line-following methods; and
(3) automatic raster-scan digitizing of complete map sheets.

7.2.1.1 *Manual line-following digitizing.* A very large number of manually-operated digitizers are available on the market, but these may conveniently be regarded as falling into one or other of two classes—mechanically-based digitizers or tablet digitizers.

Originally most manual digitizing was carried out using digitizers utilizing *mechanical slides* equipped with a measuring cursor and linear or rotary encoders to generate the rectangular (x, y) coordinate positions of the planimetric and contour data (Figs. 7.1, 7.2). The operator then carried out the measurement of the contour lines by following (tracing) each contour line in turn using a cursor equipped with a measuring mark. Either the operator decided the locations of the successive points required to best represent the contour lines—the so-called 'point mode' of digitizing—or the selection of the positions was executed automatically by the digitizer electronics operating in the so-called 'stream mode'. In this latter mode, the interval between the recording of successive points would be preset by the operator

Table 7.3 Measurement of contour lines

Mode	Measurement	Features	Characteristics of	
			(i) Manual operation	(ii) Automatic operation
(a) Line following	Points measured along contour lines only	1. Selective—only contour lines measured 2. Less data to be recorded and stored 3. Length of time required for measurement related to total length of contours	1. Low speed of measurement 2. Inexpensive hardware 3. Low-speed data recording 4. Relatively easy feature coding	1. Very high-speed measurement and recording 2. Expensive hardware 3. Operator intervention for coding etc.
(b) Raster scanning	The whole area of the map is scanned	1. Not selective—whole sheet scanned and measured 2. Very large amounts of data need to be recorded and stored. 3. Length of time required for measurement related to the size of sheet and the resolution of the scan line	1. Low speed 2. Enormous time required N.B. Not practicable to implement	1. Very high-speed measurement and recording 2. Very expensive hardware 3. Need for separate feature coding/ labelling operation 4. Very considerable post-measurement processing required

to work either on a time base (e.g., 0.1–10 s) or a distance base (e.g., 0.1–2.0 mm). The coordinate data would usually be recorded on paper or magnetic tape.

A special type of mechanically-based digitizer employing cross-slides which was widely used for digitizing contour data on existing maps in the 1970s was the *Pencil Follower* originally manufactured by d-Mac (later part of Ferranti-Cetec). In this type of device, the cross-slide and supporting rails were positioned below the surface on which the map was placed (Fig. 7.3). The measurement of contour lines was carried out with a cursor equipped with cross-hairs, around which a field coil was placed. This generated an electric field which was picked up by sensors (pick-up coils) mounted on a trolley which in turn was mounted and could move along the cross-slide. When the cursor was located directly above the trolley, the signals received by the opposite coils were equal since they were in balance. As the cursor was moved by the operator to follow the contour lines, the signals detected by the pick-up coils went out of balance.

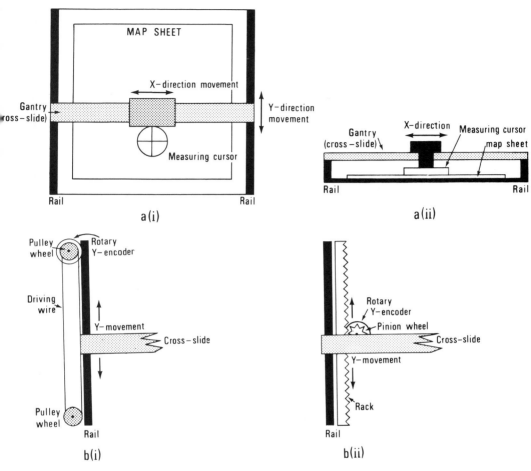

Figure 7.1 (a) Cross-slide digitizer: (i) plan view; (ii) section. (b) Linear-to-rotary converters: (i) using pulleys and wire; (ii) using rack and pinion.

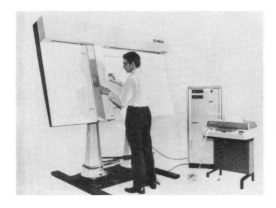

Figure 7.2 Cross-slide digitizer.

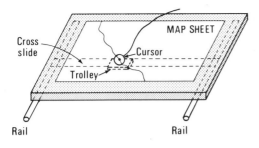

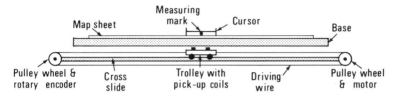

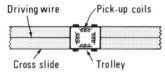

Figure 7.3 d-Mac Pencil Follower: (top) plan view; (centre) section; (bottom) plan view of trolley.

This signal then activated a motor which moved the trolley via a pulley and wire system with virtually no delay so that it was again stationed below the new position of the cursor. A rotary encoder mounted on the other pulley wheel generated the coordinate position which was passed to the output electronics. The great advantage of this arrangement was that the map surface remained completely clear of all devices (such as the cross-slide system) which might obstruct the operator's view of the contour sheet (Fig. 7.4).

Figure 7.4 d-Mac Pencil Follower.

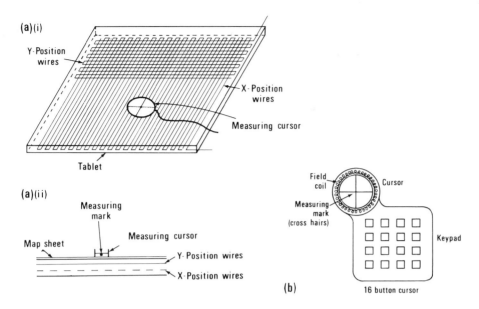

Figure 7.5 (a) Tablet digitizer: (i) plan view; (ii) section. (b) Cursor and keypad.

Relatively few such mechanically-based devices are still in use, and nowadays the process of manually digitizing contour lines for the purpose of acquiring terrain model data is almost always carried out using a solid-state *digitizing tablet* in which two sets of position-locating wires (one set in the *x*-direction, the other in the *y*-direction) forming a measuring grid, are embedded in a matrix of fibreglass, epoxy resin or plastic material (Fig. 7.5). In some recent examples, the grid is a large printed circuit board. The tracing/measuring of the map data is carried out using a cursor around which a field coil is located. This produces a signal which can be picked up by the sensing wires of the grid below. Originally most of these tablets were incrementally-based, and a loss of contact between the cursor and the map surface resulted in a loss of count in the coordinates. However, current designs are now absolutely encoded and virtually impervious to any loss of count.

The cursor may have an attached keypad (Fig. 7.5) with which contour line values or feature codes may be entered. Alternatively this may be done via a separate alphanumeric keyboard or by allocating part of the tablet surface to act as a 'menu' in which a series of individual boxes are set aside to provide feature coding or headers for specific classes of feature.

A feature of the new digitizing tablets is that most of them are 'smart': they have built-in microprocessors with PROMs which allow such functions as the time and distance modes of stream digitizing, line length and area measurement, and shift of origin to be implemented as integral features of the digitizer. Also, most have inexpensive LED or LCD displays of the measured coordinate values. Thus these tablets can be used as separate units off-line, the coordinate data being recorded on a digital cassette or cartridge drive (Fig. 7.6). This allows data to be collected without tying up a microcomputer or a substantial part of the capacity of a minicomputer. It is usually termed 'blind' digitizing, since the measured data is not displayed or plotted

Figure 7.6 Blind digitizing using GTCO digitizer.

Figure 7.7 Hewlett Packard interactive digitizing station.

Figure 7.8 LITES 2 Graphics digitizing/editing station.

out. This procedure does not permit the editing or correction of errors during measurement, which must be carried out later on a minicomputer or mainframe machine with interactive editing facilities.

The alternative procedure of providing an on-line display of the digitized contour lines and powerful interactive editing facilities is becoming increasingly popular with the advent of powerful graphics workstations equipped with a tablet digitizer as an integral part of the system. An early example is the system devised some years ago at the University of Glasgow based on Hewlett Packard equipment. This comprises an HP-9874 tablet digitizer; an HP-2647A intelligent graphics terminal programmable in Basic, and an HP-2631G plotter/printer unit (Fig. 7.7). A more recent approach is the Laser-Scan LITES-2 system (Fig. 7.8) such as that in use at the University of Glasgow which is based on the use of a DEC VAX station graphics workstation equipped with a high-resolution (960 × 840 pixel) monochrome VDU display and a large-format GTCO tablet digitizer. This has full windowing, manipulation and editing facilities.

It is a point for some debate whether it is economic to use such a powerful and still relatively expensive on-line system on a task such as the manual digitizing of contour lines. Thus there is considerable interest in the development of similar but lower-cost alternatives based on small, powerful *microcomputers* linked to tablet digitizers. An early example of such a system is that developed by Map Data Management (Fig. 7.9) based on an Apricot Xi or Xen or an IBM PC-AT microcomputer with a medium-resolution (800 × 400 pixel) screen linked to a GTCO digitizer and costing between one-third to one-quarter of the LITES system—albeit with lower functionality and capability. It is of course easy to network these microcomputer-based digitizing stations to a local file server and to transfer the edited, corrected terrain information to a mainframe or minicomputer on which the actual terrain modelling operations will be carried out.

The accuracy with which the contour lines required for terrain modelling may be measured manually is a matter of debate. The *resolution* (least count) of the commonly used tablet digitizers lies in the range $50-100\,\mu m$; the actual *accuracy* of the x/y coordinates produced by the device itself will be a little lower. To these figures must be added the errors made by the operator while measuring/tracing the contour lines and any distortion or change in shape present in the map document itself. The latter can easily amount to several millimetres across a large format sheet; much will depend on whether the map containing the contour lines is available on a stable plastic base or

Figure 7.9 Map   Data   microcomputer-based digitizing station.

on paper. To overcome some of these difficulties, measurement of the grid corners is usually undertaken, with a subsequent affine transformation to remove any regular change in scale between the x and y coordinate directions. Even with such precautions and processing, it is obvious that the accuracy (root mean square error) of the final positional coordinate values generated by manual digitizing of contour lines will be approximately ± 100–$250\,\mu m$ (0.1–0.25 mm), equivalent to 1–2.5 m at the scale 1:10 000 at which contours first become available on an OS series.

7.2.1.2 *Automatic and semi-automatic line-following digitizers.* As with manually controlled digitizers, it is convenient to classify automatic or semi-automatic line-following digitizers into two categories: mechanically-based and non-mechanically-based.

Mechanically-based automatic line-following digitizers were originally based on the use of some type of sensing device which illuminated and scanned a small area to

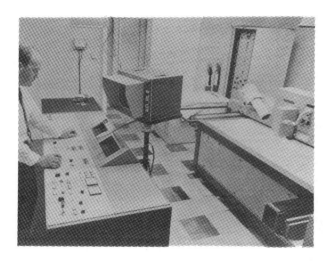

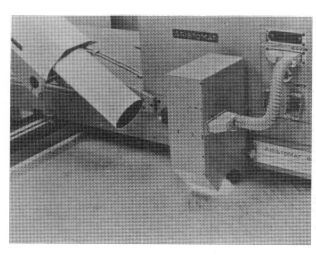

Figure 7.10 (Top) AEG/Aristomat Geameter mechanically-based automatic line-following digitizer with operator and video monitor in use at the Experimental Cartography Unit (ECU). (Bottom) AEG/Aristomat Geameter—close-up view of video camera (left) and scanning unit (right).

establish the presence and direction of the contour and so allowed the continuous tracking of a line. The sensing head was mounted on a mechanical cross-slide arrangement; high-accuracy drafting machines (flatbed plotters), e.g., from Calcomp, Gerber, AEG, etc., were often adapted to form the basis for such automatic line followers (Fig. 7.10). The tracking movement was implemented using stepping motors actuating either a lead screw or a rack-and-pinion arrangement attached to the slide defining an individual coordinate axis. Any movement of either of these slides would then be measured by linear or rotary encoders in much the same manner as in the equivalent manually-controlled digitizers. Because of the frequency of branching or crossing lines, some type of video display, e.g., using a closed circuit television (CCTV) camera to view the contour sheet, was a complete necessity to allow operator intervention—either to guide the system along the correct path in a difficult or confusing area or to accurately reposition it at a branch or junction. The expense and complication of such systems, e.g., the AEG/Aristo device used by one or two agencies in the UK and the similar Gerber 32 device used by the Royal Australian Survey Corps (RASVY) digital mapping unit in the 1970s and the need for an operator to be present throughout the digitizing process, if only for monitoring purposes, meant that such mechanically-based automatic line-following systems enjoyed little popularity among users. Well-executed manual digitizing was only a little slower and very much cheaper because of the lower capital cost of the equipment used.

This situation has however been transformed with the advent of the Fastrak and Lasertrak digitizing systems produced by Laser-Scan of Cambridge. The origin of these devices is the Sweepnik device originally developed at the Cavendish Laboratory, Cambridge, for the measurement of nuclear bubble-chamber photography (Davies *et al.*, 1970). This used a laser beam steered by a computer-controlled mirror to follow the tracks appearing on the bubble chamber film. The basic computer-controlled laser deflection technology was then used to produce the HRD-1 display/plotter (Woodsford, 1976; Bell and Woodsford, 1977). In the Fastrak device (Howman and Woodsford, 1978), the laser beam is steered to scan the contour line needing digitizing in a local raster scan (Fig. 7.11). The actual scan pattern varies in its vertical and horizontal components as the line changes direction (Fig. 7.12). The crossing of the contour line by the laser beam is detected on a reduced scale (A6 size) negative of the document being digitized at $10-15\,\mu$m intervals (Fig. 7.13). This produces x and y coordinates of points on the line at intervals of between 50 and $70\,\mu$m on the original map at speeds up to 500 points per second.

The system is highly interactive with operator intervention in the form of backtracking, redirection of the scan or fully manual digitizing if the automatic line following should falter or fail. Operator monitoring is achieved through the display of

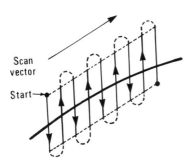

Scan
vector

Start

Figure 7.11 Laser-Scan   Fastrak—local raster scan within a single vector.

Figure 7.12 Laser-Scan Fastrak—series of successive vectors to follow a contour.

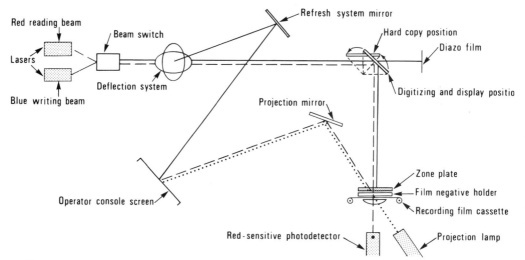

Figure 7.13 Laser-Scan Fastrak—diagram of overall system.

the lines being digitized on a large viewing screen 1 m × 70 cm in size. The operator also selects features, adds coding such as contour line values and carries out any required editing with the aid of a keyboard and a tracker ball positioning device (Fig. 7.14). The control computer is normally one of the more powerful models in the Digital (DEC) VAX-11/700 or Microvax 2 series, incorporating high-speed floating-point arithmetic and a disk-based operating system. Average rates of 500 line inches (12.5 m) per hour of fully coded and edited digitizing are achieved on contour and culture type maps. Graphics up to A1 (59.4 × 84.1 cm) in size can be handled by the Fastrak in a single pass. Graphical output for verification purposes, archival use and final plotting can be carried out using diazo-based microfilm. Digital output is normally in the form of magnetic type.

The digitizing of contour lines from existing topographic maps is particularly suited to the Fastrak and Lasertrak systems, especially if the original contour-only sheet is available for the purpose, since the crossing and branching of lines which gives so much difficulty to any form of automatic line-following device is largely eliminated. Difficulties may of course still arise in areas where the contours lie very close together or are replaced by cliff symbols, as may happen in steep terrain—in this case the

Figure 7.14 Laser-Scan Lasertrak showing display screen, control keyboard and tracker ball and graphics display terminal.

operator will have to take over the digitizing operation in this particular area of the sheet.

Large cartographic agencies such as the Ordnance Survey (OS), the Mapping and Charting Establishment (MCE) of the Ministry of Defence, and the United States Geological Survey (USGS) have the volume of graphic digitizing to justify the purchase of Fastrak or Lasertrak devices with the certainty that they will be used on a full-time production basis. For those engineers, landscape architects and planners who may have the need for the digitizing of a large amount of graphic data, occasionally or even just once, e.g., to form the topographic and locational base for their design activities, resort can be made to Laser-Scan's service bureau in Cambridge. For example, the Forestry Commission has carried out a woodland census of the United Kingdom on this basis, the data being captured by the Fastrak from the green colour separations of the Ordnance Survey's 1:50 000 scale series. There are 204 individual map sheets in the series, the throughput on the Fastrak being four to six sheets per week. Similar production rates may be expected with contour digitizing.

7.2.1.3 *Automatic raster-scan digitizers.* The raster-scan digitizers available for the digitizing of large graphic documents such as contour line sheets are fully automatic devices capable of producing and storing a file of raw coordinate data for all the lines and symbols contained in the sheet. Until recently, most of these devices were *drum scanners*, e.g., those manufactured by Optronics (used by Intergraph), Scitex (the

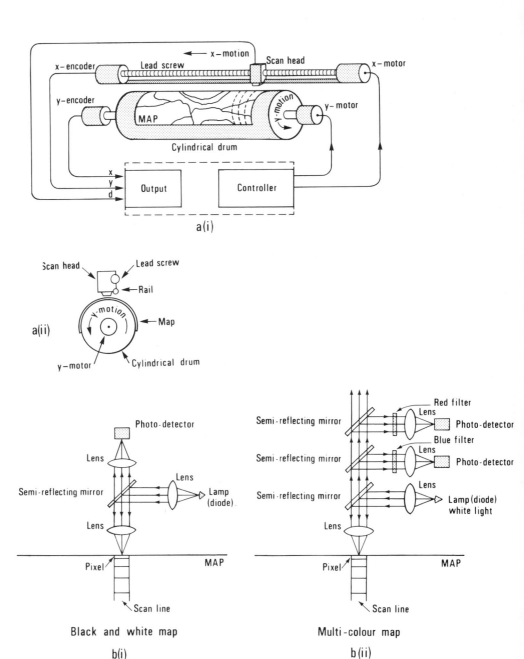

Figure 7.15 (a) Raster-scan drum digitizer: (i) perspective view; (ii) cross-section. (b) Raster-scan drum digitizer: (i) scan head for monochrome image; (ii) scan head for colour image.

Response series) and Tektronix (Model 4691) (Fig. 7.16). In each of these, the map or graphic document is wrapped round a drum which is rotated at a constant speed below a photodetector head which is moved continuously forward in steps along the axis of the drum (Fig. 7.15), the step size defining the width of the scan lines. With these devices, it is possible to digitize either black-and-white (monochrome) sheets—as in the case of colour-separated contour sheets—or full-colour maps. In the former case, a single photodetector will be used; in the latter, different filters are used to perform colour separation of the detail contained in the map, each colour separated channel being sensed for the presence or absence of line data by its own individual photo detector (Fig. 7.15).

The alternative type of raster-scan digitizer is the so-called *flatbed scanner*. Two examples of such digitizers are the SysScan KartoScan (West Germany) and the Broomall Scan Graphics System (USA). In both of these devices, the contour line sheet is laid flat on a base board and is then scanned by a cross-slide or gantry which rapidly traverses the sheet from top to bottom (Fig. 7.17). Alternatively, the gantry may be held fixed and the sheet passed below it as in the newer types of KartoScan device (Fig. 7.18). The elements which sense the presence or absence of lines on the sheet are an array of solid-state photo diodes in the case of the KartoScan device and a

Automatic Data Base Capture
4991S1 Process

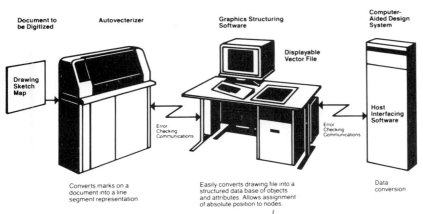

Document to be Digitized — Autovecterizer — Graphics Structuring Software — Computer-Aided Design System

Displayable Vector File

Drawing Sketch Map

Error Checking Communications

Error Checking Communications

Host Interfacing Software

Converts marks on a document into a line segment representation

Easily converts drawing file into a structured data base of objects and attributes. Allows assignment of absolute position to nodes.

Data conversion

Figure 7.16 Tektronix 4991S raster-scan digitizing system.

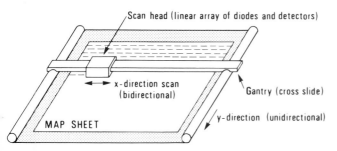

Scan head (linear array of diodes and detectors)

x-direction scan (bidirectional)

Gantry (cross slide)

y-direction (unidirectional)

MAP SHEET

Figure 7.17 Flatbed raster-scan digitizer.

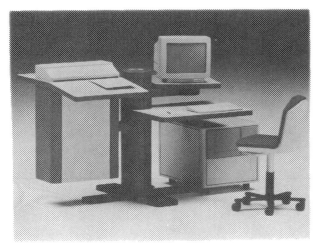

Figure 7.18 SysScan   KartoScan
CE raster-scan digitizer.

laser source travelling bi-directionally across the gantry in the case of the Scan Graphics equipment. In each case, the coordinate data is read out continuously line by line and stored on disc or tape. In the case of the Scan Graphics device, the maximum area which can be scanned is 24 × 50 inches (112 × 165 cm); with the largest KartoScan (Model FE), the maximum imaging area is 47 × 64 inches (1.2 × 1.6 m).

Recently, various alternative types of low-cost, small-format raster-scan digitizers have appeared on the market which utilise linear or areal arrays of charge-coupled devices (CCDs) to scan graphics documents. The arrangement of such a device utilizing an *areal array*, together with a lens to form a CCD camera, is shown in Fig. 7.19. Since the current state of this technology allows only small areal arrays to be manufactured, it is necessary to scan a large-format document such as a contour line plot in sections, i.e., patch by patch. This will result in an increased time in the subsequent data processing so that the individual scanned patches may be joined together to reconstitute the original contour sheet. Software must also be available to

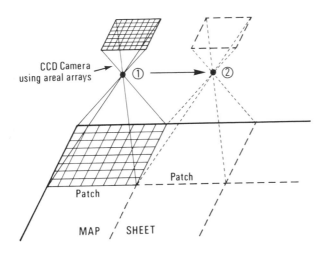

Figure 7.19 CCD   areal   array
scanner.

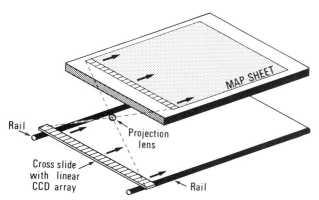

Figure 7.20 CCD linear array scanner.

reconcile the inevitable mismatches between the same lines occuring on adjacent patches.

Scanners utilizing *linear arrays* include the Topaz Picture Scanner (UK) and the Eikonix E-Z Scan Model 4434. In both cases, the map is scanned by the linear CCD array which is stepped across the image field of view behind a lens (Fig. 7.20), instead of the contour sheet being passed under the linear CCD array of detectors as in the smaller KartoScan CC and CE scanners.

The advantage of this type of raster-scan digitizing is the speed of the digitizing process, the time taken being a function of the size of the graphic document and quite independent of the length of the line to be digitized, which is the main factor in both the manual and semi-automatic line-following methods. Using a fairly fine resolution, a single large-format map may typically be scanned in one hour.

However, simple comparisons on this basis can be misleading, since line-following has the advantages of economy (in that only those lines and areas which exist are scanned and digitized) and of selectivity (since only those features required for a specific purpose need by measured). Furthermore the time and expense needed to edit and process the digitized raster data must be considered as well. With complex maps containing numerous graphic symbols, it will certainly be necessary to supplement the raster-scan digitizing to pick up all the individual point symbols or text. The display and editing of the digitized raster data and the considerable task of labelling and adding header codes and attributes and, in this specific case, the elevation values of each contour line, will also have to be carried out later on an interactive graphics workstation as an off-line process.

Consideration must also be given to the conversion of the digitized contour data from the scanned raster format to a line or vector format which must be implemented to ensure both some degree of data compaction and a more convenient format for many display or plotting devices. Software must be written or acquired to carry out the following two basic steps in this conversion operation:

(i) *line thinning*—the process of reducing the lines to unit thickness (i.e., to single pixel width) at a given resolution;

(ii) *line extraction*, involving the identification of a continuous series of coordinated points which constitute a contour line or line segment as portrayed in the original map.

The detailed procedures (including various alternative solutions) are described by Peuquet (1981) and by Baker *et al.* (1981).

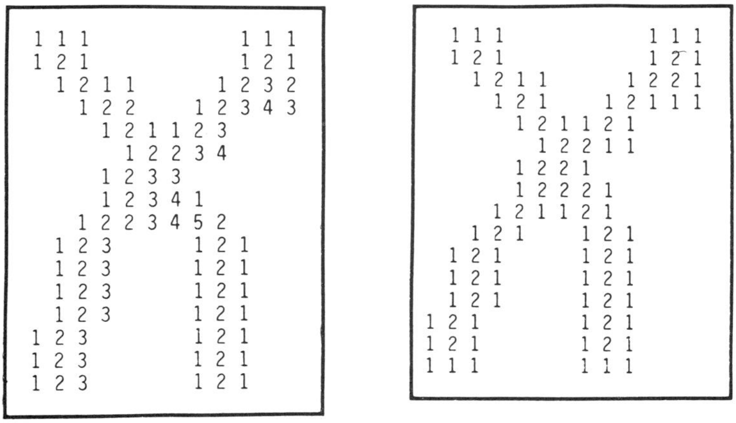

```
 1 1 1                    1 1 1
 1 2 1                    1 2 1
   1 2 1 1            1 2 3 2
     1 2 2          1 2 3 4 3
       1 2 1 1 2 3
         1 2 2 3 4
       1 2 3 3
       1 2 3 4 1
     1 2 2 3 4 5 2
   1 2 3            1 2 1
   1 2 3            1 2 1
   1 2 3            1 2 1
   1 2 3            1 2 1
 1 2 3              1 2 1
 1 2 3              1 2 1
 1 2 3              1 2 1
```

```
 1 1 1                    1 1 1
 1 2 1                    1 2 1
   1 2 1 1            1 2 2 1
     1 2 1          1 2 1 1 1
       1 2 1 1 2 1
         1 2 2 1 1
       1 2 2 1
       1 2 2 2 1
     1 2 1 1 2 1
   1 2 1            1 2 1
 1 2 1              1 2 1
 1 2 1              1 2 1
 1 2 1              1 2 1
 1 2 1              1 2 1
 1 2 1              1 2 1
 1 1 1              1 1 1
```

Figure 7.21 Line thinning operation.

One solution to the problem of line thinning is the so-called *medial axis approach*, which is based on the calculation of all those points (pixels) which lie at the maximum distance from all edges of the original contour line which may be several pixels thick (Fig. 7.21). In a first pass through the digitized raster data set, each pixel is inspected in turn in scan-line order (i.e., from left to right) starting from the upper left corner of the rasterized data. Each occupied pixel is assigned a value equal to the minimum value of its neighbours directly above and immediately to the left, plus one. In a second pass, the resulting matrix of pixel values is processed in reverse order, assigning to each pixel the minimum value among itself, its neighbour to the right plus one, and its neighbour below plus one (Peuquet, 1981). The resulting skeleton of maximum-value pixels represents the centre line of the contour at unit thickness (Fig. 7.21).

The second step, that of vectorization or line extraction, involves the scanning of the whole contour data set initially around the edges of the digitized area to pick up the starting point of each contour. Once such a contour line is found, it is then followed, typically by using a 3 × 3 pixel template to follow the line to its terminal position. Once this has been done, the data comprising the reconstructed contour line is eliminated from the data set. The starting point of the next contour line is then picked up and again followed to its terminus. This process is continued until all the digitized raster data has been picked up and assigned to a specific contour.

As discussed above, a vast amount of data processing is required both for this raster-to-vector conversion process and the supplementary manual digitizing required to edit and label the scanned contour line data. This tends to offset the very rapid initial raster-scanning of the basic map document, which preferably has to be available in the form of separation transparencies used for colour printing if high-quality results are to be obtained. Enthusiasts for raster-scanning are mainly to be found among the academic community (e.g., Boyle, 1980) and they tend to play down the difficulties and expense encountered with the method. However at the present time, given the enormous investment needed for the purchase and implementation of these fully automatic raster-scan systems, none but the largest mapping organizations with a dedicated long-term commitment to map digitizing on a huge scale are likely to contemplate their acquisition.

7.2.2 Commercial digitizing bureaux

From the point of view of the civil engineer, landscape architect or planner, manual digitizing is the only real option if the digitizing work is to be carried out in house. Comparative benchmarks show that costs compose favourably with the alternative capital-intensive, semi-automatic line-following and automatic raster-scan methods of digitizing. If however, time is very important, then the mass digitizing of the required cartographic material may be carried out using the service offered by one of the manufacturers such as Laser-Scan, Scitex and Broomall, which will scan the contour sheets, process the digitized data and generate a digital terrain model (DTM) from the digitized contours.

There has been a massive expansion of map digitizing bureau facilities in the UK as a result of the recent OS policy of putting part of its programme of digitizing existing maps out to private industry in order to cut down the massive backlog of digitizing its large-scale map series. This has encouraged several private-sector surveying and mapping firms to invest in cartographic digitizing facilities, in some cases on a considerable scale, as at Mason Land Surveys, Dunfermline, which has 30 interactive digitizing stations. Most of these firms use manual digitizing, still the most cost-effective method at the present time. While such facilities have mainly been used for planimetric map digitizing (the OS large-scale sheets are plans with no contours), they are of course equally suited to the digitizing of contours for the purposes of acquiring terrain modelling data.

7.3 Generation of gridded DTM data from digitized contours

Once the contours have been converted to strings of coordinate data—one for each contour—the actual terrain model data, which will be used to represent the terrain within the area of interest, must be generated. Different approaches may be taken to carry out this task depending on whether line-following or raster-scanning has been used for the capture of the contour line data.

7.3.1 Data processing of digitized contour data produced by line-following

An example of the procedure followed with line-based digitizing methods is that used in the PANACEA software suite devised and sold by Siren Systems (and by its licensee Laser-Scan) and used widely for this specific purpose. A fuller account of this program suite and its application is given in Chapter 9.

As Fig. 7.22 shows, the overall procedure starts with the set of digitized contours. From these, a triangulation is derived, i.e., a triangular DTM is formed, from which in turn a regular grid-based DTM can be interpolated. The various modules which carry

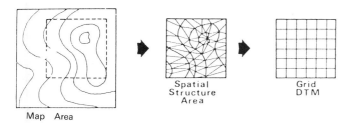

Spatial
Structure
Area

Grid
DTM

Map Area

Figure 7.22 Contours to triangulation to grid DTM.

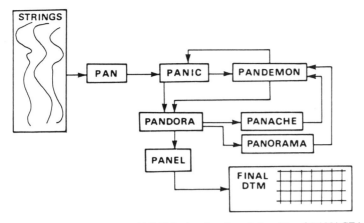

Figure 7.23 Contours to grid DTM, showing the relationship of PANACEA modules.

out these different operations and the relationships which exist between them are shown in Fig. 7.23.

- (i) The first module, PAN, is a *pre-processing program* to reduce the number of points collected during the line-following digitizing of the contours. This is less likely to be required when manual digitizing has been undertaken in point mode, when the operator will presumably have selected the most significant points along the contour in relation to the scale of digitizing. However, when semi-automatic line-following digitizing in stream mode has been employed, as with the Fastrak or Lasertrak devices, a reduction of the digitized data is necessary. This has been implemented using the Douglas–Peucker algorithm (Douglas and Peucker, 1973).
- (ii) The second module, PANIC, determines the Thiessen neighbours for each data point and then carries out the Delaunay *triangulation* over the whole data set.
- (iii) The third module, PANDEMON, is essentially a *graphics editor* which allows the editing of any erroneous data, the correction of triangles crossing break lines, the insertion of new data points, etc.
- (iv) The fourth module, PANDORA, then allows the *interpolation of the elevation* at each specified grid node with reference to the linear triangular facet in which it lies.
- (v) The fifth module, PANEL, allows the *amalgamation* of a number of contiguous grid models into a single overall grid-based DTM.
- (vi) The remaining modules are concerned with the *display* of the gridded DTM data, e.g., PANORAMA generates perspective views, and PANACHE generates contours from the gridded data.

7.3.2 Data processing of digitized contour data produced by raster scanning

The initial procedure necessary to convert the raster-scanned contour, ridge line and stream data into vector format has already been outlined. A sufficient number of the reconstituted contours must next be tagged with their numerical values so that all such contours may be identified and their height values assigned. On the other hand, there is no need to assign values to the ridge and break lines and streams, since they will normally intersect contour lines and will obtain their height values automa-

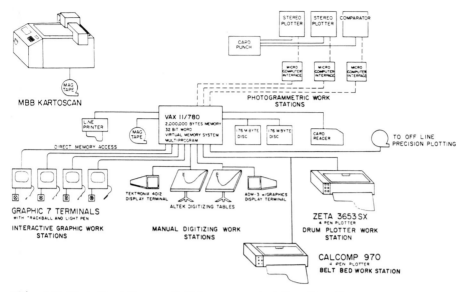

Figure 7.24 Mark Hurd Surveys digitizing systems including KartoScan raster-scanner, manual digitizing stations, photogrammetric digitizing stations and interactive editing stations.

tically, first at the intersection points and subsequently at all intermediate points by interpolation.

The overall procedure of generating DTM data from raster-scanned contour sheets has been described by Clark (1980) and by Leberl and Olsen (1982). The overall SysScan system used at Mark Hurd Surveys by Leberl and Olsen is shown in Fig. 7.24. From this it will be seen that the raster-scanning of contour sheets is carried out using the KartoScan flatbed scanner. Leberl describes a height gridding procedure which he terms a 'sequential steepest slope algorithm' (Fig. 7.25). In this

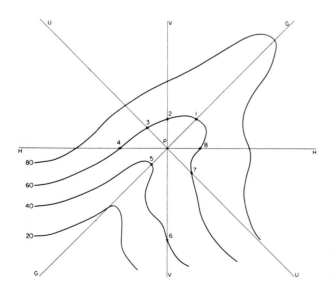

Figure 7.25 Sequential steepest slope algorithm showing cross-sections *HH*, *VV*, *UU* and *GG* and the intersection points 1–8 on the contours.

procedure, a search is made along each of four lines passing through the required grid node and oriented along the grid directions (VV and HH) and their bisectors (UU and GG). The intersection of each of the eight directions with the nearest contours is established, and the slope of each of the four lines calculated. The line with the steepest slope is then selected and the value of the elevation of the grid node established by linear interpolation along this line, e.g., in the example shown in Fig. 7.25, search line GG is the steepest and the height of the grid node P is derived from

$$H_P = \frac{H_1 - H_5}{(1,5)}(P,5) + H_5$$

7.4 DTM production in the United States

In the USA, government topographic mapping agencies—especially the National Mapping Division of the United States Geological Survey (USGS) and the Defense Mapping Agency (DMA)—are engaged in a very large programme of converting existing maps to digital form for use in digital mapping operations and geographical information systems. As an additional product, the USGS offers digital elevation model (DEM) data as part of its Mapping Program (El Assal and Caruso, 1983). Two distinctive DEM products are offered:

(i) *7.5 minute DEM data*, produced by the USGS itself, in which the data is derived from and corresponds to the standard 1:24 000 scale, 7.5 × 7.5 minute quadrangle topographic map sheets published by the Survey; and

(ii) *1:250 000 scale DEM data*, produced by the DMA and distributed by the USGS, which is derived from and corresponds in coverage to 1° × 1° blocks, which repesent one-half of a standard 1:250 000 scale, 1° × 2° quadrangle, topographic map sheet.

7.4.1 Data acquisition
In the case of the larger scale (1:24 000), the elevation data may be measured directly by photogrammetric methods using either the highly automated GPM-2 correlator-based machine or by manually-controlled profiling with a conventional stereoplotting machine equipped with encoders. Alternatively, it can be produced from digitally-encoded vector contour data measured either by a stereoplotting machine or by digitizing the contour plates of existing 1:24 000 scale maps. This vector contour data is then processed to produce, by bilinear interpolation, the DEM elevation matrix at a 30 m spacing, which corresponds to 1.25 mm at 1:24 000 scale. This data is available for only selected parts of the United States at present.

By contrast, the 1:250 000 scale DEM data is derived entirely by digitizing the contours, ridgelines and streams shown on existing 1:250 000 scale maps and deriving the elevations by interpolation at intervals of three arc-seconds, corresponding to 90 m in the north–south direction and approximately 60 m in the east-west direction at 50° N latitude. This data is available for the whole of the coterminous United States and Hawaii.

The measurement of the existing contours by the two agencies concerned has been carried out using all the different methods described above, i.e., by manual line-following (used by both agencies); the Laser-Scan Fastrak semi-automatic following device (used by USGS); and the Broomall and Intergraph/Optronics fully automatic raster-scanning systems (used by DMA).

The manual line-following digitizing approach, called Digital Graphics Recorder (DGR) in-house, has been employed by the DMA since the mid-1960s with digitizing of contours supplemented by ridges and streams. Originally it was developed to automate the production of raised-relief maps using plaster models, but the usefulness of the derived gridded DEM data was soon recognized, and the concept of a complete 1:250 000 scale DEM data base for the whole country was born. The manual digitizing approach was then supplemented by the Automated Graphic Digitizing System (AGDR), based on the automated raster scanning of stable base materials containing the contour lines and the ridge and stream bases. The digitized raster-based scan data is then vectorized back to line form, tagged and edited using an interactive digitizing station, and the final geographically-referenced DEM data interpolated from the contours.

7.4.2 Data structures

It is interesting to note the quite different data structures adopted by these two different national mapping agencies within the same country. In the case of the 7.5 minute DEM data produced by the USGS, the data is grid-based consisting of a regular array of elevations referenced to the Universal Transversal Mercator (UTM) coordinate system. The DEM data is made available as a series of parallel south-to-north profiles that are ordered from west to east. Since the spacing of the elevations both along each individual profile and between profiles is 30 m, this results in a regular matrix of points. However, since the corners and sheet-lines of the USGS 1:24 000 scale topographic map sheets are referenced to geographical coordinates (latitude and longitude), this results in a variable number of elevations within each profile and between sheets, arising from the variable angle between true north and grid north on an individual sheet (Fig. 7.26).

In the case of the 1:250 000 scale DEM data produced by DMA, the data is organized and referenced on the basis of geographical (latitude/longitude) coordinates. As in the case of the 7.5-minute data, the origin is at the southwest corner

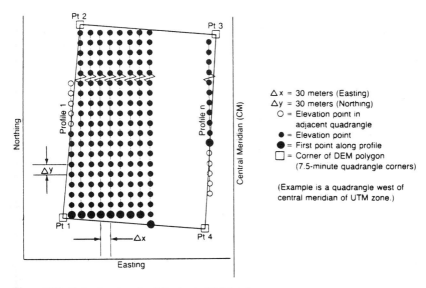

$\triangle x$ = 30 meters (Easting)
$\triangle y$ = 30 meters (Northing)
○ = Elevation point in adjacent quadrangle
● = Elevation point
⬤ = First point along profile
□ = Corner of DEM polygon (7.5-minute quadrangle corners)

(Example is a quadrangle west of central meridian of UTM zone.)

Figure 7.26 Data structure for 7.5-minute DEM data.

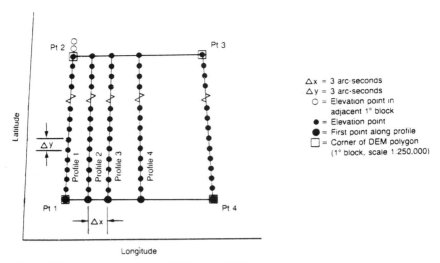

Figure 7.27 Data structure for 1:250 000-scale DEM data.

of the block being covered, and the data is organized in a series of profiles, each ascending from south to north and running in an ordered series from west to east (Fig. 7.27). However, the interval used is 3 arc-seconds both in latitude and longitude for the whole of the main continental extent of the United States. For the state of Alaska, located closer to the North Pole and hence having a strong convergence of the meridians, the spacing remains at 3 arc-seconds along the profiles, i.e., in the latitudinal direction, but varies between six arc-seconds (corresponding to 601 profiles per $1° \times 1°$ DEM) in the south of the state to 12 arc-seconds (corresponding to 151 profiles per $1° \times 1°$ DEM) in Northern Alaska in the longitudinal direction.

7.4.3 DEM accuracy
With regard to the accuracy of each type of DEM, this will of course depend on the source materials from which the data has been obtained. In the case of the 7.5-minute DEM data, where this has been acquired photogrammetrically, a root mean square error (RMSE) of ± 7m is usually cited since this is consistent with stereomeasurements from the high-altitude ($H = 13\,000$m), 1:80 000 scale aerial photography from which the 1:24 000 scale series has been compiled. If the 7.5-minute DEM data has been acquired via digitized contours and subsequent interpolation, a lower accuracy figure will apply (an RMSE of ± 7–15 m). In the case of the 1:250 000 scale DEM data, a much lower accuracy must be expected and this will in fact vary with the claimed accuracy of the contours (El Assal and Caruso, 1983): 50 ft (15 m) in flat terrain, 100 ft (30 m) in moderate terrain, and 200 ft (60 m) in steep terrain.

7.5 DTM production in the United Kingdom

The generation of a national UK database providing terrain elevations for the whole country is somewhat different to that described above for the United States. The national mapping agency, the Ordnance Survey, has been very active in digital mapping since the early 1970s, but this has been devoted almost entirely to large-scale mapping at 1:1 250 and 1:2 500 scales which are purely planimetric series. Furthermore, the OS does not produce any photomap series from which terrain elevation information would be gathered as a byproduct of the photogrammetric scanning of stereomodels. Nor was the opportunity taken to directly digitize the photogrammetric contouring carried out for the new 1:10 000-scale topographic map series covering the whole country, although feasibility experiments were conducted (Sowton, 1970). Thus the OS does not produce digital contour and terrain model data at present.

However, the needs of military users have led to an active and extensive programme of digitizing existing UK cartographic material to generate digital elevation data. This has been carried out by the Mapping and Charting Establishment (MCE) of the Ministry of Defence, which corresponds to the DMA in the United States. As with the DMA programme in the USA, the initial work has been to digitize the contour plates of the existing 1:250 000 scale topographic series of the UK. This work has involved both manual digitizing and the use of the Laser-Scan Fastrak semi-automatic digitizer, followed by interactive editing using the LITES system. Thus the line-following method of digitizing has been used throughout. From this data, various program packages supplied by Laser-Scan, Scicon, etc., have been used to generate a regular grid model of elevation data entitled the Digital Land Mass Simulation (DLMS), which has been interpolated for the whole country. Obviously, given the scale, form and accuracy of the input data, this height data has a limited resolution and accuracy, but it has proven to be entirely suitable for its intended purposes of providing data for radar and aircraft simulators, radar visibility studies, etc. Many spectacular examples of perspective views, slopemaps, layered contour plots, etc., have been produced using both the Terrain Visualisation and Exploitation Software (TVES) from Laser-Scan and the Viewfinder package from Scicon, both of which are designed specifically to exploit the DTM data generated by MCE.

A still further development has been the initiation of a project by MCE to generate

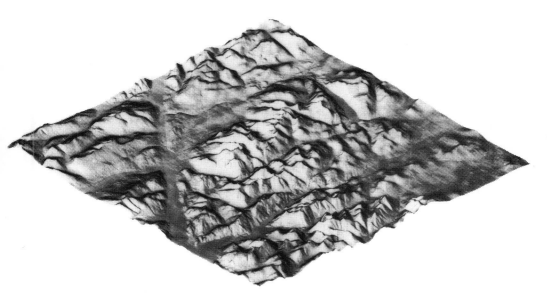

Figure 7.28 Perspective view of Ben Nevis produced by PANACEA program (PANORAMA module) from grid DTM derived from digitized OS contours.

DTM data at an improved resolution and accuracy for the whole of the UK via digitizing of the existing contours on the 1:50 000 scale OS map series. Already the whole of Wales has been covered together with several more limited areas of southern England and the Ben Nevis area in Scotland. Digitized contour data is converted to gridded elevation data using Laser-Scan's DTM and DTM Plus software or the PANACEA program suite mentioned above. Figure 7.28 shows a representative example, a perspective view of Ben Nevis generated by the latter system.

At present, it is still not clear whether the DTM data of the UK produced by MCE will be made available to a wider range of users, e.g., on a commercial basis via the OS. This would be of widespread interest to engineers, landscape architects, planners, geologists, geophysicists, etc., all of whom are concerned with the planning of projects or the presentation and visualization of data on a regional scale.

7.6 DTM production in Sweden

Sweden is another country where there has been a concerted plan (since 1975) to build up a nationwide database of mapping information and of terrain elevations under the so-called AutoKa system of the National Land Survey (NLS) of Sweden (Moren, 1983). As in the United States, there are two levels of provision of the terrain elevation database, with quite different densities and accuracies associated with each.

The first of these databases has been derived by digitizing the contour data present on the 1:50 000 and 1:100 000 scale topographic map series. This data is used as the base for the interpolation of a grid of spot height of 500 m spacing which now covers the whole country. The individual spot height values have an expected accuracy (RMSE) of $\pm$ 10 m. As in the USA and the UK, the main impetus for the provision of this DTM data has come from the Swedish defence authorities for use in aircraft simulators, radar visibility studies, etc.

The second terrain elevation database aims at a much higher density, with a spacing between grid nodes of 50 m, and greater height accuracy, estimated about $\pm$ 2.5 m. In this case, the availability of this height data has been greatly aided by the fact that the standard medium-scale (1:10 000 and 1:20 000) topographic map coverage of Sweden (the so-called Economic Map series) is based on photogrammetrically-produced orthophotographs. The terrain height data is thus available as a byproduct of the profile scanning of the stereomodels generated from aerial photographs taken at 1:30 000 scale (for the 1:10 000 scale Economic Map series) and at 1:60 000 scale (over more remote areas where the basic Economic Map series is produced at 1:20 000 scale). This higher-density, higher-accuracy DTM of the whole country was due to be completed in 1988.

7.7 Conclusion

The field of activity covering the production of digital elevation data from the contours available on existing topographic maps is an extremely active and fast-developing one. The technology employed ranges from the simplest manual type of digitizer to the most sophisticated types of automatic line-following and raster-scan digitizers. The accuracy of the resulting elevation data will be much lower than that achievable with field survey or photogrammetric instrumentation, but it is quite acceptable to a considerable body of users concerned with the modelling of large areas of terrain. The consequent demand has led to the institution of programmes of digital data acquisition on a regional or national scale by several large government mapping agencies in the highly developed countries of North America, Western Europe, etc. Already these programmes have been completed for smaller scales; e.g., at 1:250 000 scale in the USA and UK, and at 1:100 000 scale for Sweden, to provide low-resolution/low-accuracy DTM data on a national scale. The generation of higher-accuracy DTM data at a higher density from the contours available on medium-scale topographic maps promises to keep the national mapping agencies busy for many years to come. Digital elevation data will perhaps eventually be available off the shelf on a national basis in the same manner as topographic maps are available at present.

References

Baker, D.J., Gogineni, B., Shepphird, G.R., Williams, L.M. and Bethel, J.S. (1981). Linetrac—a hardware processor for scanning and vectorizing contours, and associated systems software for automatically generating digital terrain elevation data. *Technical Papers of the American Society of Photogrammetry, ASP-ACSM Fall Meeting*, 542–557.

Bell, S.M.B. and Woodsford, P.A. (1977) use of the HRD-1 laser display for automated cartography. *Cartographic Journal* 14(2), 128–134.

Boyle, A.R. (1980) The present status and future of scanning methods of digitization, output drafting and interactive display and edit of cartographic data. *International Archives of Photogrammetry* 23(B4), 92–99.

Clark, R. (1980) Cartographic raster processing programs at USAETL. *Technical Papers, 40th Annual Meeting of the American Congress on Surveying and Mapping, St. Louis*, 110–125.

Davies, D.J.M., Frisch, O.R. and Street, G.S.B. (1970) Sweepnik. A fast semi-automatic track-measuring machine. *Nuclear Instruments and Methods* 82, 54–60.

Douglas, D. and Peucker, T. (1973) Algorithm for the reduction of the number of points required to represent a digitized line or its caricature. *Canadian Cartographer* 10(2), 112–122.

El Assal, A.A. and Caruso, V.M. (1983) *Digital Elevation Models*. US Geological Survey Circular **895-B**, 40 pp.

Howman, C. and Woodsford, P.A. (1978) The Laser-Scan Fastrak automatic digitizing system. *Presented Paper, 9th International Conference on Cartography (ICA), Maryland, USA*, 13 pp.

Leberl, F.W. and Olsen, D. (1982) Raster scanning for operational digitizing of graphic data. *Photogrammetric Engineering and Remote Sensing*. 48(4), 615–627.

McCullagh, M.J. (1983) Transformation of contour strings to a rectangular grid based digital elevation model. *Presented Paper, Euro-Carto II*, 18 pp.

Moren, A. (1983) AutoKa, applications at the NLS. *Landmäteriverket, Professional Papers, LMV-Rapport* 1983, **13**, 19 pp.

Peuquet, D.J. (1981) An examination of techniques for reformatting digital cartographic data—Part I: The raster-to-vector process. *Cartographica* 18(1), 34–48.

Sowton, M. (1971) Automation in cartography at the Ordnance Survey using digital output from a plotting machine. *Bildmessung und Luftbildwesen* 39(1), 41–44.

Woodsford, P.A. (1976) The HRD-1 laser display system. *Computer Graphics (SIGGRAPH/ACM)* 2(10), 68–73.

Part B
Theoretical basis of DTM formation and display

8 Modelling, interpolation and contouring procedures

G. PETRIE

8.1 Introduction

A very large number of program packages have been devised and written for terrain modelling applications in surveying and civil engineering. Each program package has been written independently, has usually been optimized for a specific application or for use on a specific computer, utilizes a different programming language, or has a distinctive user interface. However, in spite of this diversity, when their characteristics are analysed, it can be seen that basically they follow one of two main approaches (Fig. 8.1):

(i) they are based on, or make use of, height data which has been collected or arranged in the form of a regular (e.g., rectangular or square) grid: or

(ii) they are based on a triangular network of irregular size, shape and orientation, and use randomly-located height data.

8.2 Grid-based terrain modelling

In the first approach, in many ways the simplest, the data comprising the terrain model is measured or collected in the form of a regular grid. This is in widespread use, e.g., in the grid levelling of building sites by surveyors and civil engineers and in the acquisition of profile data measured photogrammetrically in stereoplotting instruments, in the latter case especially when derived automatically or semi-automatically

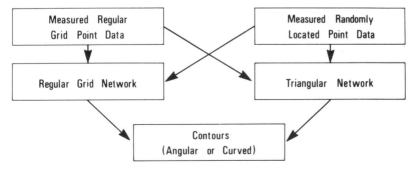

Figure 8.1 Overall relationship between measured point data, networks and contours in terrain modelling.

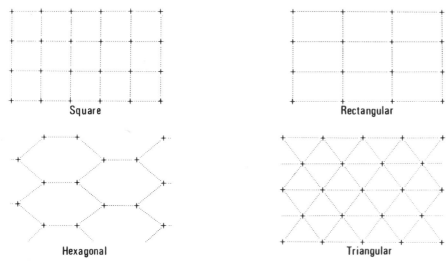

Figure 8.2 Regular grid patterns.

during the production of orthophotographs or under computer control using an analytical plotter. While the square grid is the most common form, rectangular, hexagonal and triangular-based grids are also encountered (Figure 8.2).

The shortcoming of this regular grid-based approach is that the distribution of data points is not related to the characteristics of the terrain itself. If the data-point sampling is conducted on the basis of a regular grid, then the density must be high enough to accurately portray the smallest terrain features present in the area being modelled. If this is done, then the density of data collected will be too high in most areas of the model, in which case there will be an embarrassing and unnecessary data redundancy in these areas. In this situation, filtering of the measured data may need to be carried out as a pre-processing activity before the terrain model can be defined.

8.2.1 Progressive sampling

Since the grid-based approach is most easily implemented in computer-controlled photogrammetric instruments, a solution to the shortcoming mentioned above has come from the photogrammetrists in the form of progressive sampling and its development, composite sampling, originally proposed by Makarovic of the ITC in the Netherlands (Makarovic 1973, 1975, 1977). Instead of all points in a dense grid being measured, the density of the sampling is varied in different parts of the grid, being matched to the local roughness of the terrain surface.

Starting with a widely spread (low-resolution) grid which will give a good general coverage of height points over the whole area of the model the density of sampling (or measurement) is progressively increased on the basis of an analysis of terrain relief and slope using the on-line computer attached to the photogrammetric instrument. Thus the basic grid is first densified by halving the size of the grid-cell in certain limited areas, based on the results of the preceding terrain analysis. Measurements of the height points at the increased density are carried out under computer control only in these predefined areas. A further analysis is then carried out for each of these areas for which the measured data has been increased or densified. Based on this second data analysis, an increased density of points may be prescribed for still smaller areas.

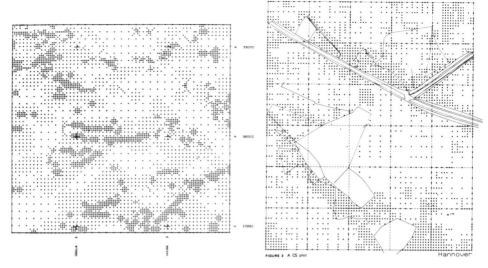

Figure 8.3 Examples of progressive sampling.

Normally, three such runs or iterations are sufficient to acquire the terrain data necessary to define a satisfactory model. Figure 8.3 illustrates two examples of progressive sampling applied to different data sets.

In this way, the progressive sampling technique attempts to optimize automatically or semi-automatically the relationship between specified accuracy, sampling density and terrain characteristics. Recently the method has been widely implemented in grid-based terrain modelling packages devised by photogrammetrists, e.g., HIFI (Height Interpolation by Finite Elements, Ebner *et al.* 1980, 1984); SCOP (Stuttgart Contour Program, Assmus, 1976; Stanger, 1976; and Kostli and Wild, 1984); the package devised in the ITC by Huurneman and Tempfli (1984), and the Graz Terrain Model (GTM) package implemented at the Technical University of Graz.

8.2.2 Random-to-grid interpolation
Especially in land surveying operation, terrain data will be collected by optical or electronic distance-based methods combined with the necessary angular measurements. Since the surveyor or civil engineer must visit every measured point to set up the measuring staff or retro-prism, the opportunity exists to ensure their positioning on points which are important in terms of terrain morphology or representation, e.g., on hilltops, in pits, hollows or saddles, along breaklines such as ridges, breaks of slope, rivers, etc., besides giving an appropriate general spread of height points. If a grid-based terrain modelling package is to be used with such specific, but randomly located, points, then a preliminary interpolation must be carried out which converts the measured data to a suitably dimensioned regular grid.

The distinguishing feature between different random-to-grid interpolation procedures is the range of the interpolation function which is employed in the interpolation. Usually the following are distinguished:
(a) *pointwise methods* which involve the independent determination of different functional parameters and height values for each and every grid node being interpolated (Fig. 8.4a);

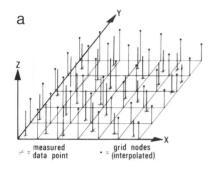

POINTWISE INTERPOLATION

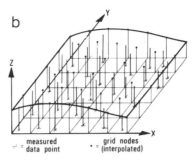

GLOBAL INTERPOLATION

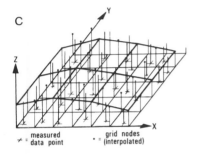

PATCHWISE INTERPOLATION

Figure 8.4 Random-to-grid inter-polation methods.

(b) *global methods* which involve the establishment of a single complex three-dimensional surface through the complete data set of measured height points, with the subsequent interpolation of terrain heights at all the required nodes or points on the regular grid (Fig. 8.4b); and

(c) *patchwise methods* in which a series of local 3D surfaces or patches are established from which the elevations of a series of neighbouring grid points lying within each patch can be interpolated (Fig. 8.4c).

(i) Pointwise methods. These involve the interpolation of the values of the height at a specific grid node from its neighbouring randomly-located measured height points. Since each point, or node, on the final grid is determined independently of any other node it has no effect or impact on the adjacent points in the terrain model, and a continuous surface may be generated through all the derived grid nodes without any discontinuities or boundary problems.

Almost all the algorithms used for the determination of the height of each individual grid point are based on a search for a set of *nearest neighbours*, followed by the averaging of their heights weighted inversely by some function of their respective distances d from the position of the grid node. This weight $w = 1/d$, where m is the power used, and is typically in the range 0.5–4. The various search methods are illustrated in Fig. 8.5.

The search for the nearest neighbours may involve a simple area search (Fig. 8.5a) where a circle of predefined radius, or a box of predefined size, is used to select the data points from which the grid-point value is determined. A variant of this technique, in which the n nearest neighbours are searched for (Fig. 8.5b), may also be used, n being

SEARCH METHODS

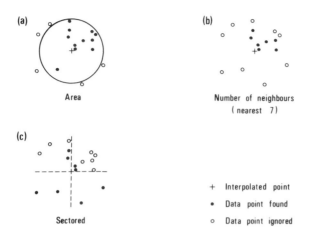

Figure 8.5 Pointwise interpolation: area search techniques.

user-defined but usually in the range 6–10. Since these two methods offer no control over the distribution of the points used for the interpolation of the grid node point, a further variant is the *sectored nearest neighbour* technique in which the area around the grid point, or node, is divided into equal sectors, usually four (quadrants), or eight (octants), with the nearest two or four neighbours being searched for in each case (Fig. 8.5c).

(ii) Global methods. These involve the fitting of a single three-dimensional surface, defined by a high-order polynomial, through all of the measured randomly located terrain height points existing within the model. Once this global surface has been defined, and the specific values of the polynomial coefficients have been determined, the values of the heights for each grid node can then be interpolated. Difficulties may arise if many terms are used (as will be the case with high-order polynomials), and if a large data set is required to form the model, resulting in large amounts of computation time. Also, the unpredictable nature of the oscillations produced by such high-order polynomials may produce poorly interpolated values for the grid node points.

(iii) Patchwise methods. These lie in an intermediate position between the pointwise method and the global method. The whole area to be modelled is divided into a series of equal-sized patches of identical shape. The shapes of each patch are regular in form, typically square or rectangular. Quite separate mathematical functions are then generated to form the surface for each patch. Thus separate sets of fitting parameters are computed for each individual patch.

Two distinct methods of patchwise interpolation exist (Fig. 8.6):

(a) *Exact fit patches* (Fig. 8.6a) may be defined in which each patch abuts exactly on to its neighbours. Such patches may between, result in sharp discontinuities along their junctions, which show up markedly when the isolines or contours are finally produced. The patches are larger than the cells defined by the grid nodes, so several grid nodes will fall within a single patch.

(b) The alternative is an arrangement of overlapping patches (Fig. 8.6b), in which case there will be common points lying within the overlap which will be used in the computation of the parameters for each patch.

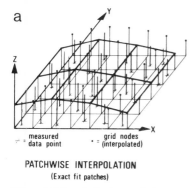

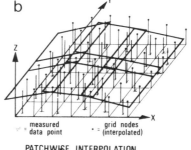

Figure 8.6 Patchwise interpolation.

The advantages of using patchwise methods over global methods are (i) that quite low-order terms (parameters) can satisfactorily be used to describe each patch, so only a few unknowns have to be solved via simultaneous equations using least-square methods for each patch; (ii) once the unknowns have been solved for, it is easy to calculate the derived points, i.e., the grid nodes by back-substitution in the functions or equations describing the patch.

However, there are also some disadvantages of the patchwise method: it needs much more organization of data and of its processing than pointwise or global methods, and the subdivision of the model surface into patches needs to be carried out with care—if the data is poorly distributed towards the patch corners, then this affects the computed parameters and, in turn, the accuracy of the subsequent heights determined for the grid node points.

(iv) Polynomials used for surface representation. As noted above, polynomial equations are used to represent the terrain surfaces in the global and patchwise methods of interpolation. The basic *general polynomial equation* for surface represent-ation is shown in Table 8.1, where z_i is the height of an individual point i, x_i and y_i are the rectangular coordinates of the point i and $a_0, a_1, a_2 \ldots$ are the coefficients of the polynomial. One such equation will be generated for each individual point i with coordinates x_i, y_i, z_i occurring in the terrain model.

In the first step, the values of x, y and z are known for each measured point present in the overall data set or patch. Thus the values of the coefficients $a_1, a_2, a_3 \ldots$ can be determined from the set of simultaneous equations which have been set up, one for each data point. Once the values of the coefficients $a_1, a_2, a_3 \ldots$ have been determined,

Table 8.1 General polynomial equations used for interpolation. For explanation, see text.

Individual terms	Order of term	Descriptive term	No. of terms
$z = a_0$	Zero	Planar	1
$+ a_1 x + a_2 y$	First	Linear	2
$+ a_3 x^2 + a_4 y^2 + a_5 xy$	Second	Quadratic	3
$+ a_6 x^3 + a_7 y^3 + a_8 x^2 y + a_9 xy^2$	Third	Cubic	4
$+ a_{10} x^4 + a_{11} y^4 + a_{12} x^3 y + a_{13} x^2 y^2 + a_{14} xy^2$	Fourth	Quartic	5
$+ a_{15} x^5 + \ldots$	Fifth	Quintic	6

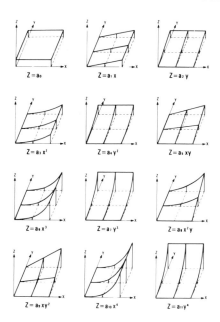

Figure 8.7 Surface shapes produced by individual terms in the general polynomial equation.

then for any given grid node point with known coordinates x, y the corresponding height value z can be calculated.

To make a correct selection of the terms which will best represent or model the terrain surface, the surveyor or civil engineer must keep in mind the shape of the individual terms produced in the polynomial equation, as illustrated by Fig. 8.7.

Typical of the simpler types of surface used to model patches are:

(a) the 4-term *bilinear polynomial* with the form

$$z = a_0 + a_1 x + a_2 y + a_3 xy$$

(as used in BIPS, HIFI, etc.);

(b) the 9-term *biquadratic polynomial*

$$z = a_0 + a_1 x + a_2 y + a_3 xy + a_4 x^2 + a_5 y^2 + a_6 x^2 y + a_7 xy^2 + a_8 x^4 y^2$$

(as used in SCOP)

(c) the 16-term *bicubic polynomial*

$$z = a_0 + a_1 x + a_2 y + a_3 xy + a_4 x^2 + a_5 y^2 + a_6 x^2 y + a_7 xy^2 +$$
$$a_8 x^2 y^2 + a_9 x^3 + a_{10} y^3 + a_{11} x^3 y + a_{12} xy^3 + a_{13} x^3 y^3 +$$
$$a_{14} x^3 y^2 + a_{15} x^2 y^3,$$

(as used in HIFI, Rijkswaterstaat DTM).

In spite of the disadvantages of the grid-based terrain models discussed above, they are still widely used. Furthermore, the development of progressively sampled data sets using computer-controlled photogrammetric instruments has ensured that the principal disadvantage of grid-based models, lack of the variation in sampling density with terrain type, has been overcome to a considerable extent. Also, the incorporation of terrain breaklines in program packages such as HIFI and SCOP has given further credibility or life to grid-based terrain modelling methods, especially when based on photogrammetric data.

8.3 Triangulation-based terrain modelling

This method is being used increasingly in terrain modelling, and many recently devised program packages are based on it. The reasons for this are firstly that every measured data point is being used and honoured directly, since they form the vertices of the triangles used to model the terrain, to determine the heights of additional points by interpolation and to carry out the construction of contours, and secondly, that the use of triangles offers a relatively easy way of incorporating breaklines, faultlines and so on.

Originally the triangulation method suffered from the fact that it proved difficult to ensure that the same network of triangles would be generated from a single set of randomly-located measured data points, no matter from which point in the data set the triangulation started. Also, automatic triangulation often took an exorbitant time to execute in a computer. However, these problems have been overcome (Fig. 8.11a, b), hence the present popularity of the triangle-based modelling of the terrain.

Any triangular-based approach should attempt to produce a unique set of triangles that are as equilateral as possible and with minimum side lengths (McCullagh, 1983). One or other of two main algorithms are used to implement these requirements: (a) the *Delaunay triangulation* method; and (b) the *radial sweep* algorithm.

8.3.1 Delaunay triangulation

Associated with the Delaunay triangulation (Delaunay, 1934) is the *Thiessen polygon* (Fig. 8.8), which endeavours to define geometrically the region of influence of a point on an areal basis. This is done by constructing a series of perpendicular bisectors on each of the triangles formed around that specific point. These intersect at the Thiessen vertices. The polygon so defined is the Thiessen polygon. Because of their distinct pattern when viewed on a computer graphics terminal, they are often referred to as

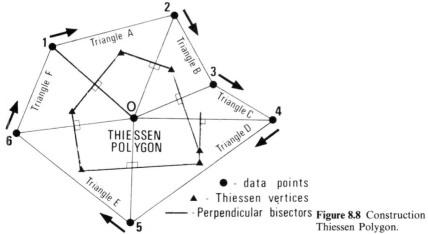

● - data points
▲ - Thiessen vertices
—— - Perpendicular bisectors

Figure 8.8 Construction of the Thiessen Polygon.

a

b

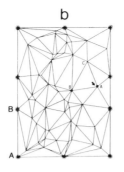

Figure 8.9 Delaunay triangulation (McCullagh, 1983). (a) Data points; (b) triangulated surface and boundary points.

'tiles'. The data points surrounding a specific data point (e.g., 0) are known as its 'Thiessen neighbours.'

A preliminary step before triangulation begins is to define a set of (artificial) boundary points to form a perimeter around the edges of the data set area. This is necessary to create a frame to the terrain model and a set of boundary triangles which allow contours to be extrapolated outside the area of the data set itself. Once these boundary points (which may have arbitrary values) are defined and have been added to the data set, the whole area can then be triangulated, starting with a pair of the artificial boundary points known as the initial known neighbours (A and B, located in the bottom left-hand corner of the area illustrated in Fig. 8.9).

The search for the next neighbour is then made by constructing a circle with the base AB as diameter and searching to the right (i.e., clockwise) to find if any point falls within this circle (Fig. 8.10). This search can be carried out quite quickly by computer. If no data point lies within the circle, it is increased in size to perhaps twice the area of the original circle, with AB now a chord in the larger circle (Fig. 8.10). Any data points lying within the new circle are tested to discover which meets the criteria set for the nearest Thiessen neighbour.

Once this has been achieved, the search for the next neighbour then continues to the right (i.e., clockwise), then on to the next neighbour, and so on till the next boundary point is reached. The triangles so formed constitute a so-called *shell*. The process of the triangulation then continues, each point in the shell being used in turn as the starting point for the search until the next set of Thiessen neighbours. This continues

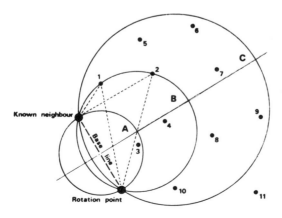

Figure 8.10 Expanding circle search for nearest neighbour (McCullagh, 1983).

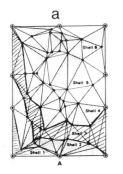
a

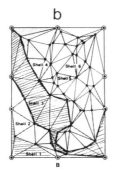
b

Figure 8.11 Advancing shells of a Delaunay triangulation (McCullagh, 1983). (a) Orientation as in Fig. 8.9; (b) rotated through 180°.

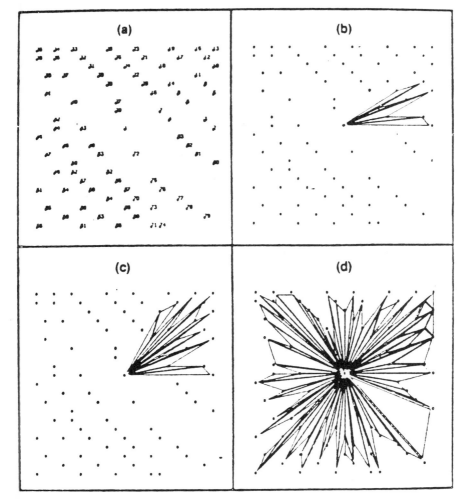

Figure 8.12 Sorting and triangle generation. (a) Nodes sorted by heading, distance and pitch. (b) Beginning of radial sweep. (c) Radial sweep past two points with same heading. (d) Beginning of concavity infill. (Mirante and Weingarten, 1982.)

in a systematic manner until the neighbours for all the points existing in the data set have been found and the corresponding triangles formed (Fig. 8.11).

Now that it is better known and understood, the Delaunay method is that used to form triangles in the majority of terrain modelling packages based on the triangulation method.

8.3.2 Radial sweep algorithm

The radial sweep algorithm is an alternative to the Delaunay triangulation devised and first published by Mirante and Weingarten (1982).

As before, the input data is in the form of randomly located (as distinct from systematically located) points with x, y and z coordinates. The points will have been located on summits, on ridges and breaklines, etc., in the usual way. The point or node which is located nearest to the centroid of the data is selected as the starting point for

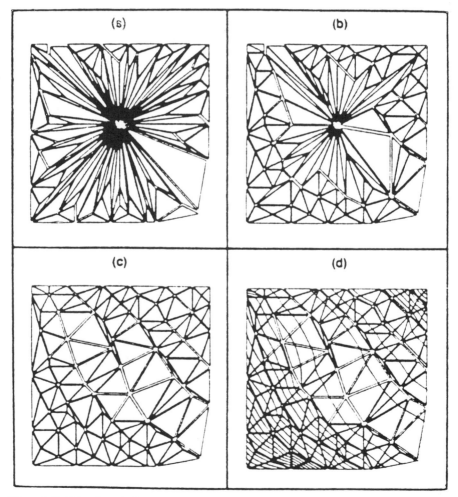

Figure 8.13 Triangle switching and contouring. (a) Mesh before first switching pass. (b) After first switching pass. (c) After all switching passes. (d) With contour cuts. (Mirante and Weingarten, 1982.)

the triangulation. From this central point the distance and bearing to all the other points in the data set is calculated and the points are then sorted and placed in order by bearing.

Once this has been done, the radiating line to each point is established, and a long thin triangle formed by connecting a line between the new point and the previous point (Fig. 8.12). If two points have the same bearing, then they are used to form a pair of triangles on each side of the common line. As each point is accessed, it is added to a linked list which forms the other boundary of the triangulated network. Once the initial sweep has been achieved, the concavities created by the initial radial sweep triangulation must be filled by new triangles. Each point or node on the boundary list is combined and compared with the next two nodes on the list and checked to see if an inside triangle can be formed. If so, then the new triangle is added to the database, and the second (inside) node is removed from the list of outer boundary points. After the process has been completed, the list will comprise the points or nodes forming the convex edge of the terrain model. As can be seen from Fig. 8.13, all the data points are now triangulated with non-overlapping triangles. However, the shapes and connections are far from desirable. To optimize the shapes, each triangle is now tested against each of its neighbours.

A quadrilateral is formed by a pair of triangles. This is tested by calculation of the two distances by the opposite pairs of nodal points. If the distance between the two common points is greater than the distance between the two unique nodes, then the triangle indices are switched and the database pointers updated. This process is repeated with successive passes through the database till an entire pass through the database produces no changes.

8.4 Contouring

Considering a single grid cell as illustrated by Fig. 8.14, a simple linear interpolation is carried out along each of the four sides in turn based on the values at the nodes. The positions of all the contour values are determined for each side. Next they are connected by straight lines or vectors, since for every entry point there must also be an exit point.

However, ambiguities may also occur with alternative solutions and impossible situations. Taking the data points given in the example in Fig. 8.15, there are two possible solutions which give quite different positions for the contour, and a third (impossible) alternative. A solution to the problem is to revert to a centre-point figure, i.e., split the grid cell into four triangles, assigning the average of the four nodes to the

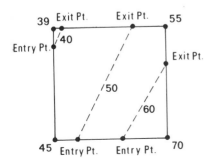

Figure 8.14 Linear contour interpolation in a single cell.

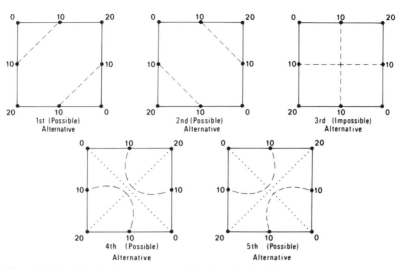

Figure 8.15 Ambiguity in contour threading in a single cell.

central point and then implementing an arbitrary rule such as 'keep the high ground to the right of the contour'. This is, of course, quite an arbitrary solution.

In view of these potential difficulties many contouring packages based on regularly gridded data make no attempt to carry out direct threading of the contour. Instead, the interpolation of the contour is based on some type of function fitting, e.g., typically using the bicubic polynomial already discussed in the context of random-to-grid interpolation. Thus a new patch is formed based on a group of grid cells. If the full 16-term bicubic polynomial is employed, then the minimum size of patch that can be used is one made up of $4 \times 4 = 16$ nodes with 9 cells (Fig. 8.16). If a smaller number of coefficients is used, e.g., as in the 9-term biquadratic polynomial, then there will be redundancy, and a least-squares solution will be employed in the determination of the parameters. This can help to give extra smoothing and a better fit or continuity between cells. The interpolation of the contour through an individual grid cell is then carried out with reference to the whole patch rather than by simple linear interpolation in each individual grid cell.

Thus, first of all the values of the coefficients, $a_0, a_1, a_2 \ldots$ of the local surface patch are determined using the 16-grid node values. In some contouring packages, e.g., Calcomp's GPCP (General Purpose Contouring Package), additional height values

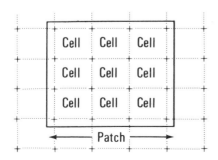

Figure 8.16 Patch formed by 16 grid nodes (9 grid cells).

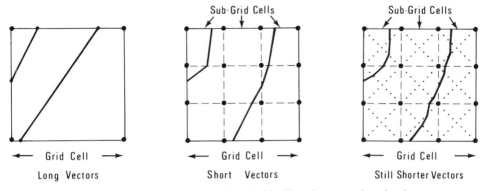

Figure 8.17 Contour threading through grid cells, subgrid cells and centre point triangles.

are then determined using the known values of the coefficients for all the points located on a *sub-grid* which is generated within each grid cell. The contours are then interpolated linearly and threaded between the points on the sub-grid cell.

Each of the straight line vectors will be quite short and the quite abrupt changes in direction which would be seen with contours derived from linear interpolation on the whole cell are lessened considerably with evident benefit to the visual appearance of the contours. It is possible to further divide the sub-grid cell itself into centre point triangles and conduct the contour threading on the cutting points on the diagonals, in which case, the vectors will be still shorter and the appearance even more acceptable (Fig. 8.17).

Some contour packages using grids try to achieve more pleasing contours by an alternative approach, contour smoothing, which involves fitting a series of cubic splines between the entry and exit points. This does produce smooth contours which are visually more acceptable than the straight-line type produced from the same data. However, a separate curve or spline will be fitted to each segment of each contour line to ensure its smoothness. A difficulty which can then arise in an area of steep terrain with close contouring is that the curved smoothed contours may then cross, which of course is not permissible in terms of actual terrain.

8.4.1 Contours from triangulated data

As with contouring of regular gridded height points, so with randomly located triangulated height data, there are two main options for the contour threading: (a) simple linear interpolation of the contours; and (b) generation of curved smoothed contours using some type of function.

Unlike the situation with gridded data, the use of direct linear interpolation for the contour generation is very common when the terrain model is based on triangulated data. The potential difficulty with ambiguities regarding the direction that the contours might take are either not present or can be resolved. So direct linear interpolation gives a simple and robust solution. The contour threading usually begins at the boundary triangles. Using linear interpolation, all the entry points along the perimeter are located and the corresponding exit points found on the interior sides of each boundary triangle. These then of course act as entry points for the next internal triangle (Fig. 8.18).

Just as with the grid-based method, so in the triangle-based method, the area of an individual triangle can be subdivided into smaller triangles (subtriangles) in order to

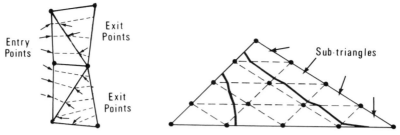

Figure 8.18 Contour threading through triangles and subtriangles.

get over the difficulty of long vectors with abrupt changes of direction along the common line between two adjacent triangles. The values at the vertices of each sub-triangle can be calculated by simple linear interpolation. Then the cutting points are determined for each required contour along the sides of each subtriangle. Joining up the cutting points gives the required contour; again this looks smoother since it consists of a series of short vectors, instead of long vectors spanning the whole triangle. Obviously the threading decisions in any subtriangle are also simplified to a choice between the two remaining sides of the triangle as against the three possible sides within the grid cells.

If curved smoothed contours are required, then this can again be achieved using a series of cubic splines or polynomials fitted through the string of interpolated cutting points along the triangle boundaries. In areas of very close contours in steep areas, crossing contours may occur if this procedure is adopted.

The alternative approach is again to fit some type of curved three-dimensional surface patch to each triangle so ensuring a smooth transition from one triangle to the next, instead of having a series of planar triangular facets with abrupt changes in the direction of contours between facets along the lines common to adjacent triangles. This will entail forming a 'polygon patch' using the central point and its Thiessen neighbours (Fig. 8.19). There is no difficulty in generating the polygon patch, but it will be irregular in shape (since there is no regular grid), so the data points will have a rather intractable format. The distances from the central point will all be unequal and the fitting function could be weighted inversely with increasing distance. So an inverse distance-weighted, low-order polynomial surface could be fitted through the irregularly located data points which comprise the patch.

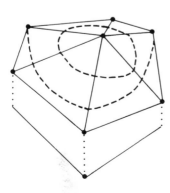

Figure 8.19 Surfaces and contours on a centre polygon comprising several triangles.

8.4.2 Terrain model data derived from contours

The present discussion has concentrated on the situation most commonly encountered in surveying and civil engineering, namely that the individual points comprising the digital terrain model have been measured either by land surveying or by photogrammetric methods. The contours (and any perspective views) representing the model surface are then constructed or derived from measured spot height data.

However, the reverse situation may also occur, in which the terrain model comprising a set of individual points is formed from existing contours already recorded on a topographic map. This is unlikely at large scales where the accuracy requirements of the terrain model for road and dam design and construction, volume determination, etc., are such that the spot height data must be directly measured to a high degree of accuracy. However, in other applications where a wide area has to be modelled or represented at small scales, e.g., in landscape visualization, aircraft simulators, landform or geological representation, geophysical work, etc., the situation is quite different. Accuracy requirements are much lower and less stringent and data has to be acquired for large terrain area. In such situations, the terrain modelling will often be conducted on the basis of data derived from contours on existing maps. Several alternative methods of creating terrain models from existing map sources are possible. The relative merits of each of these are discussed in more detail elsewhere in this book (e.g. in Sections 7.3 and 9.8).

References

Assmus, E., 1976. Extension of Stuttgart Contour Program to Treating Terrain Break Lines—Theory and Results. *Presented Paper. Commission III, XIII Congress of the I.S.P.*, Helsinki, 13 pp.

Delaunay, B. (1934) Sur la sphère vide. *Bulletin of the Academy of Sciences of the USSR, Classe Sci. Mat.*, 793–800.

Ebner, H., Hoffman-Wellenhof B., Reiss, P. and Steidler, F., 1980. HIFI—a Minicomputer Program Package for Height Interpolation by Finite Elements. *Presented Paper. Commission IV, XIV Congress of the I.S.P.*, Hamburg. 14 pp.

Huurnemann, G. and Tempfli, K. (1984) Composite sampling using a minicomputer supported analogue instrument. *ITC Journal 1984–2*, 129–133.

Kostli, A. and Wild, E. (1984) A digital terrain model featuring varying grid size. *Contributions to the XVth ISPRS Congress, Rio de Janeiro, Institute of Photogrammetry, University of Stuttgart;* **10**, 117–126.

Makarovic, B. (1973) Progressive sampling for digital terrain models. *ITC Journal 1973–3*, 397–416.

Makarovic, B. (1975) Amended strategy for progressive sampling. *ITC Journal 1975–1*, 117–128.

Makarovic, B. (1977) Composite sampling for DTMs. *ITC Journal 1977-3*, 406–433.

McCullagh, M.J. (1983) Transformation of contour strings to a rectangular grid based digital elevation model. *ITC Journal II*, 18 pp.

Mirante, A. and Weingarten, N. (1982) The radial sweep algorithm for constructing triangulated irregular networks. *IEEE Computer Graphics and Applications* **2**(3), 11–21.

Stanger, W., 1976. The Stuttgart Contour Program SCOP—Further Development and Review of its Application. *Presented Paper, Commission III, XIII Congress of the ISP.* Helsinki, 13 pp.

9 Digital terrain modelling and visualization

MICHAEL J. McCULLAGH

9.1 Introduction

This chapter acts as a pathway through part of the field of computer mapping that relates to digital terrain modelling. Coverage is limited to grid interpolation, triangulation procedures, contouring, and image display systems, with a short section on databases. Three computer software systems are considered: PANACEA, PANTHEON, and MIRAGE. PANACEA provides a general surface mapping system that reaches a compromise between honouring the data points of any data set, and having general application in a number of fields. Thus far PANACEA has been used for topographic digital terrain model work, seismic interpretation, demographic studies, and various standard geographical thematic projects. MIRAGE, originally designed as a satellite image processing system, has found applications in surface modelling as a display and analysis tool for either the model results from PANACEA or derived mapping from PANTHEON. Previously surface mapping and image processing have been considered separate areas, but increasingly there is a need to marry both together to achieve the most interpretable results.

9.2 Grid-based interpolation

Topographic models are commonly developed from strings of data—contours rather than seismic lines. In both cases, spot heights or well data can be thought of as single-point strings. There are three ways conversion from data point set to a map can be achieved: by simulating the manual approach using methods involving partitioning of the area into some form of triangulation of the data points; by using direct contour determination (e.g., Schagen, 1982); and lastly by using the computer's own ability to store two-dimensional data sets in the form of grids (tables, arrays, networks, or dimensions, depending on local terminology). This last process is the one in widespread use today, although the manual mimic methods are catching up fast and perhaps have significant accuracy advantages in the long term.

The term grid implies a network of values arranged in a rectangular mesh and calculated in such a fashion that the values at the grid nodes (where a given row and column pair intersect) are accurate samples from the terrain surface that is being contoured. The grid is so arranged that its rectangular area coincides with the map area to be contoured. The estimation of the values at the nodes is the major problem associated with the first phase of the contouring process—interpolation. The second phase is the lacing, in a smooth fashion, of the contours through the regular grid that has been created, followed by automatic labelling, and the generation of commands to drive the plotting device.

(i) *Choice of grid size.* There are two constraints on the accuracy of the final contour map: the size of grid itself, and the accuracy of interpolations at the grid nodes. It should be clear that a small grid of 10 rows and 10 columns cannot possibly represent accurately a data set of 1000 points, although it may be adequate for a 50-point data set. It is therefore necessary to consider the likely requirements of grid size related to the number of data points in the area of concern. Information theory would lead to the conclusion that the number of grid interpolations should be roughly equivalent to the number of data points. Unfortunately, even for randomly distributed data and especially for data collected in strings (seismic, contours), cosmetic and accuracy requirements of the final map preclude the use of such small grids. The criterion used, to stand any chance of honouring the data points, has to be at maximum half the gap between the closest data points in the area. This is a horrifying conclusion, as the standard grid interpolation process has a speed proportional to the number of interpolations being performed and not to the number of data points being contoured!

A simple solution to this problem would be to use a rectangular grid, but to vary the mesh size across the map to accommodate the different data densities in different parts of the map area. This has been tried (Hayes and Halliday, 1974) but was found cumbersome from both a computational and contour-following or smoothing viewpoint. In many contouring packages using the gridding technique, the size is set automatically by the program, and the user never knows the inaccuracies he is enduring as a result. Grid size determination for a data set is at best difficult, and at worst leads to significant errors in over-generalizing the surface and removing data fluctuations. As a result, the process does not honour the data points unless they occur exactly at the grid locations.

(ii) *Function requirements.* There is an immense literature on methods of interpolating values at given locations from scattered data points (Lam, 1983; Rhind, 1975). This alone indicates that there is rarely an optimum technique, and that the choice of method is very much a function of user whim and the desired characteristics of the surface to be produced. We are looking for an interpolating function that:

(a) will provide a continuous surface (at least in the first derivative—visually smooth) from a scattered and possibly linear data set;
(b) is easy to calculate—because where we have N data points we are calculating N^2 interpolations to create the grid;
(c) preferably has the mathematical properties in which we may be interested (this is sometimes possible for some surfaces, for instance gravity surveys).

(iii) *Choice of interpolation method.* There are two major classes of interpolation methods: global and local. The global methods fit some function to the entire data set in such a way that all the data points exist on the function surface. All data points are honoured, but not necessarily on the grid sampling of the surface that will be used for contouring. These techniques include the multiquadric approach used for many weather models (Hardy, 1977), and bicubic spline systems applied to aerial seismic line data (Bhattacharyya, 1969). They sound ideal, but have one major drawback— on most computers they take up an amount of memory that is always at least the square of the number of data points, and require the solution of a matrix equation that is the same size. This is not a trivial task for situations where the data points may be in excess of a few hundred. For seismic analysis, where the number of mapped shot points can easily run into the tens of thousands, or digitized contour maps where a 150 000-point set is not unusual, such techniques are of limited value unless the data set can be partitioned in some fashion. Nevertheless, for small data sets the global approach is a very successful one, but is only of limited value in most applications.

Local interpolation functions produce a continuous surface because they assume that there is an auto-correlative effect present in the surface that decreases with distance away from the location where the interpolation is made. Therefore, some distance must be specified beyond which no data point will have any effect on the interpolated value of the surface at a particular grid node. The process by which the influence of a data point diminishes with distance away from an interpolation is referred to as weighting. The weight of a point will be greatest when it is near to the location of an interpolation, and will diminish to zero at some preset distance away. The choice of function to represent the decline in weighting is as variable as is the choice of interpolation function itself. The advantage of a locally applied function is that it has only to consider local data points, and can be computationally as small and simple as required (within the continuity constraint). It is therefore fast to compute, providing that the necessary local points can be found quickly from the entire data set.

(iv) Local point search strategies. An efficient search strategy for local points is a prerequisite of the PDWA technique. The most efficient sorting technique is probably a bucket sort (Knuth, 1973) that partitions the data into geographically defined boxes. The process of discovering local points within a given range in the interpolation location is then reduced to the problem of determining the boxes covered by the search circle, a simple operation, and then opening the boxes to take the data out.

(v) Weighting function determination. The function chosen for weighting the local points that will be used to calculate the interpolation is usually kept simple, as the autocorrelation function of the surface involved is often not known; a simple inverse power function is frequently used. The function for calculating the interpolation can also be kept simple; some form of inverse distance weighted average may be employed, where the distance weighting is provided by the weighting function previously discussed. Sometimes far more complex functions are used, particularly local forms of the global procedures using partially banded matrix solutions relying on the local operation to enable the global procedure to deal with the large data set problem.

An alternative local function is the process known as kriging (Olea, 1975; Journel and Huijbregts, 1978). The autocorrelation function (variogram) is defined before-hand, and the distributional properties of the local data are taken into account by means of a matrix equation system expressing their relationships to each other, to the point where an interpolation is to be made, and to regional trends in the surface as a whole. This system of equations is solved and the coefficients used to calculate both the interpolation and an error estimate. The latter is of variable reliability, depending on the accuracy of the autocorrelation function and other estimates necessary to the calculation.

(vi) The projected distance weighted average (PDWA). The function in most widespread use is the projected distance weighted average (PDWA). It has the advantages of simplicity, speed, and, when used correctly, of achieving a very good continuous surface. Any problems in use are largely due to incorrect application or understanding of the properties of the function, and over-reliance on a grid mesh size sufficiently coarse to allow any discontinuities that result to fall between the grid nodes and hence be only rarely discernible on the final map. A range is first defined to

determine which points are considered local to the calculation. Points falling within the range are weighted according to the inverse of some function (often the square) of their distance from the interpolation location. The interpolation could then be calculated from the weighted values as a simple average, but this has problems. Averages never extrapolate beyond the range of the original data, and therefore the tacit assumption is made (erroneously) that all highs and lows on the map have data points located at their highest and lowest locations—a most unlikely event! Secondly, a straight average assumes a zero derivative at the data point and thus the final surface, although smooth, is covered with 'flat spots' at each data point location. The solution is to use an average of the projected values at the interpolation location based on estimated derivatives for each data point.

(vii) Problems of data distribution. Although the function now has the correct properties, it has limitations in applicability owing to data distribution variations, for instance in low gradient areas between contours. Enlarging the range would solve these problems but the range must be small in order to keep the searching time for local points and subsequent interpolation to a reasonable minimum in areas where the data is densely clustered. Rather than increase the range in some continuous fashion, an octant search is often used. The area around the interpolation point is divided into eight sectors and each sector is considered as a separate search problem. The program looks for the nearest two (sometimes more or less depending on whim) data points in that sector. They may lie either in the basic range or outside, but will always be the nearest. Usually a limit is set (often double the range) as to how far beyond the range a search may continue if too few points have been found. In this way a stable distribution of points is built up from which a PDWA can be calculated. This method however, makes, the tacit assumption that the weighting function does not go to zero at a single range but declines asymptotically to zero with increasing distance. The rest of the PDWA calculation is identical. By using sectors and only the nearest so many points in each one, the continuity criterion for the PDWA is violated.

The reason why discontinuity often does not show on the final map is related to the grid mesh size. Where the mesh is large, the discontinuities are averaged out by the subsequent smooth contours that are laced through the grid mesh. This criticism of the sector method used in PDWA interpolation can become critical in areas of high data density. The only solution is not to use the sector method, but to use an alternative where the range fluctuates smoothly over the whole map area dependent on data density. In this latter case, continuity of the surface is ensured.

9.3 Grid contouring

Once a grid of interpolated values has been generated that reflects the data to the required level of accuracy, it is still necessary to contour it to provide the visual interpretation of the surface variation. This can be done by one of three methods, depending on the output medium to be used for final plotting:

(a) Flatbed or drum plotter using an ink pen on paper, mylar, or some special surface. This is a vector plotter: a series of vectors are sent together with pen commands and it moves in the appropriate directions. The resolution (quality of line obtainable) is very good on the higher-quality plotters, usually 0.005″ or better.

(b) Raster plotters have become an important output medium. Information is transmitted to a raster device as a series of dots on a printer (either electrostatic or matrix mechanical). The contour image is built up as a series of scans across the grid at very fine resolution, with a black dot switched on every time a contour is crossed. Most raster graphics manufacturers also supply software to convert from vector to raster format without user intervention. Resolution of an electrostatic device is typically about 200 dots per inch.

(c) Colour graphics terminals are now common, and some colour printers (usually ink jet, or laser photoplotted for the best quality) are now found in many computing establishments. The standard resolution used to be about 512×512 dots over a screen of smaller than one foot square, but now may be in excess of 1000×1000. Colour plotters on the other hand exhibit the 200 dots per inch characteristics if they are electrostatic, but 1000 or more dots per inch for a laser photoplotter. Prices of both are, however, commensurate with resolution.

(i) Contour lacing. The most general technique used for contouring grids is the contour lacing approach where each contour in the grid is found, threaded through the grid, and sent as vectors plus any appropriate labelling and annotation to the hardware device which may or may not perform a vector/raster conversion before display.

(ii) Fault depiction. A perennial problem with automated contouring is the representation of faults, cliffs, and other discontinuities in the surface being contoured. It is not possible on the one hand to stipulate that a surface continuous in the first derivative is essential, and on the other to specify that breaklines must also be included. There have been many attempts to achieve some form of compromise between these opposed objectives, but they have usually resulted in the need to segment the data set into a number of areas that are both gridded and contoured independently and then plotted as one map. A few versions have attempted to generate weighting functions for a PDWA interpolation that are effectively almost discontinuous in the area of a fault by introducing the concept of barriers into the data. The calculation of distance weightings allowing for cliff position in a complex topographic map, although possible, takes an inordinate amount of computer time on anything but a large mainframe. Some work has been done (Pouzet, 1980) in geology with 3D seismic (already gridded) data, where the distances can be formalized owing to the regular nature of the grid. These introduced barrier effects without great time penalty. For grid-based systems the problem of developing faulted structures is related to:

(a) cliff recognition from the basic contour and spot height data—this is not really solved;
(b) computer time; and
(c) the problem of making the contours run up to a cliff and no further, without the need for elaborate masking techniques. This is an essentially unsolved problem, but definitely one of interest!

9.4 Smooth contour creation from grids and networks

The smoothness of contours representing a surface created from an irregular data set is not always of paramount importance. It is often perfectly acceptable to represent the surface by a series of linear approximations within the grid or network being used. Methods have been outlined by Heap (1972, 1974). There are two common methods of creating a contour map from an irregular data distribution. The first uses some local function to interpolate a regular grid that can then be contoured. The second method, triangulation, relies on the development of a network of triangles based on the data points to subdivide the area. Modifications can easily be made to the triangulation to allow for the insertion of ridge lines (Yoeli, 1977) or the insertion of such discontinuities as fault patterns or similar barriers. Both grids and triangulations use lacing to thread the contours through the area or, possibly, direct heap-sort techniques in more complicated systems. In addition to these, there are direct contouring from data approaches using iterative search techniques to locate contours without precalculating any sort of grid or network (Hibbert, 1977; Schagen, 1982).

In both triangular and rectangular cases, linearly interpolated contours appear angular to the eye, although the surface is continuous in value. In general such a map based on a large grid cell size with considerable surface variability and many data points has a very unappealing appearance and is only marginally interpretable. The solution to this problem can take one of two courses: either the grid mesh resolution can be increased by regridding until every cell of the grid has a side length less than a few millimetres, thus creating reasonably short vectors giving the appearance of a smooth curve; or the line has to be smoothed in some way. Smoothing a line *in vacuo* can be very dangerous: it is possible to make the lines individually smooth, but it is not possible to ensure absolutely that they do not cross. The solution is to create a smooth surface 'patch' that fits every grid cell in such a manner that the course of a contour through a grid cell can be determined by reference to interpolations on the patch rather than by linear interpolations to the grid cell edges. The patch function (either local or global) will generate a surface that honours exactly every grid node value and also estimated first derivatives, thus ensuring continuity over the whole area and hence a smooth surface. As smoothing is now performed with reference to the surface itself, a coherent pattern of contours is presented to the eye.

(i) Contouring surface patches. The increased quality of smooth contours exacts a price in terms of computation. Whereas a linear contour lacing is cheap to compute, the smoothed surface represents an investment in computer time at least proportional to the level of smoothness required. The subdivision of a patch cell into a series of subcells enables contours to be laced smoothly through the cell. Usually, interpolations within the patch area are only made at subgrid intersections bounding the contour as it passes through the patch cell. The same procedure can be applied to rectangular grids or triangles. For a triangle, three coordinates specify intersections in the subtriangle network.

The patch for a rectangular cell can be calculated using either globally or locally calculated exactly fitting functions. Some authors have favoured the global approach, as, once the coefficients for the surface are calculated, a fast contouring can be achieved compared with local patch-fitting operations (Holroyd and Bhattacharyya, 1970). The arguments against the global approach are constraints imposed by available computer memory. In some cases, the size of grid that adequately represents the detail in a data distribution of 10 000 points would be considerably larger than the memory available. It is necessary in a large grid to be able to contour and continuously join a number of pieces of the grid independently of each other. The local surface patch, generated quickly and only where needed, comes into its own in the contouring process where space is at a premium. It will only become inefficient where, on average, several contours pass through the same cell. The process of lacing contour lines through an arbitrarily-shaped triangle divided into subcells is analogous to the rectangular case. The main differences are that the coordinate system is more complex, and lacing decisions in any subcell are simplified to a choice of one of the two remaining sides of the subtriangle.

(ii) Derivative estimation. Whatever patch is employed, it is essential to have an estimate of the first derivatives at the vertices of the patch. The shape of the calculated surface, and hence the pattern of contour lines upon it, depends critically on these estimations. In the rectangular case the process is simpler, in that a regular grid is available and some method can be used to determine the derivatives in X and Y at each grid node using the immediate surrounding nodes in a stencil form to provide estimators of the true values. The estimate is often made using a divided difference approach, or a polynomial surface-fitting method. In either case, reasonably stable estimates can be achieved. All this is predicted on the availability of a grid of values. If these have to be interpolated from a randomly located data set, special, usually quite lengthy, methods will be needed to create the grid.

The triangular case is more complex, although the vertices of the triangles are data points and need no interpolation. They lie on no regular grid, giving them a very intractable format. The Thiessen neighbours of a point can be used as the equivalent of the stencil. Owing to their varying distances from the central location, they must be allocated weights according to some function representing variability increasing with distance. An inverse distance weighted low-order polynomial surface-fitting process seems a good method of estimating the first derivates in X and Y. The choice of order becomes critical where one or more of the neighbours is close to the estimate location, as the surface slope could be near-vertical in the extreme case where neighbours have very disparate values. Considerable care has to be taken to maintain a good distribution of neighbours (if necessary, several layers) in order to provide a stable solution.

(iii) Rectangular surface patches. The literature on surface patches for rectangular grids, either global or local, is immense. A summary is provided by Lancaster and Salkauskas (1977) of some of the major methods involving analytic solutions. The interest in surface patches has been a result not only of the desire to create smooth contours, but also of the need to produce a surface having certain physical properties. For instance, in geophysics the type of variation in the surface is predictable, and the function chosen for local interpolation should bear the same properties. In this chapter there are no assumptions as to the type of variability.

The surface patch computation must maintain a balance between coefficients computed on entry to the grid cell and the amount of extra calculation necessary for the creation of discrete individual, arbitrarily located interpolations. A commonly used patch is the Hermitean bicubic fitted to the corner points of the rectangle and their first derivative estimates. This equation has 16 terms, but, owing to the discrete regularly spaced subcell locations where interpolations are to be made, many of the coefficients can be partially precalculated at the start of the program, and others on entry to a given cell. For storage reasons the derivatives are usually recalculated every time a contour traverses a given grid cell. As grid cells are adjacent, it is not necessary to calculate all estimates for a cell every time. Other possibilities for the rectangular case include the use of blending functions (Barnhill, 1977; Lancaster and Salkauskas, 1977; Tipper, 1980) and the combination method involving a combination of a standard Hermitean cubic interpolation along the sides of the cell, and then a PDWA of these values for the final interpolation.

(iv) Triangular surface patches. The number of available surface patch methods for triangles is large, reflecting their importance in numerical analysis in engineering. The problem is that, although many methods are available (reviewed extensively by

Ritchie, 1978, and Zienkiewicz, 1971), the majority have made assumptions about the types of triangle that will be used. Most methods assume either acute or right-angled triangles and do not operate on an arbitrary triangle shape without subdivision or transformation to standard form. Any shape of triangle may be encountered in mapping and particularly in the development of a Delaunay triangulation. Many variations of the Clough-Tocher (1965) element exist (extensively reviewed by Ritchie, 1978), where the basic triangle is split into three sub-areas. Powell and Sabin (1977) suggest alternative subdivisions into four, six and 12 triangles and the use of piecewise quadratic approximations. Kluceiwicz (1978) developed Barnhill and Gregory's (1975) Boolean sum (blending) interpolant by mapping on to a standard triangle.

Coefficient calculation for the triangular surface patch is much more complex than for the rectangular case because of arbitrary triangular shape. This means that time in calculation becomes a major factor, and the user has to make some decisions about the smoothness of his surface. Two patches are considered here, first the Birkhoff-Mansfield (1974) 9-parameter rational interpolant based on a quintic polynomial that would normally require 21 coefficients, and secondly Akima's (1978) triangular quintic interpolant. The former reduces the quintic interpolation problem by making certain assumptions about the condition of the surface along the edges of the triangles, particularly by reducing the normal derivatives on the edges of the triangle to linear functions rather than cubic polynomials. Both interpolants operate without subdivision and guarantee continuity of the first derivative, and in Akima's interpolant the second as well. Triangle shapes can be extremely non-equilateral, leading to a problem of sharp surface changes occurring in the interior of a triangle in response to particularly violent changes in local gradient at the vertices.

The user of interpolated patches should, however, be aware of the different visual impressions created by the different functions. Imagine the same data set triangulated and then contoured using either a PDWA to interpolate heights within triangles based on a sufficiently large range to give good continuity, or alternatively a local patch such as Akima's. All data points would exist on either surface and be honoured, and the estimated derivative at every data point would be identical in both cases. But the surfaces drawn would differ, reflecting the mathematical formulation of the patch as either global or local. The results, while similar, would lead to the same regional structural interpretation. On a local level, the exact shape and possible interconnections of highs and lows represent an area of uncertainty.

9.5 Mapping using the triangulation alternative

The objections to the grid based methods outlined previously were:

(a) considerable computer time needed to interpolate a detailed regular grid to represent few data points;
(b) lack of flexibility in responding to variable data densities in different parts of a map;
(c) non-honouring of data points caused by an insufficiently fine grid chosen in order to keep computer time down to reasonable levels; and
(d) difficulty in representing cliff and breakline information adequately on a continuous surface.

This has led to the intensive development in recent years of alternative methods of generating contour maps.

The most widely known method is based on triangulation of the data set. The human cartographer, given a scattered data set to contour, will visualize a set of triangles in the area in which he is working that help him locate the contour he is tracing. These triangles have no real existence, but provide a structure for use in estimating the position of the contour line relative to other data points.

It seems likely that an automated approach doing the same would have considerable benefits, particularly as it would always honour all data points, because the data points would form the vertices of the triangulation. There used to be little interest in the automatic triangulation of terrain and other data sets however, because:

(a) it appeared impossible to generate the same triangulation, and hence the same map, from the same set of data, independent of the starting point of the triangulation process; and
(b) the time taken for automatic triangulation was excessive—the time taken for gridding is related to the square of the number of data points, but in some methods of automatic triangulation, it is at best related to the cube!

Improvements in the last ten years have now produced reliable triangulation procedures that produce the 'most equilateral' (and therefore unique) set of triangles possible—the Delaunay triangulation—in a time linearly related to the number of data points, and without large computer memory requirements. Some of the multitude of alternative names for the same procedure (or its dual) found in the literature are Thiessen, Voronoi, Dirichlet, and Deltri. For any given data set, it is now much faster to generate unsmoothed contours from an automatic triangulation procedure than from a grid interpolation approach. At the same time, all data points are honoured, the resolution of the map varies with the data density, and maps can be joined together without error at the margins. A major reason for the upsurge of interest in triangulation techniques has been that they are ideally suited to cliff and breakline insertion. If the locations are entered as a set of data points, the triangulation process will include them and will automatically relate them to the rest of the data set. Then, the triangulation can be 'unzipped' for a cliff, or marked for a breakline so that there is no direct interpolative connection in the data structure between the two halves. Contouring can then take place, and the result will be a perfect edge to the fault depending only on the input resolution of the fault line. Similarly, geological fault planes can be both inserted and contoured.

Any triangulation that is to be used for the basis of isarithm map production must have as a goal the properties of stability, equilateralness, and non-intersection. It is desirable that the triangulation resulting from any data distribution should be independent of the starting point of the triangulation process inside the distribution. This is particularly important in cases where ambiguity of triangulation might be expected to occur, for instance where four near-equidistant data points could be divided into two triangles in either of two ways. From the triangulation viewpoint there would be little difference, but in terms of the contoured surface there could be dramatic changes, related not to real variation but purely to the imposed triangulation. Any triangular approach usually attempts to achieve a set of triangles that are as near equilateral as possible with minimum line length. In the past this has often meant iterative processes such as those in SCA (1975) and GTN (1977) that attempted refinement of an initial triangulation. This did not produce a unique solution, and was expensive in computer time.

Brassel and Reif (1979) published a paper concerning the subdivision of a two-dimensional area into Thiessen polygons—the dual of the Delaunay triangulation—based on the location of a set of random data points. Their method is related to work by Rhynsburger (1973), Shamos and Bentley (1978), Green and Sibson (1978), Gold *et al.* (1977) and Elfick (1979). The Delaunay triangulation (Delaunay, 1934) has all the desired properties for use as a base for automatic contouring. The problem of calculating the triangulation is closely akin to that of Thiessen polygon generation, but certain modifications can be made which increase the speed of computation, help the algorithm reach linearity, and allow certain calculations to be omitted. The only major problem is that considered by Yoeli (1977), relating to the representation of known topographic structure where 'breaklines' may have to be included to maintain ridges and valleys. More recently there has been a great increase of interest in triangulation approaches to mapping, and a number of algorithms and reviews have appeared (Sabin, 1980; Peucker 1980; Sibson, 1981; Watson, 1982). Workers in different fields do not wish to achieve the same objectives. An interesting approach has been attempted in land surveying using a 'radial sweep algorithm' to generate a triangulation that approaches the Delaunay, but has apparent benefits for representing buildings on landscape (Mirante and Weingarten, 1982).

(i) Algorithmic organization. The Brassel and Reif algorithm approaches the problem of Thiessen polygon formation by choosing an arbitrary starting point and, as necessary, creating a set of imaginary guaranteed neighbours outside the data are to be polygonized. Once a known neighbour has been determined by this arbitrary starting method, each neighbour of a given point can be found by rotation about that point in a clockwise direction. While the neighbours for the rotation point are being discovered, it is possible to update indexes of the neighbours for all connected points, greatly reducing calculation effort at later stages. In a random data set, about two-thirds of all calculations will have already been made by the time any given data point is investigated.

Features of the Brassel and Reif algorithm are:

(a) It uses a one-dimensionally sorted data list and has to check a considerable number of points that may be the correct next neighbour out of the neighbours surrounding the data points—a two-dimensional sort structure would minimize searching time for any given point;

(b) much calculation is involved in determining the new circumscribing circle centre and radius—for Delaunay triangulation only, this is not necessary;

(c) it proceeds in a logical spatial manner through the data set, never covering the same ground again, thus giving considerable economy in storage—a major strength of the approach.

(ii) Delaunay triangulation for mapping. In all triangulation systems there is a boundary problem that must be solved in some manner. The points lying within the data window may well not be isolated, but only part of a larger data set. If so, then any arbitrary triangulation around the outside of the present data area must be incorrect. As it is impossible to know what lies outside the data window, some boundary condition must be set up to act as a frame and to provide a set of boundary triangles so that isarithms can be extrapolated outside the present apparent data area. Many possibilities exist for this, but one of the most efficient is to place a set of imaginary points around the outside of the area, just outside the data window. Once these imaginary points have been added to the original data set, the whole area can be triangulated, starting with any pair of imaginary points as initial known neighbours. The question of how many imaginary points should be used is moot.

(iii) Grid/triangular comparison. The justification for using a triangular network approach for contour mapping lies mainly in questions of accuracy of representation

of the surface and speed of calculation. The best comparison is with a regular rectangular grid covering the same area to an equivalent level of detail. Imagine 1000 randomly located data points over a square area. The size of grid used to represent this area would probably be of the order of 100 rows by 100 columns. The creation of the grid preparatory to contouring would involve the calculation of 10 000 interpolations, each one requiring searching and numeric effort. The Delaunay triangulation, on the other hand, requires the same initial X and Y sorting procedures to provide a box structure, followed by 1000 sets of neighbour calculations. If the average Delaunay computations for one point are considered approximately equivalent to the amount of effort in a grid interpolation, it can be seen that the saving through producing a triangular structure rather than a square grid can be great, particularly as the number of points rises. Added to this is the advantage of having the original data points located exactly on the triangular network and on the surface on which they should be found, as opposed to the rectangular grid approach which only occasionally honours the data points.

In order to give a fair comparison, it is also necessary to consider the contour lacing time and surface patch-fitting necessary to produce a final smooth contour map in both the rectangular grid and triangular network cases. If linear surface patches are considered adequate, then there is little difference in speed of execution for contour lacing between the rectangular grid and the triangular network approach when the triangular relational data structure formation time is taken into account. If, however, a continuity condition is to be imposed on the surface, then surface patch creation times can be considerably different. It has been found that the creation of a rectangular surface patch using bicubic hermitean interpolants is substantially faster than the time using an irregular triangular surface patch (the reduced quintic interpolant of Birkhoff and Mansfield, 1974, or the patch of Akima, 1982). On the other hand, the definition in the triangular case will be equal to the data density in all areas and thus accurately represent the information content at different points in the surface.

9.6 Contouring cliffs and breakline data sets

The number of data points used in constructing a contour map from such data will vary very considerably, from a small set of a few tens of points for some geological maps, up to more than 100 000 points for one map in the case of some high-accuracy digital terrain models based on contour input. Both these examples will entail the need to insert barriers, either vertical or sloping, in the form of discontinuities in the surface. In the first case, faults and fault planes are essential features of the map, and in the latter case, road cuttings, dams, and cliffs are important features of the landscape.

Some methods used to contour data sets are usually particularly quantity-sensitive. For instance, the multiquadric approach used by Hardy (1978) requires the inversion of an $N \times N$ matrix which quickly becomes prohibitively expensive. For small data sets of much fewer than 100 points, however, this approach may be very cost-effective. Faulted data sets are very difficult to represent using this technique. Other contouring systems based on direct searching for contours without any intervening grid interpolation or network structuring stages (Lodwick, 1970; McLain, 1974; Schagen, 1982) always honour the data but usually involve quite prodigious amounts of cpu time, are only applicable to reasonably small data sets, do not allow volumetric estimates easily to be made, and become impossibly slow if faulting or barriers are introduced. Grid-based systems have problems in representing barriers accurately without a 'step' appearance, even if they can be made to honour the data itself. An attractive solution to the problem of honouring data points without using an excessively fine grid has been proposed by Tobler (1979), using a lattice-tuning procedure, but its applicability to large data sets has yet to be demonstrated. On the other hand, breakline or cliff representation in such a scheme might well be possible.

In rectangular grid terms, the insertion of a fault is not easy, but the automatic recognition of the presence of a fault from the data set itself is witchcraft—far beyond the scope of this paper! Many approaches to fault insertion have been tried, with varying success, and requiring varying amounts of computer time. Examples of common methods are given by Bolondi *et al.* (1976) and Pouzet (1980), where a rather large amount of computer power is used to determine the fault boundary effect on grid surface values in the area of the fault pattern. The problems are twofold: recognition of the presence of a boundary, and the estimation of upthrown and downthrown elevations along the margins of the fault. The latter is compounded by the fact that faults die out and the surface must be continuous around the end.

(i) Vertical cliff/fault insertion by point replication. All the problems mentioned above are ameliorated when an irregular triangular structure is used instead of a grid. The data points for the nodes of the structure, and any fault pattern, can then be superimposed upon that triangulation. Vertical fault traces, defined as a set of coordinate pairs, can be considered 'special' data points, usually with indeterminate values. These can either be triangulated along with the real data to produce a set of triangles that include the exact path of the fault along the triangle edges, or can be inserted as a post-process. The PANZER program in PANACEA allows the structure of the triangulation to be transformed to turn potential cliffs into actual ones. The contouring process needs no extra guidance over a normal contouring job, as the cliff forms a hole in the map surrounded by an edge, no different from the edge of the map area as far as the program is concerned.

It is not really feasible to represent breaklines in the same fashion. The solution to this problem is to consider a breakline as a discontinuity marker rather than as a hole in the data. There is then no need for duplication of any points, providing that the surface patch function used on the triangles can deal with breaks of slope on some but not necessarily all of the triangle vertices. The one used in PANACEA does this.

9.7 Hybrid mapping systems

Incoming string data usually consists of digitized x,y data locations, each with a specific associated z or height value. An alternative for isarithmic data would be strings of x,y locations with a single implied surface value. Single point (spot height) data could also be used. Data could be of various types:

(a) smooth continuous variation in slope around the point;
(b) slope discontinuities at the point (ridge lines, river courses, sea lake or embankment edges, non-vertical geological faults); or
(c) discontinuities in the thematic variable (cliffs, vertical faults, boundaries).

All three types are still represented by location and value, but in addition every string must also possess a type identifier corresponding to the classes given above.

The output from the transformation will be a data structure binding all the points together according to some predefined set of rules: for PANACEA a triangular one that exactly represents the data. The data point locations form nodes of the structure. The manipulation of such a network does, however, require more sophisticated programming methods than for the regular grid. A review of some polygonal methods is given by Peucker (1979).

(i) Mixed methods. Where both speed in execution and convenience of use are required, a combination of grid and triangular methods may be preferred. The PANACEA system is an attempt to achieve this result by calculating, editing, and storing a triangular digital terrain model (DTM), but providing as final output a grid DTM based on the triangular structure but at the resolution required by the user. The same triangular DTM can then be used, without further recalculation, to generate as many rectangular grids as required from all or part of the data area.

9.8 PANACEA

The PANACEA system is a surface generation system typically taking in contour strings, spot heights, and cliff and river lines, and producing as output high-accuracy regular grid DTMs with associated block diagram and contour output. The complete system contains modules arranged to perform six logical functions:

 (i) string preprocessor (PAN) to reduce string complexity and size to a level consistent with the final display cell resolution;
 (ii) spatial structure generation program (PANIC) that defines internode relationships, and string integrity restorer (PANZER);
 (iii) graphic editor (PANDEMONium), including structure display and contouring, to adjust the automatically derived relationships to suit special purposes, and to allow insertion of extra special purpose strings (such as faults);
 (iv) derivative estimation program (PANDER) operating on the data files, and a separate grid generation program (PANDORA) operating on data structures created by PANIC and PANDEMON;
 (v) display suite to permit direct checking of the grid by sampling using either isometric views (PANORAMA) or contouring (PANACHE) from regular rectangular grids, direct contouring of the triangular structure (PANTECH);
 (vi) amalgamation module (PANEL) that takes grids generated by the system (possibly at differing resolutions and possibly covering overlapping areas) and combines them into one grid.

(PAN creates PANIC and PANDEMONium, but PANDORA generates with PANACHE a PANORAMA to be PANELled together, thus providing a PANACEA).

PANACEA is written in standard Fortran 77 with the usual exceptions of system dependent I/O operations and assignments. File opening and closing are separated into a few small routines to allow easy installation of the system on most machines. The system is presently operational on a number of microcomputers, minicomputers, and mainframe machines.

(i) PAN—string pre-processing. It is desirable that excess points should be removed from line input leaving only those that best define the line at the chosen resolution. Thus if a contour line has far too high a density of data, some weeding out has to take place. Many methods have been proposed (Peucker, 1975) for achieving this goal: in this module a modification of the Douglas–Peucker (1973) algorithms has been

adopted, although most users have their own systems of data reduction with direct entry to the PANIC module.

(ii) PANIC—data structure formation. PANIC determines the Thiessen neighbours for each point in the complete data set. The PANIC module can be dimensioned to deal with whatever size of problem the user considers to be a reasonable maximum. In the case of a virtual memory computer such as the VAX, there are no theoretical limits, but speed will decline as more virtual memory is used.

In the case of topographic mapping, the procedure does not rely on full contour integrity—gaps are allowed, as are spot heights, breaks of slope lines, and clifflines. Thus one can use partially digitized maps where the digitizing has attempted to maintain a given density of information rather than represent every contour. Contour strings can be supplemented by data strings for ridge, river, and valley lines to provide slope discontinuities, always bearing in mind that all locations will have had to be tagged with height values.

Associated with PANIC is PANZER, an option to restore any breaks in logical string path that may have occurred. If, for instance, a river string marking the bottom of a valley is broken during the Delaunay triangulation phase, then PANZER will, on command, restore the continuity of the logical river or contour connections thus ensuring that the path of the river is well defined and does not break into a series of 'puddles' down the valley. The same applies to cliffs where a break in the logical cliffline would lead to the equivalent of a landslide, and to a lake boundary where a break could lead, logically, to its emptying!

(iii) PANDORA—grid generation. PANDER reads the files output from PANIC, estimates values for the imaginary points around the edge of the data area, then derivatives for the whole data set, and writes out the derivative set. PANDORA can then generate a grid covering any area of the data structure. Grids (DTMs) of virtually unlimited size can be created from the spatial data structure using PANDORA. The largest possible segment of memory is reserved as a balance between storing all data points and holding the DTM. The program automatically chops the DTM into sections that hold as much as possible in memory, and can therefore handle virtually any size of grid. Either smooth surface or linear facet interpolation can be performed.

(iv) DTM—display systems—PANORAMA and PANACHE. Once a grid has been calculated, it must be checked. The simplest method is usually to inspect it visually for errors using either a contour plot (PANACHE) or a false three-dimensional view such as an isometric (PANORAMA) or perspective block diagram. Each approach has its own merits and each should be used for checking different elements of the final DTM. Both PANORAMA and PANACHE are menu-driven, giving default options but allowing selection of vertical exaggeration, specific area grid coverage, colour output, and, for PANORAMA, optional colour coded contoured isometric views. An example of PANACHE (Fig. 9.1) and PANORAMA (Fig. 9.2) output is shown below for a part of the Cairngorms in Scotland.

In addition, PANTECH can be used to derive contours directly from the triangular data structure without first creating a grid using PANDORA. This has the advantage of being exactly faithful to the original data, whereas the standard PANACEA grid will not be to some extent, depending on its cell resolution.

(v) PANDEMON—graphical structure display and editor. When data is entered into a system it often contains errors caused by insufficient or incorrect data in certain areas of the map, leading to the incorrect definition of particular characteristics of the shape. Examples of this kind of flaw could be inaccurate location of point data, or

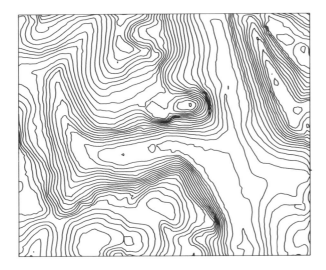

Figure 9.1 A mountainous area of the Cairngorms, 6 km × 5 km, was used to create the model displayed below. Contours at 100 ft are shown on PANACHE output from a 121 column × 101 row grid.

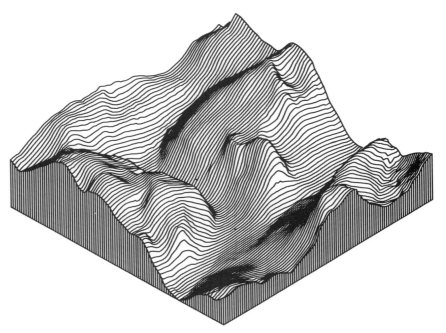

Figure 9.2 Isometric PANORAMA view of the grid displayed in Fig. 9.1, in monochrome, using scan lines parallel to the x axis.

inaccurate height values for specific locations. These can be modified on a flexible basis using the graphic editor, working either on single points or on logically connected strings, e.g., contour segments, river lines, etc. The other activity to be performed by the editor is the insertion of extra strings into the existing network. This is particularly important for cliff lines, breaks of slope edge (as in quarries, or lake/sea

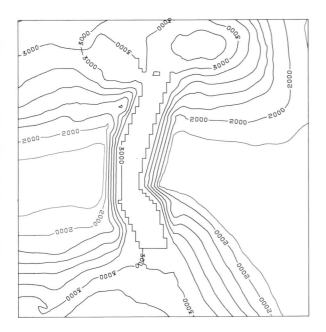

Figure 9.3 PANACHE contoured dam inserted using PANDEMON commands.

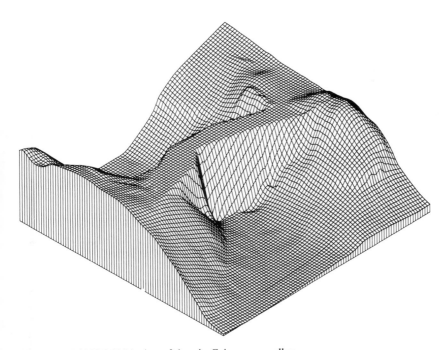

Figure 9.4 PANORAMA view of dam in Cairngorms valley.

F

areas), and highway engineering (embankments, cuttings). PANDEMON has options allowing flexible file or keyboard input of any such strings to superimpose on the original triangulation. The process of data insertion maintains permanent connection between points in all logical input strings in much the same fashion as PANZER. Figs. 9.3 and 9.4 show the result of using PANDEMON to insert a rather primitive dam across the valley displayed in the centre of the example given for PANACHE and PANORAMA (Figs. 9.1, 9.2). The black-and-white isometric and contoured views show the effectiveness of using the breakline option for insertion of the top of the dam in combination with smooth patches on the valley sides.

(vi) PANEL—grid amalgamation and output. Not all grids need be created in a single pass, particularly for large DTMs where it may be sensible to grid different terrain areas separately, but allow some overlap between the areas. PANEL can then amalgamate these areas into one large final DTM. Every grid will be held in a file with its own unique filename. PANEL reads a file containing these filenames, opens them in the order which they are needed, and pieces together the final grid.

9.9 Derived mapping

So far, only the direct applications of computer mapping have been considered where the product has been a contour map representing some thematic variable whose distribution must be represented visually. Once that variable has been determined over an area, in the form of grid or triangulation, it is then possible to carry out further analyses on the same data or combinations of data.

The PANTHEON subsystem—a collection of derived mapping programs based on PANACEA DTMs as input—has been used to generate planimetrically correct grids representing slope and aspect (PANSLOPE); relief shading and insolation information (PANSHADE); a halftone mapping system (PANTONE), intervisibility (PANVIS), and other information that can be derived from the DTM and other relevant information. PANTHEON also allows an interface (PANTOMIR) to be created between the topographic DTM and any areally associated imagery such as Landsat/TM or SPOT scenes in classified, level sliced, or raw mode, and cartographic line/area data. Similarly, PANCHROME is an interface from the companion image-processing system (MIRAGE) that allows isometric representation of terrain with any associated derived mapping or imagery to be draped over it to give a 3D view of an area colour fill rather than wire frame as in PANORAMA.

Halftone output (Fig. 9.5) is very useful as a base map on which other information can be printed. Another example of PANTONE output is given in McCullagh *et al.* (1985) where halftones were used to produce hazard maps for landslide prediction.

Other possibilities abound: the conversion of seismic velocity data to depth information is one example, and another geological example could be the creation of isopach maps from subsurface horizon maps. In civil engineering, derived mapping is a daily occurrence for site investigations, and also in the planning field where environmental impact assessment can be greatly enhanced using DTMs and derived mapping. An esoteric example is the use of Fourier spectra for the analysis of landforms. Another is the Wakarusa Quadrangle Kansas Atlas jointly produced by the Kansas Geological Survey and the Experimental Cartography Unit (Campbell and Davis, 1979). Using information concerning cultural features, topography, geology and soils, these authors were able to create maps defining the landsurface elements, the depth to bedrock, recreational areas, the permeability of surface

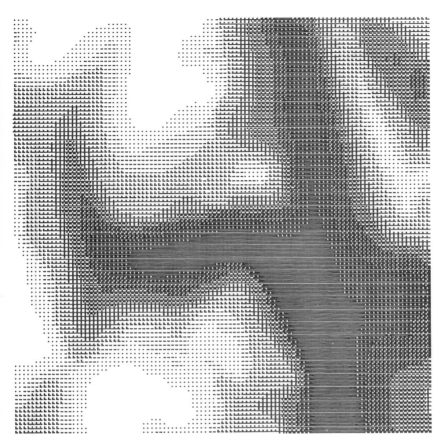

Figure 9.5 Halftone example produced by PANTONE of the Cairngorms test area. Low areas are blackest, and high areas are white, using a total of 10 classes of shading pattern based on a 3 × 3 grid of dots. Output can be on almost any hardware device, although preference should be given to some form of raster device such as an inkjet printer.

materials, foundation support for low buildings, soil capability groups, slope maps, and the suitability of surface materials for use as roadfill. It can be appreciated that one has rarely finished when one produces the first map of the original variable under consideration, be it topographic, seismic, or any thematic variable!

9.10 Image processing

The MIRAGE system, currently in its final development phase, consists of a set of modules to perform a wide range of functions involved in the acquisition, registration, display, enhancement and analysis of monochrome or false-colour digital images. There are three basic types of module in the system:

(a) tasks initiated by the user to enter and format imagery from outside the system via magnetic tape;

(b) image merging, combination, geometric rectification and registration processes; and
(c) real-time image manipulation, classification, and display.

All tasks are menu-driven from a fully interactive question/answer sequence. A series of utility programs allows the user to read a range of satellite computer-compatible tapes, such as Landsat, Tiros-NOAA, CZCS, and HCMM. Landsat formats presently readable include Kiruna, Fucino, EROS, EDIPS, South African, and NRC Thailand.

Interaction is the keynote of image processing systems. All image display functions are performed in real time and involve no disk I/O delay, as the images are stored in main memory or graphic memory if space permits. Thus the only limitation on update speed for this fully interactive phase is that imposed by the processor and memory and the parallel link to the colour hardware. All the basic operations are available:

(a) windows are implemented whereby the user can allocate different parts of his actual screen to hold different images, and hence operate on several images at once;
(b) image display—the images may be displayed either in monochrome or in false colour;
(c) zoom/unzoom—full incremental zoom and unzoom facilities are provided with automatic windowing;
(d) roam facility—if the image in memory is larger than the displayed image, it is possible to 'roam' the image at will. If the image is in zoom mode then the roaming will maintain the zoom, thus allowing detailed inspection of the complete memory-held image. Any other images on display will remain 'locked' to the present focus of attention.

Arithmetic facilities include:

(a) linear or histogram based contrast stretch methods using uniform function;
(b) standard spatial convolution techniques are available using either a fast median filtering algorithm where the high-pass filtered image is formed as the subtraction of the low-pass filtered image, or by incorporating user defined filter weights as the filter function (e.g., Robert's gradient, Laplacian);
(c) edge detection options enable the user to define edge detection criteria adaptively over the image area;
(d) frequency domain filtering using the fast Fourier transform can be undertaken—a range of bandpass and directional filters can be used, including user defined filter functions;
(e) arithmetical facilities include addition, subtraction, functional transformations, multiplication and division (ratioing);
(f) principal components analysis can be undertaken to replace the original images with the component score images;
(g) a fast classifier is included, based on the parallelpiped method or, optionally, a one-pass k-means algorithm. This is particularly useful for first look and exploratory studies, as it operates in near real time. Alternative, but slower, methods are available optionally as well. These include ISODATA, and maximum likelihood algorithms, and contain options to allow user specification of class prior probabilities.

(i) Hardware/software design concept. As Fortran-77 allows an almost unlimited

address space, an image processing task can occupy a great deal of memory without any need for disk access to replenish its image memory, and all current images can be held in memory. MIRAGE takes advantage of this memory availability. Many of today's supermicrocomputers are based on 68 000 or superior processors capable of addressing up to 16 Mb of memory of roughly 32 512 × 512 images! Not all these computers allow the full 16 Mb memory to be used, but all do allow any process to access at least 1 Mb. Thus, even at the lower end of the hardware scale a reasonably sized image can be held totally in memory. Typically, for a minimum system, three 512 × 512 images can be handled in memory without the need for disk transfer. After initial loading, all operations are simply memory transfers and can be performed very quickly. Similarly the displayed image can be completely updated to reflect the results of these changes at memory access speed. Alternatively, the user can specify at run-time that certain images are to be retained on disk, and others held in memory.

The usual hardware configuration assumed available for MIRAGE is:

(a) 68 000 processor (or better) with 1 Mb memory, possibly UNIX multi-user operating system, and Fortran 77;
(b) magnetic tape drive (or perhaps floppies—slow!), controller, interface, and supporting software;
(c) preferably more than 40 Mb of Winchester disk for system and image storage;
(d) graphics display hardware—colour with either 8- or 24-bit planes, parallel or DMA interface, supporting basic system software—colour RGB monitor capable of resolving a 512 × 512 image;
(e) console terminal—VT100 or equivalent.

A smaller system could be implemented, perhaps by using a minimum 10 Mb Winchester, floppy disks, no tape drive, economizing on the colour hardware, but providing the potential for upgrading at later date. A larger system need not be much more expensive, as the main additions would be memory to allow larger images to be handled, and possibly a higher-resolution colour graphics hardware system.

Bibliography—general

Brassel, K.E. (1984) Strategies and data models for computer-aided generalisation. *Euro-Carto III*, ed. W. Kainz, Int. Cart. Assn. and Austrian Computer Society, 19 pp.

Brodlie, K.W. (ed.) (1980) *Mathematical Methods in Computer Graphics and Design*. Academic Press, New York, 63–86.

Campbell, J.B., Davis, J.C. and the Experimental Cartography Unit (1979) Northern Part of the Wakarusa Quadrangle, Kansas: Atlas, No. 5, Series on Spatial Analysis, Kansas Geological Survey, Kansas, US, 29 pp.

Crain, I.K. and B.K. Bhattacharyya (1967) Treatment of non-equispaced two-dimensional data with a digital computer. *Geoexploration* 5, 173–194.

Davis, J.C. (1973) *Statistics and Data Analysis in Geology*. John Wiley and Sons, New York, 550 pp.

Davis, J.C. and McCullagh, M.J., (eds.) (1975) *Display and Analysis of Spatial Data*. John Wiley and Sons, London, 383 pp.

Douglas, D.H. and Peucker, T.K. (1973) Algorithms for the reduction of the number of points required to represent a digitised line or its caricature. *Canadian Cartographer* 10(2) (Dec), 112–122.

Eidenbenz, Ch. (1984) The evaluation of cartographic systems: experiences of the evaluation phase in project DIKART. *Euro-Carto III*, ed. W. Kainz, Int. Cart. Assn. and Austrian Computer Society, 14 pp.

Foley, J. and Dam, van A. (1984) *Fundamentals of Interactive Computer Graphics*. Addison-Wesley, Mass.

Franke, R. and Neilson, X. (1980) Smooth interpolation of large sets of scattered data. *International Journal for Numerical Methods in Engineering*, 15, 1691–1704.

Hardy, R.L. (1971) Multiquadric equations of topography and other irregular surfaces. *Journal of Geophysical Research*, 76(8) 1905–1915.

Hardy, R.L. (1977) Least squares prediction, *Photogrammetric Engineering and Remote Sensing*, **47**(4), 475–492.

Hayes, J.G. and Halliday, J. (1974) The least squares fitting of cubic spline surfaces to general data sets. *J. Inst. Maths. Applic.* **14**, 89–103.

Journel, A.G. and Huijbregts, C.J. (1978) *Mining Geostatistics*. Academic Press, London, 600 pp.

Kasvand, T. (1983) Computerised vision for the geologist. *Journal of IAMG* **15**(1) 3–23.

Lam, N.S. (1983) Spatial interpolation methods: a review. *American Cartographer* **10**(2) 129–149.

Lancaster, P., and Salkauskas, K., (1977) *A Survey of Curve and Surface Fitting*. Published by the authors, 3052 Conrad Drive, Calgary, Alberta, Canada.

McCullagh, M.J. (1982) Mini/micro display of surface mapping and analysis techniques. *Cartographica, Perspectives in the Alternate Cartography* (*Monograph 28, Euro-Carto I*), **19**(2) 136–144.

McCullagh, M.J. and Bennett, A.M. (1983) An interactive surface modelling system for thematic data. *Monograph of Euro-Carto II?*

McCullagh, M.J. and Doornkamp J.C. (1984) Hazard mapping in engineering geology. *Euro-Carto III*, W. Kainz, ed., Int. Cart. Assn. and Austrian Computer Society, 13 pp.

McCullagh, M.J., Cross, M. and Trigg, A.D. (1985) New technology and supermicros in hazard map production. *Surveying and Mapping 85*, 2nd UK National Land Surveying and Mapping Conference, 16 pp.

McLain, D.H. (1974) Drawing contours from arbitrary data points. *Computer Journal* **17**, 318–324.

Olea, R.A. (1975) Optimal mapping techniques using regionalised variable theory. *Kansas Geological Survey Series on Spatial Analysis*, No. 2, 137 pp.

Peucker, T.K. (1975) A theory of the cartographic line. *Proceedings of Auto-Carto II*, ACSM and Bureau of the Census, 508, 518.

Peucker, T.K. (1979) Digital terrain models: an overview. *Proceedings of Auto-Carto IV*, ACSM and ASP, 97–107.

Peucker, T.K. (1980) The impact of different mathematical approaches to contouring. *Cartographica* **17**, 73–95.

Peucker, T.K. and Chrisman, X (1975) Cartographic data structures. *The American Cartographer* **2**(1) 55–69.

Pouzet, J. (1980) Estimation of surface with known discontinuities for automatic contouring programs. *J. Int. Assn. Math. Geol.* **12**(6) 559–576.

Rhind, D. (1975) A skeletal overview of spatial interpolation techniques. *Computer Applications* **2**(3 and 4) 293–309.

Sabin, M.A. (1980) Contouring—a review of methods for scattered data. In *Mathematical Methods in Computer Graphics and Design*, Brodlie, K.W., (ed.), Academic Press, New York, 63–86.

Sampson, R.J. (1975) The Surface II Graphics System. In *The Display and Analysis of Spatial Data*, Davis, J.C. and McCullagh, M.J., eds, John Wiley, New York, 244–266.

Schagen, I.P. (1982) Automatic contouring from scattered data points. *Computer Journal* **25**(1) 7–11.

Tobler, W.R., (1979) Lattice tunning. *Geographical Analysis* **11**(1) 36–44.

Traylor, C.T. and Watkins, J.F. (1985) Map symbols for use in the three dimensional display of large scale digital terrain models using micro-computer technology. *Proceedings of Auto-Carto VII, Digital Representations of Knowledge*, ASP and ACSM, 527–531.

White, M.S. (1984) Technical requirements and standards for a multi-purpose geographic data system. *American Cartographer* **11**(1) 15–26.

Zienkiewicz, O.C. (1971) *The Finite Element Method in Engineering Science*. McGraw-Hill, London.

Bibliography—triangulation

Brassel, K.E. and Reif, D. (1979) Procedure to generate Thiessen polygons. *Geographical Analysis* **11**, 289–303.

Delaunay, B. (1934) Sur la sphère vide. *Bulletin of the Academy of Sciences of the USSR, Classe Sci. Mat. Nat.*, 793–800.

Gold, C., Charters, T. and Ramsden, J. (1977) Automated contour mapping using triangular element data. *Computer Graphics* **11**, 170–175.

Green, P.J. and Sibson, R. (1978) Computing Dirichlet tessellations in the plane. *Computer Journal* **21**, 168–172.

Heap, B.R., (1972) Algorithms for the production of contour maps over an irregular triangular mesh. *National Physical Laboratory Report* NAC 10.

Hodges, D.J., Alderson, J.S. and Johnson, S.M. (1987) Computer graphics for mine planning enquiries. *Mine and Quarry*, February, 46–48.

McCullagh, M.J. (1980) Triangulation systems in surface representation. *Proceedings of Auto-Carto IV*, International Symposium on Cartography and Computing. Vol. 1, 146–153.

McCullagh, M.J., and Ross, C.G. (1980) Delaunay triangulation of a random data set for isarithmic mapping. *Cartographic Journal* **17**(2) (Dec) 93–99.

McCullagh, M.J. (1982) Contouring of geological structures by triangulation. *Geological Society of London, Miscellaneous Paper No 15–Computer Applications in Geology III*, 47–60.

Mirante, A. and Weingarten, N. (1982) The radial sweep algorithm for constructing triangulated irregular networks. *IEEE CG and A*, May, 11–21.

Rhynsburger, D. (1973) Analytic delineation of Thiessen polygons. *Geographical Analysis* **5**, 133–144.

Sibson, R. (1981) A brief description of natural neighbour interpolation. In *Interpreting Multivariate Data*, Barnett, V. (ed.)., John Wiley, New York, 21–36.

Watson, D.F. (1981) Computing the *n*-dimensional Delaunay tessellation with application to Voronoi polytopes. *The Computer Journal* **24**, 167–172.

Watson, D.F. (1982) ACCORD—Automatic contouring of raw data. *Computers and Geosciences* **8**, 97–101.

Bibliography—patches

Akima, H., (1974) A method of bivariate interpolation and smooth surface fitting based on local procedures. *Communications of the ACM*, **17**(1) 18–20.

Akima, H. (1978) A method of bivariate interpolation and smooth surface fitting for irregularly distributed data points. *ACM Transactions on Mathematical Software*, **4**(2) 148–159.

Akima, H. (1978) Algorithm 526: Bivariate interpolation and smooth surface fitting for irregularly distributed data points (E1). *ACM Transactions on Mathematical Software*, **4**(2) 160–164.

Barnhill, R.E. (1975) Blending function interpolation: a survey and some new results. *International Series of Numerical Mathematics* **30**, Birkauser Verlag, Basel.

Barnhill, R.E. (1977) Representation and approximation of surfaces. *Mathematical Software III*, J.R. Rice (ed.), Academic Press, New York, 69–118.

Barnhill, R.E. and Farin, G. (1981) C^1 quintic interpolation over triangles—two explicit representations. *International Journal for Numerical Methods in Engineering* **17**, 1763–1778.

Bhattacharyya, B.K. (1969) Bicubic spline interpolation as a method for treatment of potential field data. *Geophysics* **34**(3) 402–423.

Bolondi, G., Rocca, F. and Zanoletti, S. (1976) Automatic contouring of faulted sub-surfaces. *Geophysics* **34**(3) 402–423.

Ellis, T.M.R., and McLain, D.H. (1977) Algorithm 514: A new method of cubic curve fitting using local data (E2). *ACM Transactions on Mathematical Software* **3**(2) 175–178.

Franke, R. (1977) Locally determined smooth interpolation at irregularly spaced points in several variables. *Jnl Inst. Maths Applics.* **19**, 471–482.

Franke, R. (1982a) Smooth interpolation of scattered data by local thin plate splines. *Computers and Mathematics with Applications* **8**, 273–281.

Franke, R. (1982b) Scattered data interpolation—tests of some methods. *Mathematics of Computation* **38**, 181–200.

Klucewicz, I.M. (1978) A piecewise C^1 interpolant to arbitrarily spaced data. *Computer Graphics and Image Processing* **8**, 92–112.

Lodwick, G.D. and Whittle, J. (1970) A technique for automatic contouring of field survey data. *Australian Computer Journal* **2**(3) (Aug), 104–109.

McCullagh, M.J. (1981) Creation of smooth contours over irregularly distributed data using local surface patches. *Geographical Analysis* **13**(1) (Jan), 51–63.

McCullagh, M.J. (1983a) Automatic contouring of faulted seismic data sets. *Computer Graphics Forum, Journal of Eurographics Association*.

McCullagh, M.J. (1983b) Transformation of contour strings to a rectangular grid based digital elevation model. *Monograph of Euro-Carto II*.

Powell, M.J.D. (1976) Numerical methods for fitting functions of two variables. *IMA Conference on the State of the Art in Numerical Analysis*, University of York, April 1976.

Ritchie, S.I.M. (1978) Surface representation by finite elements. Unpublished Master's thesis, University of Calgary, Alberta, Canada.

Tipper, J.C. (1980) Surface modelling techniques. *Series on Spatial Analysis* **4**, Kansas Geological Survey, University of Kansas.

Bibliography—image processing

Gonzalez, R.C. and Wintz, P. (1977) *Digital Image Processing.* Addison-Wesley, Mass.
Lillesand, T.M. and Kiefer R.W. (1979) *Remote Sensing and Image Interpretation.* John Wiley, New York, 612 pp.
Pavlidis, T. (1982) *Algorithms for Graphics and Image Processing.* Computer Science Press, Rockville.
Rosenfeld, A. and Kak, A.C. (1976) *Picture Processing by Computer.* Academic Press, New York.
Saebo, H.V., Braten, K., Hjort, N.L., Llewellyn, B. and Mohn, E. (1985) Contextual classification of remotely sensed data and development of a system. *Report No 768, April 1985*, Norwegian Computing Center, Oslo. 101 pp.

Bibliography—data structures and information systems

(Many references are from Auto-Carto VII, where there was much interest in GIS, and also some from the International Symposium on Spatial Data Handling, Zurich, 1984.).
Abel, D.J. (1984) A B$^+$-tree structure for large quadtrees. *Computer Vision, Graphics, and Image Processing* **27**, 19–31.
Aronson, P. (1985) Applying software engineering to a general purpose geographic information system. *Proceedings of Auto-Carto VII–Digital Representations of Knowledge*, ASP and ACSM, 23–31.
Brassel, K.E. (1984) Strategies and data models for computer-aided generalisation. *Euro-Carto III*, W. Kainz (ed.), Int. Cart. Assn. and Austrian Computer Society, 19 pp.
Cebrian, J.A., Mower, J.E. and Mark, D. (1985) Analysis and display of digital elevation models within a quad Tree-based geographic information system. *proceedings of Auto-Carto VII-Digital Representations of Knowledge*, ASP and ACSM, 55–64.
Chen, Z.-T. and Peuquet, D. (1985) Quad Tree spatial spectra guide: a fast heuristic search in large GIS. *Proceedings of Auto-Carto VII–Digital Representations of Knowledge*, ASP and ACSM, 75–82.
Kent, W. (1983) A simple guide to five normal forms in relational database theory. *Comm. of ACM* **26**(2) 120–125.
Mark, D. and Lauzon, J.P. (1985) Approaches for Quadtree-based geographic information systems at continental or global scales. *Proceedings of Auto-Carto VII–Digital Representations of Knowledge*, ASP and ACSM, 355–364.
Morehouse, S. (1985) ARC/INFO: a geo-relational model for spatial information. *Proceedings of Auto-Carto VII–Digital Representations of Knowledge*, ASP and ACSM, 388–397.
Peuquet, D.J. (1983) A hybrid structure for the storage and manipulation of very large spatial data sets. *Computer Vision, Graphics, and Image Processing* **24**, 14–27.
Peuquet, D.J. (1984) Data structures for a knowledge based geographical information system (KBGIS). *Proceedings, International Symposium on Spatial Data Handling, Zurich*, August 1984, Vol. 2, 372–391.
Pfaltz, J.L. (1977) *Computer Data Structures.* McGraw-Hill, New York, 446 pp.
Roessel, van, J.W. and Fosnight, E.A. (1985) A relational approach to vector data structure conversion. *Proceedings of Auto-Carto VII– Digital Representations of Knowledge*, ASP and ACSM, 541–551.
Saalfield, A.J. (1985) Lattice structure in geography *Proceedings of Auto-Carto VII–Digital Representations of Knowledge*, ASP and ACSM, 482–489.
Saalfield, A.J. and O'Reagan, R.T. (1985) Streamlining curve-fitting algorithms that involve Vector-to-raster conversion. *Proceedings of Auto-Carto VII–Digital Representation of Knowledge*, ASP and ACSM, 491–497.
Samet, H. (1984) The Quadtree and related hierarchical data structures. *Computing Surveys* **16**(2) 187–260.
Shapiro, L.G. and Haralick, R.M. (1980) A spatial data structure. *Geo-Processing* **1**, No. 3.

Bibliography—artificial intelligence

Davis, R. and Lennat, D.B. (1982) *Knowledge-based Systems in Artificial Intelligence.* McGraw-Hill Series in Artificial Intelligence, McGraw-Hill, New York, 490 pp.
McCorduck, P. (1979) *Machines Who Think.* Freeman, San Francisco.
Rich, E. (1983) *Artificial Intelligence.* McGraw-Hill Series in Artificial Intelligence, McGraw-Hill, New York, 436 pp.

Smith, T.E. (1984) Artificial Intelligence and its applicability to geographical problem solving. *Professional Geographer*, **36**.

Smith, T.E. and Pazner, M. (1984) Knowledge-based control of search and learning in a large GIS. *Proceedings, International Symposium on Spatial Data Handling, Zurich*, Vol. 2, 498–519.

Waltz, D. (1983) Artificial intelligence: an assessment of the state of the art and recommendations for future directions. *AI Magazine* **4**, 55–67.

Part C
Systems and Software

10 Surface modelling systems

E. MALCOLMSON

10.1 Introduction

Early computers lacked the memory to cope with the considerable amount of data used for surface definition. Square and rectangular grids of levels, where a simple easting and northing register and grid interval could be used to position an array of levels, were the first serious attempts at storing the surface of the ground in a computer. However, the technique breaks down where more intimate changes in topography occur. The string ground model technique of storing a series of easting and northing coordinates with an attached level enabled better definition of the ground by relating the computer data to features of the topography, and cross-sections taken through this type of data represented more accurately the changes of shape. An early attempt at producing triangular surface models, which not only show the features of change in topography, but represent the surface as well, required the observer to specify which points were linked to form triangles. This very tedious method of creating a surface model has been superseded by algorithms for automatically generating triangular surface representations from different types of data.

10.2 Data collection

It is often difficult to match data collection methods to the type of surface model required for a particular analysis. Methods of data collection and the most suitable type of model will be discussed initially, followed by methods of analysis and the feasibility of converting between different types of model.

Theodolite surveys, whether based on manual, electronic, or total station systems, present information most related to features in the topography and can include random points in open ground in order to represent the elevation of featureless terrain. With direct graphical editing of the analysed data, the surveyor can arrive very quickly at mapping that represents his detail as well as an elevation survey. Sophisticated graphics software facilities such as fitting curves to sections or features, squaring buildings, creating parallel lines, using ties and lines of sight by reference to the surveyor's field book will enable the full map to be completed efficiently. This type of model, because of its subjective choice of surveyed points, is ideal for producing a true surface model of a triangular plate type. Its disadvantages in use as a string-type ground model relate to the directions of the strings and the interpretation of open ground strings when producing cross-sections, the inevitable method used for any analysis of string-type ground models. Photogrammetry, via stereophotographs, can provide detailed information and surface level data in a variety of forms. Current practice from stereophotogrammetry includes the production of contours and three-dimensional strings in true string format, with a large number of points representing the curvature of features. Due to the high volume of data inherent in capturing such information, the data is usually only suited to being handled in the form of a string ground model type. With specific instructions, the data can be extracted from stereophotographs in a way more suited to surface representation such as square or rectangular grids and triangular mesh models.

Table digitizing has often been resorted to when the user has existing maps requiring input to a computer system in order to be used for surface modelling. Contours and three-dimensional information result from this method of data collection and, in common with photogrammetric techniques, produces a considerable number of points representing the strings. Video digitizing in a raster format, supplemented by raster-to-vector conversion techniques, produces similar results to table digitizing but with a higher throughput in terms of data preparation (see Chapter 7). Occasionally the user will have access to sections from existing surveys or from analysis of other computer data. This type of ordered data can be extremely efficient in creating surface models of the triangular element type.

10.3 Surface model types

Square and rectangular grid models suit open topography or other surfaces that do not change abruptly. Applications abound in this area, but the data collection problem of precision in locating grid points detracts from the overall cost effectiveness of this type of data storage. It would be hypocritical to take data from actual feature observations and interpolate a grid of levels unless there were specific reasons for lessening the accuracy of data. (Some 'smoothing' techniques involve grid interpolations.) String models generally look good, because the strings are usually features that the user can relate to. Often a string-type data collection is the most convenient. However, processing overheads are extremely expensive, because of the necessity to 'section' the data in order to simulate a 'solid' surface, and serious discrepancies can occur if the section extraction process is not sophisticated enough to detect side-effects. Section data can be used where analyses are related directly to the original sections, but creates its own problems in a changing environment, when representation of data may become necessary each time a new analysis is required.

Triangular models are formed such that each line of a triangle is common with another (except at the edge of the model). They are formed from original data points, and can have variable amounts of data to represent the surface to different degrees of 'fit' or accuracy. Generally the triangular plate is regarded as a plane surface, and so more elements give a better definition of the surface.

Surface validation It is important to provide users with easy-to-use procedures which will encourage the validation of data representing the surface. Typically, a string model will undergo a planimetric check but little in the way of an elevation check, since the data representation requires extensive and costly computing to produce sections, or to give perspective views of the data. So the checking is done only when obvious errors are produced in the actual processing of the model.

With square and triangular models, analysis is fast enough to produce contouring, which is an efficient means of checking elevation data. Perspectives also provide a useful checking method. Any system should allow the user to adjust the surface data at any point which is in error, and not have to resort to raw data editing and re-processing.

Data model analysis. Several analyses may have to be carried out on models for various purposes, but they can be categorized into the following broad headings:

(a) distance and area calculations;
(b) volume calculations;
(c) contouring;
(d) plan drawing;
(e) isometric and perspective projections.

The type of model utilized plays an important part in determining the efficiency of the above analyses and it is worthwhile discussing the methods used by the different types of models in the computation of results. In *distance* and *area calculations*, and the interpolation of spot points, only the square and triangular model types offer a reasonable basis for computation. In these types, the required points will be interpolated on a plane, or mathematical surface represented by surrounding data, within the structure of the model itself. For all other types of model where the data essentially represents features or lines, the difficulty lies in locating the surrounding points on which to base an interpolation. Some systems utilize multiple or secondary interpolation to achieve reasonable results, but if the data prepared in the first place is not conducive to this type of analysis, then the process is questionable.

Volume calculations can be categorized similarly. However, the restrictions of this analysis when the data model type is formed from strings is less of a problem, since sectioning the data can be employed to provide sufficient representation of the surface, albeit at the expense of long computational procedures. Where the surface is represented by grids and triangles, volumes are computed by column analogy, that is, the actual data points are 'averaged' and the area of the column multiplied by the height difference between surface and datum. This technique is very much faster and more accurate than the interpolation of sections and the use of average-end-area or other methods of computing volumes.

Contouring once again displays the difference in computational requirements between the different model types. Again in principle, a section has to be used to produce a representation of a surface through a string type model in order to assess where contours occur. With square and rectangular models, direct interpolation of contours across each surface can be used. Even if adjacent 'elements' are not stored such that one continuous contour is extracted, but instead each line element representing a contour is produced for each surface element, at the end of all such computations joined contours result. It is a simple matter to take the individual straight-line elements interpolated through each element and string them together to form continuous string-type contours.

In *plan drawing*, it is obvious that a string-type model can represent information related to detail, and that the presentation of this detail on the screen or plotter is at its most efficient. With square or rectangular meshes, it is impossible to carry detailed information that would be useful for plan plotting. With triangular meshes, if the triangular elements were originally formed from features, the sides of the triangles can carry coding to allow their representation in plan drawing form. However, since triangular elements that could have features attached to their sides will generally have been formed from a string type of model in the first place, the output of plan information would be taken from the original data, not the surface model.

10.4 User interfaces

With the new graphics workstations (IBM PC and compatibles, Perq, Sun, Apollo, Whitechapel, Microvax II, etc.) a new breed of software became necessary to exploit the advantages these systems provided over the old technology graphics terminals.

Implicit in the operation of this new architecture of software and hardware is 'interactive' working. It is of paramount importance for a user to feel that he is at all times in charge of his data, and not merely remotely connected to the data.

Any graphics system must therefore have built-in interactive working, and a good system will follow this principle from initial manipulation of raw data, through editing and the creation of new details, to the final ability to adjust surfaces. This is all accomplished on-screen with the display reflecting all operations and their effects, as they are made.

Summary. A comprehensive surface modelling facility should include:

Input	Data from digitizers, data-loggers and survey instruments, aerial surveys, data from other computer systems, OS digital data
Validation	User-defined tolerances, fast 2D and 3D data viewing and correction
Editing	Fast location of data and graphical and attribute data editing; selection of data by boundaries
Modelling	String, square/rectangular, section and triangular forms supported
Design	Creation of new features, new surfaces in all forms of modelling type; modular approach to allow special cases such as highway design to be incorporated easily
Analysis	Distance, area, volume, contouring, isopachyte, sectioning, interpolative facilities
Graphical output	User-defined design of characters and symbols, line type, colour; user selection of output for 'layering'.

11 Software packages for terrain modelling applications

T.J.M. KENNIE

11.1 Introduction

Previous chapters have focused on the alternative procedures and algorithms for both the regular grid and triangular-based techniques together with the various methods employed for the generation of contours. This chapter considers how these techniques are used by a selection of the software packages currently available internationally and provides a framework for the future discussion of the individual packages.

A number of publications have discussed the various methods used for the formation of DTMs and also the interpolation procedures which can be adopted (such as Grist, 1972; Jancaitis and Junkins, 1973; Schut, 1976). However, these papers do not report some of the most recent developments in this field. In order to provide a framework for discussion of the range of program packages available, a classification scheme (Fig. 11.1) will be used.

11.2 General-purpose modelling/contouring packages

The most general, and often least flexible, approach to adopt for terrain modelling is to use one of the many general purpose modelling/contouring packages. Most of these were written in the early 1970s. Almost all are essentially grid-based and usually contain random-to-grid routines to convert non-grid data to the required grid format. Most organizations which possess a large computer installation will normally find that this software is readily available. These packages can be used to model any regularly or irregularly distributed scientific data set. The characteristics of some of the most common grid-based packages are illustrated in Table 11.1.

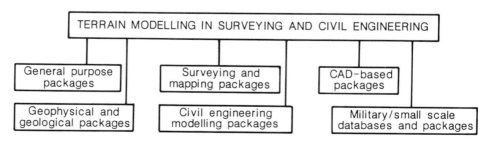

Figure 11.1 Classification of terrain modelling software packages in surveying and civil engineering.

Table 11.1 Comparison of the interpolation and contouring procedures used in a selected number of general-purpose modelling/contouring packages (adapted from Grassie, 1982)

Name of package	Method of interpolation from random points to grid nodes	Method of contouring
GHOST	Pointwise (distance weighted average)	Triangle (diagonal method)
SURFACE II GRAPHICS	Pointwise (distance weighted average)	Rectangle (linear)
GINOSURF	Pointwise (paraboloid)	Triangle (centre point method)
GPCP (General Purpose Contouring Package)	Pointwise (projected weighted average)	Polynomial (bicubic)
SACM (Surface Approximation and Contouring Mapping)	Patchwise	Rectangle
SURFACE II GRAPHICS (TREND)	Global (up to 6th order)	Rectangle (linear)
GPCP (TREND)	Global (up to 10th order)	Polynomial (bicubic)

As mentioned above (Section 8.2.2), any one of three interpolation procedures can be utilized if randomly positioned data are being used: (a) *pointwise*, involving the determination of different functional parameters for each point, using either the distance weighted average, paraboloid or projected weighted average technique; (b) *patchwise*, involving the establishment of a series of local surfaces; and (c) *global*, based on a single polynomial function (including up to 10th order terms). Similarly various contouring functions can be used to thread contours through the interpolated grid of data points. An evaluation of the accuracy of some of these general-purpose packages for cartographic applications has been presented by Grassie (1982). Grassie carried out an empirical study using various topographic data sets. Using contours measured photogrammetrically as a reference, he compared the accuracies obtained using different packages. Triangulation methods produced consistently better results than random-to-grid methods of which the patchwise and global techniques often produced poor results. Gridded or evenly distributed data sets were more satisfactory than poorly distributed linear data.

11.3 Surveying and mapping modelling/contouring packages

While the packages outlined in the previous section were designed to handle digital data from a large number of different technical fields, there are also several programs which have been designed specifically for handling topographic and terrain data.

11.3.1 Height Interpolation by Finite Elements (HIFI)

HIFI (Ebner *et al.*, 1980; Ebner and Reiss, 1984) is a suite of programs developed at the Technical University of Munich, which is available commercially from Carl Zeiss (Oberkochen) as part of the software associated with their Planicomp C100 analytical plotter.

The finite element method has been applied to many problems in civil and mechanical engineering, but to date it has not been used extensively in the surveying and mapping industry. In essence, however, HIFI is a grid-based package and was developed especially to deal with the grid-based data sets generated by photogrammetric methods, e.g., during profile scanning for orthophotograph production and under computer control using an analytical plotter. Thus in some respects it is not too different from the packages described in 11.2. If non-gridded data is being used, it performs firstly a random-to-grid interpolation, followed by contouring based on the interpolated heights at the grid nodes. In contrast to the pointwise methods used by most of the programs discussed in 11.2, it uses a patchwise interpolation technique. Furthermore, it is able to include the effects of breaklines which are important in terrain modelling but are not generally included in the general-purpose programs. Breaklines are terrain discontinuities (such as ridges and valley floors) measured by the surveyor or photogrammetrist. The inclusion of such data prevents a model forming, for example, across a valley (Fig. 11.2). Two versions of HIFI are available: the first uses *bicubic finite elements*, or patches, whereas the second, which allows for the inclusion of breaklines, uses *bilinear elements*. In areas where breaklines are not considered necessary, the use of bicubic patches enables a significant reduction in computational effort to be realised. The form of the two surfaces is illustrated by Fig. 11.2. The software is written in Fortran and is designed to run on Hewlett-Packard, DEC and Prime hardware.

11.3.2 Contour Interpolation Program (CIP)

CIP (Steidler *et al.*, 1984 and Chapter 15), is the current terrain modelling/contouring package from Wild Heerbrugg (previously they offered the SCOP package). It is written in Fortran and is designed to run on the Data General Nova series (used by the Wild AC-1 and BC-1 analytical plotters) and DEC-VAX 11 series (used by the Wildmap system).

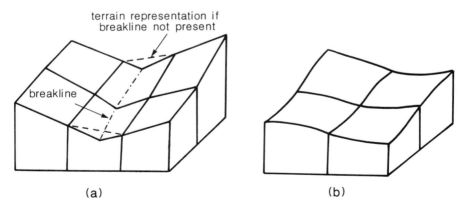

Figure 11.2 HIFI-interpolation surfaces (adapted from Ebner *et al.*, 1980). (a) Bilinear finite elements. (b) Bicubic finite elements.

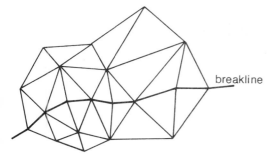

breakline

Figure 11.3 CIP-triangular mesh including breaklines (Steidler *et al.*, 1984).

In contrast to HIFI, the program is based on the use of triangles as the finite elements or patches (Fig. 11.3).

The triangular method fits the measured data exactly and it also lends itself to the incorporation of breaklines.

The interpolation procedure involves the use of a 10-term cubic polynomial which forms a smooth surface across each triangular patch. Continuity is ensured at the boundaries, except across breaklines. Contouring is carried out using 16 subtriangles. The cutting points along the edges of the subtriangles are then determined. Other options include facilities for generating the heights of single points, a regular grid DTM, profiles, perspective views and the volume relative to a reference surface.

11.3.3 Wild System 9-DTM

Wild Heerbrugg have recently introduced a new contouring and terrain modelling package which forms part of the software for the System 9 digital mapping and geographic information system (McLaren and Brunner, 1986; Sykes and Craine, 1986). The package, System 9-DTM (Steidler *et al.*, 1986), is written in the programming language 'C' and runs under the UNIX operating system.

Data can be input in either random or regularly spaced format together with information about the location of breaklines and fault lines. The model is created using a Delaunay triangulation, ensuring that lines or polygons which have known three-dimensional coordinates always form the edges of triangles. The surface normals at each node are then computed and the surface defined by a series of finite elements using Zienkiewicz functions (Bazeley *et al.*, 1965).

Editing of the triangle net and the interpolated contours can also be performed, each change automatically effecting a change to the point and any adjacent points in the database. Contouring is performed by subdividing the result of the Delaunay triangulation into a series of subtriangles. The degree of subdivision ranges from 0 to 100. Further options available within the package include:

(a) the creation of perspective views using either a wire frame hidden line removal technique or flat shading in which the triangles are given a colour which is dependent on the slope of the terrain;
(b) interpolation of a regular grid DTM or the height of single points of specified X, Y coordinates;
(c) the production of intervisibility maps;
(d) the production of sections through the model together with volumes, mass haul diagrams and isolines indicating the differences between models.

11.3.4 Intergraph

The well-known US systems supplier Intergraph has developed several terrain modelling packages, generally as part of an integrated digital mapping system. One particular system, developed in 1982 for the US Defense Mapping Agency (DMA), specifically for viewing topographic surfaces and mapping operations, is the Terrain Edit System/Elevation Matrix Processing System (TES/EMPS). The hardware installation, probably one of the largest in the world dedicated to terrain modelling, consists of two Intergraph 780 systems each equipped with an array processor connected to 12 graphics workstations, with twin 1280×1024 pixel high-resolution displays.

Data can be input in either grid, random point or string (contour) form. In the latter two instances the data are reformatted into a grid or matrix of elevations. Input in vector or raster form is also possible with the facility to convert between both forms of data structure. Up to 8 scenes of the terrain can be viewed at any one instant, together with full interactive editing of the data.

Intergraph has also developed a digital terrain modelling nucleus (DTMN) as part of its civilian mapping system. Again, various forms of data input can be accommodated including random point data from field instruments, grid data from photogrammetric stereoplotters and vector or raster cartographic data acquired from contours. A combination of the Interactive Coordinate Geometry (COGO) System (ICS) with the DTMN enables users to carry out horizontal and vertical road design and site planning operations. More details are given in Sections 6.5 and 6.6.

11.4 Geological and geophysical modelling/contouring packages

While general-purpose modelling/contouring packages can be used for examining geological and geophysical phenomena, again they are limited in their ability to deal with surface discontinuities such as geological faults. Consequently a number of packages have been developed specifically to deal with problems of this type.

11.4.1 Contour Plotting System (CPS)

CPS is a widely established modelling and contouring package which was originally developed in 1972 by UNITECH and is now available from the Radian Corporation. It is written in Fortran and was developed primarily for users in the oil, gas and mining industries. It is available in three forms: CPS-1, which is designed to operate on mainframe and superminicomputers; CPS/PC which operates on IBM/At and P5/2 range of personal computers; and CPS-1/G, an interactive version for the display and editing of projects which have been run on CPS-1.

Unlike the majority of the packages discussed so far which are generally restricted to one or two methods of interpolation, CPS-1 offers users a choice of various interpolation algorithms including global trend surfaces using polynomials and local interpolation by a number of methods. However, the major benefit of packages such as CPS is their ability to handle geological faults and other surface discontinuities. Other features of CPS-1 include the facility to generate 'ortho-contour maps' in which regions of maximum gradient are illustrated. This feature is of particular use in water run-off and oil migration studies. Three-dimensional raised contour plots can also be created.

11.4.2 UNIMAP
Another package which has been used for the depiction of geological and geophysical surfaces is UNIMAP. UNIMAP, and its subsidiary products GEOINT and GEOPAK, enable the user to access a range of interpolation procedures for the creation of datasets from gridded input, the formation of gridded datasets from randomly distributed observations, and for the production of contoured maps. Interpolation can be either of a global nature, or patchwise using bicubic polynomials and local (pointwise) using distance weighted average methods. The ability to deal with faults is also a feature of the package.

11.4.3 Further packages
In addition to the packages mentioned above, geological and geophysical modelling and contouring has also been performed using the PANACEA system (McCullagh, 1983; and Chapter 9). Another package which is in common use is Z-Map from the Zycor Corporation.

For some users the opportunity to tailor a suite of programs to meet the in-house requirements of an organization may be important. Thus, in addition to the packages already mentioned, several libraries of subroutines are also available. Two examples of these libraries are those available from Precision Visuals (DI-3000 subroutines) and from Northwest Digital Research (Interactive Surface Modelling, ISM). The latter, available in Fortran 77 and Pascal, has been used for landfill operations (Garvis and Gee, 1985).

11.5 Civil engineering modelling, design and drafting packages

Civil engineers have been using digital terrain model data for commercial operations since the mid-1960s. The early systems, developed primarily for highway design, enabled the engineer to replace the traditional manual design process by an automated system. Thus, several different road alignments could be examined within a narrow corridor of interest, a process which was not practical before the advent of these Computer-Aided Highway Design (CAHD) packages. One of the most popular of these early systems was the primarily grid-based British Integrated Program System (BIPS) for highway design. By producing links to other suites of programs such as the Highway Optimisation Program System (HOPS) for cost minimisation (Davies, 1972), the engineer had at his disposal a powerful design tool.

These early systems were, however, limited in several respects:

(a) restricted terrain modelling capabilities, with data normally held in a square grid or cross-section form (consequently terrain definition was poor);
(b) inability to deal with complex geometrical conditions at junctions and interchanges;
(c) the terrain model and the civil engineering design could not be merged to produce a composite model;
(d) limited drafting facilities (although automatic output could be generated, it was generally not possible to produce either cartographically acceptable drawings, or final contract design drawings, without considerable manual intervention); and
(e) no interactive graphics capability for interrogating and editing the database.

One of the first systems to address many of the limitations of these early systems was the MOSS Modelling System (Craine, 1985; Houlton, 1985, 1986).

11.5.1 Mainframe and minicomputer-based packages

(i) MOSS. MOSS was developed from 1974 onwards by a consortium of UK county councils (Durham, Northamptonshire and West Sussex) acting on behalf of the County Surveyors Society. The system continued to be developed by this consortium until 1983 when the privately owned MOSS Systems Ltd was formed. Its use is widespread in the civil engineering industry. It is particularly dominant in local authorities and consulting engineers' offices, and has over 150 users worldwide.

The concepts of 'point' and 'string' data are fundamental to an understanding of the MOSS system. Both survey and design data are considered to consist of either single points (such as lamp-posts and trees), or strings, which are a series of linked three-dimensional points (kerb lines, contours, breaklines, etc.). The use of this string data together with triangular modelling techniques enabled a much better terrain surface representation to be achieved than with cross-section or grid-based techniques. Furthermore, defining both the terrain and the design surface (e.g., road, railway, dam, quarry) in this way, within a common database, overcame the limitations outlined in (b) and (c) to a considerable degree. The drafting and interactive problems have also been overcome to a large extent with the advent of graphics workstations, such as those based on the Apollo Sun and DEC Microvax range. With these a new range of interactive facilities termed Interactive MOSS have been added to the current system. These facilities allow the user to create graphics windows and display all, or part, of a model on a graphics screen, and by means of mouse-driven screen menus to alter the way in which graphic information is drawn.

Visualization of the impact of new structures by using photomontage techniques (Siddans, 1980), has been a feature of MOSS for several years. A new development is the production of colour-shaded perspective views using image-processing techniques (Cobb, 1985). Systems such as VOSCO from the CAD Centre and VIGIL from the Transport and Road Research Laboratory (TRRL) enable a more realistic impression of the relationship between the terrain and the structure to be obtained. Futhermore, it is possible to stimulate the effects of varying lighting and weather conditions (e.g., fog), and also the examine the visual intrusion caused by the proposed structure on the terrain.

Although MOSS was initially developed for highway design projects, it has developed into a general purpose package for civil engineering modelling, design and drafting. It has been applied to a wide range of projects including the design of roads, railways, airports and dams, together with applications in land reclamation and mining operations (see 11.8). Further information is given in Chapters 3 and 16.

(ii) AXIS Surface Modelling System (SMS). AXIS-SMS is a modelling design and drafting package similar to MOSS. Written in Fortran-77, the package has been installed on a variety of mainframes (IBM, ICL), minicomputers (DEC PDP 11/23) and engineering workstations (ICL Perq and Apollo Domain). Recently it has also been rewritten to run on the IBM PC/XT, AT, and Olivetti M24 range of microcomputers.

The facilities offered by the package are similar to those available from other manufacturers and include triangular modelling (using the radial sweep method), area and volume analysis, horizontal and vertical road design, together with various output options including perspective viewing. The system can be linked to a separate mapping package.

Closely associated with the SMS package is the AXIS Mapping System (AMS) which, as the name suggests, is an interactive digital mapping system designed to handle field survey, photogrammetric and cartographic data specifically for topographic and utility mapping purposes.

(iii) ECL/Computervision. Elstree Computing Ltd (ECL) have recently developed three packages which are particularly relevant to civil engineering. The packages were initially developed as part of an in-house system; however, the packages are now available commercially. Three distinct packages are available: Digital Ground Modelling (DGM) is a conventional triangular modelling system based on the Delaunay algorithm, Highway Design System (HDS) offers facilities for the interactive design of horizontal and vertical alignments together with 3D visualization of projects, and finally, the Site Layout and Engineering System (SLES) enables designers to perform building and drainage layout designs interactively.

Currently the software is available only on the Computervision CGP 200X processor running the CADDS graphics database softwaree. It is, however, proposed to offer the software on the computers.

11.5.2 Personal computer (PC) and desktop computer-based packages

During the past five years or so, a variety of terrain modelling packages have become available on PC and desktop computers. Milne and Motlagh (1986) recently reported details of many of the packages available in the UK which run on IBM PCs, Apricot and Hewlett Packard microcomputers. Some of the more common systems currently in use are listed below.

(i) PANTERRA. PANTERRA (previously known as ECLIPSE) is marketed by Ground Modelling Systems Ltd (Chapter 13), and was initially developed in the late 1970s as a workstation system based on the Wang 2200 series of desktop computers. Recently, however, the system has also been installed on a variety of personal

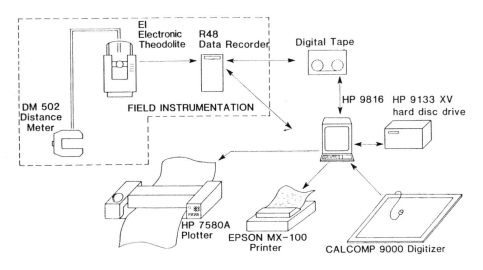

Figure 11.4 HASP-DIGITAL system flowchart (Hodges and Alderson, 1985). Reproduced, with permission, from *Mining Department Magazine*, University of Nottingham.

computers, e.g., the Wang PC and APC, together with the IBM PC/XT and AT
(Gould, 1986). Like MOSS, it is widely used, particularly for surveying operations,
and has over 100 users worldwide.

PANTERRA is a highly interactive package, written largely in Basic, which enables
the user to input survey data from a wide variety of sources (field data recorders,
stereoplotters and digitizers), process the data using standard surveying computational techniques and create a 3D database. This database can be used subsequently
for the creation of digital mapping data or be integrated with road, building layout or
drainage design information. Triangular terrain models can be created, and area and
volume reports generated. Output can be cartographically enhanced, and models can
be transformed to MOSS, GDS, RUCAPS (an architectural package) and
AUTOCAD formats. ECLIPSE is now named PANTERRA and is described in more
detail in Chapter 13.

(ii) HASP DIGICAL 11S. The HASP-DIGICAL 11S system consists of a series of
program modules developed for surface modelling, volumetric calculations, road and
building design, digital mapping and cadastral surveying (Hogan, 1984 and
Chapter 17).

The modules, although originally written in HPL, are now available in Pascal.
They are designed to run on the Hewlett Packard (HP) 200 and 300 series of desktop
computers and the HP 9816 and 9825 series of personal computers. Modelling is
carried out using the triangular irregular network technique using the *radial sweep
algorithm* developed by Mirante and Weingarten (1982). Breaklines can be incorporated either at the data collection stage or during the data processing. A
typical hardware configuration using the HP 200 series for processing is illustrated
in Fig. 11.4.

11.6 Computer-aided design and drafting (CADD) modelling systems

The cost-effective benefits of using CADD systems are being increasingly realized by
the civil engineering industry. Many organizations now possess a CADD facility for
the production of engineering drawings. For many of these users, the opportunity to
use a terrain modelling package within the context of a more general-purpose CADD
system has some obvious advantages, particularly in terms of data compatibility and
speed of data transfer.

Two distinct types of CADD system are available to the engineer; those based on
surface modelling and those which form a solid model. Terrain modelling packages
are currently available for both types of system.

11.6.1 Surface modelling
Surface modelling CADD systems are the most common type in use in the civil
engineering industry. A terrain modelling capability operating within a conventional
drafting system has been developed by McDonnell Douglas, the interactive SIte
TErrain System (SITES) to operate within its General Drafting System (GDS).

(i) GDS-SITES. GDS is a command-driven drafting and modelling system which
was launched in 1980 and enables the interactive creation, recall and editing of
drawings to be carried out. Two complementary packages, SURVEY and SITES,
have recently been added to the system for the creation and analysis of terrain data.

SURVEY (Bray and McGarry, 1986) provides a means of entering raw field-
surveyed data into the modelling part of the system. Control data is processed by

standard methods, an interesting feature being the inclusion of an interactive pre-analysis package for the optimization of control networks. Detail survey can be entered by various methods, including data from field data recorders. One product of the SURVEY package is a database of spot heights which forms the terrain model.

SITES (McGarry, 1985) is a separate element of the system which carries out the interpolation and analysis of the model. The basic algorithms used were developed for an earlier package known as CIIS (Contouring, Interpolation and Integration of Surfaces; Grayer, 1980). Essentially it consists of a modified Delaunay triangulation to allow for the inclusion of breaklines. Contours are determined by a linear interpolation along each triangular face: smoothed contours can also be created by fitting a quadratic b-spline through the points.

At present the system does not include any design capabilities, for example for road design, although volumetric analysis can be carried out. One feature of particular interest to hydrologists is the facility to display watershed lines and to indicate, using cross-slope vectors, the direction of run-off. This ability to extract ground geometry, and interrogate the terrain for physical attributes other than simple spot height information, has led to the coining of the term *Intelligent Ground Models* (IGM) (Strodachs and Durrant, 1986) to describe systems which offer these additional features.

(ii) ACROPOLIS—Building Design Partnership (BDP). A further example of a surface modelling system which forms part of an integrated CADD system is the ACROPOLIS system from BDP Computing Services. Although developed primarily for architectural operations, the system can also generate ground models for visualization purposes.

11.6.2 Solid modelling
Solid modelling systems have been developed primarily for the display and manipulation of mechanical engineering components. Systems of this type can be classified, according to Pratt (1984) and Requicha and Voelker (1982), into Constructive Solid Geometry (CSG), Boundary Representation (B-Rep) and simple sweep types. In the first case, shapes are represented in terms of simple volumetric primitives such as the cuboid, cylinder, cone, sphere and torus. The limitations of this type of modelling system for terrain modelling operations are discussed in Strodachs and Durrant (1986). The second approach, which involves the storage of all the elements which make up an object's boundary (the faces, edges and vertices together with topological information about relationships between elements) is currently under investigation by researchers at Loughborough University of Technology (Strodachs and Durrant, 1986; Mayo et al., 1986). The system which they are examining is based on the MEDUSA CADD system developed by Cambridge Interactive Systems (CIS). The third type of solid modelling system has not been assessed for terrain modelling.

(i) MEDUSA terrain modelling (MTM) package. The MTM package is currently a development system. Facetting techniques are used by the system to develop a series of triangular elements (using the Delaunay triangulation) to form a solid model of the terrain. At present, however, the system does not allow for the inclusion of breaklines, although this has been suggested as an improvement for future versions of the package. Boolean or set operations can be invoked to enable new structures (such as roads and dams) to be superimposed on to the terrain model. From this composite model, volumetric data could subsequently be determined. Figure 11.5 illustrates the use of these techniques to simulate the effect of a proposed highway on an existing

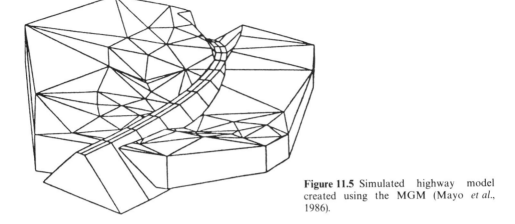

Figure 11.5 Simulated highway model created using the MGM (Mayo *et al.*, 1986).

ground model. Although the system uses a different modelling technique, the final results are not too dissimilar from those obtained from many of the packages discussed earlier. More information is given on the Medusa-based systems in Chapter 18.

11.7 Military and small-scale databases and packages

Terrain modelling has been widely used for many years by the military to create elevation databases over large regions. Data of this type, particularly when integrated with terrain feature information, can be used not only for basic military planning but also for radar simulation, weapons guidance and for flight simulation. Central to all systems which are used by the military is a series of databases created by the United States Defense Mapping Agency (DMA).

DMA is probably the largest mapping agency in the world and employs about 10 000 personnel in over 50 locations throughout the world. A substantial number are engaged in the production and revision of the global Digital Land Mass System (DLMS). Sections 7.4.2 and 7.4.3 provide information on the DEM data for the USA.

A wide range of special-purpose software has been developed by military organizations to deal with the specific problems of providing realistic terrain views over large geographical areas. Examples include the UK Directorate of Military Survey Terrain Validation and Exploitation System (TVES), the US DMA Terrain Edit System/Elevation Matrix Processing System (TES/EMPS), the US Army Topographic Laboratory (USATL) Digital Image Analysis Laboratory (DIAL) and Field Exploitation of Elevation Data (FEED) systems. Several systems, however, have been developed by non-military organizations for this type of operation. Three examples are EAMACS, Lasercheck/Viewfinder and PANACEA, which are discussed in Chapters 21, 22 and 23 respectively.

Finally, Turnbull *et al.* (1986) describe the use of small-scale DTMs as part of a computer-aided visual impact analysis (CAVIA) system. A typical application of the system has been the integration of terrain models with land use information in order to assess the visual impact of the route for a series of electricity transmission towers. The use of video techniques to mix computer-generated and natural landscapes, and to simulate different climatic conditions, is also described. Further details are provided in Chapters 19 and 20.

11.8 Applications

The main impetus for the development of many of the packages outlined in section 11.5 was the demand for major highways during the late 1960s. Since then, however, the use of these packages has diversified into many other fields of surveying, geology and civil and military engineering. The following list, while unlikely to be complete, provides an indication of the wide range of projects where terrain modelling methods have been used.

11.8.1 Surveying and mapping
Terrain models can be created both by field surveying and photogrammetric techniques and generally form the basis for the applications outlined in sections 11.7.2 and 11.7.3. In addition to the standard production of square grid, triangular or string terrain models for engineering projects, terrain models have also been used to:

(a) model the sea bed from depths derived from echo soundings and positions obtained from electronic position-fixing equipment;
(b) rectify satellite remote sensing imagery (for example as created from the Landsat or Seasat series of satellites); and
(c) enable the superimposition of satellite imagery on to a realistic 3D surface.

11.8.2 Geology and geophysics
Modelling systems have been developed (section 11.4) to deal with the specific problems of representing heavily faulted geological structures and of modelling geophysical data. Some applications in this area include:

(a) oil reservoir simulation studies;
(b) the determination of reservoir capacity of an oil field from geological and geophysical measurements; and
(c) the modelling of aeromagnetic and seismic data.

11.8.3 Civil engineering
(i) Highway engineering. Modelling systems can be used to solve a variety of highway engineering problems, including:

(a) the measurement of earthworks from borrow pits or spoil areas or to produce volumes of different strata for precise costing;
(b) the design of major roads;
(c) the design of interchanges and junctions or complex urban redevelopment schemes; and
(d) the visualization of structural elements (for example, bridge decks or concrete foundation slabs).

(ii) Other linear civil engineering projects. Many of the techniques developed for highway engineering projects can also be applied to other linear projects, including:

(a) the design of major railways and rapid transit systems;
(b) the design of canals and river restraining schemes;
(c) airfield design; and

(d) monitoring and visualization of tunnels using tunnel profiling systems.

(*iii*) *Areal civil engineering projects*. Large areal projects can also benefit from the use of terrain modelling techniques. Some recent examples include:

(a) land reclamation and rehabilitation operations associated with mineral extraction, flattening of spoil tips or the filling of gravel pits;
(b) catchment/watershed analysis to assess the storage capacity of reservoirs; and
(c) dam design and monitoring of tailings dams.

11.8.4 Military engineering

In addition to the use of terrain modelling techniques for military (as opposed to civil) engineering projects the techniques have also been applied to specific military problems, including:

(a) the siting of ground-based radar stations;
(b) line-of-sight and intervisibility analyses for artillery planning;
(c) analysis of exposure and fallout regions associated with nuclear, biological or chemical attack;
(d) flight simulation over restricted airspace or ground corridors; and
(e) simulation of the performance of missile systems which use terrain comparison guidance techniques.

11.9 Conclusions

Terrain modelling has been applied to a wide range of problems in the fields of surveying and mapping, geology and civil and military engineering. For the future, the use of terrain modelling methods will continue to develop and expand into other areas of activity. Although trends are difficult to identify, it seems possible that the creation of a national terrain model of the UK is a likely product for the future, such a model being created from existing 1:50 000 scale contour maps formatted into a grid, the grid dimensions depending on the variability of the terrain. While the technical problems of creating such a database are considerable a much greater problem is the cost, revision and general management of such information. While such a derived model would be sufficiently accurate for general visualization, or flight simulation, it would be unlikely to meet the needs of surveyors and civil engineers. For such users, the primary source of terrain data is likely to continue to be directly measured spot heights and the creation of grid- and triangular-based models.

Acknowledgements

The author wishes to acknowledge the assistance given by many manufacturers and individuals. They have supplied a great deal of relevant and interesting material, much of which has had to be condensed considerably to fit within this chapter.

References

Assmus, E. (1976) Extension of the Stuttgart contour program to treating terrain break lines. *Presented Paper, Commission III, 13th Int.Soc. of Photogrammetry (ISP) Congress,* Helsinki, 13 pp.

Bazeley, G.P., Cheung, Y.K., Irons, Y.K. and Zienkiewicz, B.M. (1965) Triangular elements in bending conforming and non-conforming solutions. *Proc. 1st Conf. on Matrix Methods in Structural Mechanics.* Air Force Inst. of Technology, Wright Patterson AFB.

Bray, D. and McGarry, M. (1986). From total station to digital model—the role of CAD. *Proc. CADCAM'86, Birmingham*, 249-254.

Cobb, J. (1985). A new route for graphics. *Proc. of Computer Graphics'85*, On-Line Publications, 173-183.

Craine, G.S. (1985). MOSS. *Civil Engineering Surveyor* 10 (10) 17-21.

Davies, H.E.H. (1972). Optimizing of highway vertical alignments to minimize construction costs: program Minerva. *TRRL Report, LR463*, 60 pp.

Ebner, H., Hoffman-Wellenhof, B., Reiss, P. and Steidler, F. (1980) HIFI—a minicomputer program package for height interpolation by finite elements. *14th Congress of the Int. Soc. of Photogrammetry, Hamburg*, Commission IV, 14 pp.

Ebner, H. and Reiss, P. (1984) Experience with height interpolation by finite elements. *Photogrammetric Engineering and Remote Sensing* 50 (2) 177–182.

Gould, S. (1986) ECLIPSE. *Civil Engineering Surveyor* 11(1) 12–16.

Grassie, D.N.D. (1982) Contouring by computer: some observations. *British Cartographic Society Special Publication* 2, 93–116.

Grayer, J.L. (1980) A site modelling system. *4th Int. Conf. on Computers in Design Engineering (CAD 80)*, 724–732.

Grist, M. (1972) Digital ground models—an account of recent research. *Photogrammetric Record* 7 (40) 424–441.

Hirst, D. (1986) SURVPAK. *Civil Engineering Surveyor* 11(2) 15–16.

Hodges, D.J. and Alderson, J.S. (1985) Automated volumetric surveys. *University of Nottingham, Mining Dept. Magazine* 37, 71–76.

Hogan, R.E. (1984) HASP Digital survey design system. *Proc. of the ASP-ACSM Fall Convention, San Antonio*, 468–475.

Houlton, J.M. (1985) Computer aided drafting in highway engineering. *Current Issues in Highway Design. Proc. of Conference of Highway Design, Inst. of Civil Engineers*, Telford Publ., 149–162.

Houlton, J.M. (1986) MOSS integrated modelling, design and drafting for civil engineers. *Proc. CADCAM'86, Birmingham*, 235–242.

Jancaitis, J.R. and Junkins, J.L. (1973) Modeling irregular surfaces. *Photogrammetric Engineering* 39(4) 413–420.

Jarvis, G.K. and Gee, B.C. (1985) Computer graphics plot the lie of the land. *Surveyor*, 2nd May.

Mayo, R.H., Durrant, A.M. and Finniear, L.J. (1986) The application of the new Medusa ground modeller. Preliminary findings of the beta-test. *Medusa User Group Meeting*, Cambridge, 8 pp.

McCullagh, M.J. (1983) 'If you're all sitting comfortably then we'll begin'. Workshop Notes on Terrain Modelling, *Australian Computing Society*, Siren Systems, 158 pp.

McGarry, M. (1985) Surface modelling in the context of a general drafting system. *2nd Int. Conf. on Civil and Structural Engineering Computing*, 1, 195–199.

McLaren, R.A. and Brunner, W. (1986) The next generation of manual data capture and editing techniques: the Wild System 9 approach. *Proc. ACSM-ASPRS Spring Meeting*, Washington, Vol. 4, 50–59.

Milne, P.H. and Motlagh, K.C. (1986) Land surveying software review. *Civil Engineering Surveyor* 11(7) 14–21.

Miller, C. and Laflamme, R.A. (1958) The digital terrain model–theory and applications. *Photogrammetric Engineering* 24(3) 433–442.

Mirante, A. and Weingarten, N. (1982) The radial sweep algorithm for constructing triangulated irregular networks. *IEEE Computer Graphics and Applications*, 2(3) 11–21.

Miyazaki, Y., Tsukahara, K. and Hoshino, Y. (1986) Digital map information in Japan. *Autocarto London*, Vol. 1, 25–33.

Pratt, M.J. (1984) Interactive geometric modelling for integrated CAD/CAM. *Eurographics Tutorial Notes*, Springer-Verlag, Berlin-Heidelberg-New York.

Requicha, A.A.G. and Voelker, H.B. (1982) Solid Modelling: a historical summary and contemporary assessment. *IEEE Computer Graphics and Applications* 2(2) 9–24.

Schut, G.H. (1976) Review of interpolation methods for digital terrain models. XII Congress of the Int. Soc. of Photogrammetry, Helsinki, Commission II, 23 pp.

Siddans, D.R. (1980) Photomontages from oblique aerial photographs. *The Highway Engineer*, March, 14–19.

Stanger, W. (1976) The Stuttgart Contour Program (SCOP)—further development and review of its applications. *Presented Paper, Commission III, 13th ISP Congress*, Helsinki, 13 pp.

Sykes, D. and Craine, S. (1986) System 9, the new concept from Wild Heerbrugg, *Civil Engineering Surveyor* 11(3) 20–23.

Steidler, F., Dupont, C., Funcke, G. and Vuattoux, W.A. (1986) Digital terrain models and their applications in a database system. Wild Heerbrugg publication, 10 pp.

Steidler, F., Zumofen, G. and Haitzmann A. (1984) CIP—a program package for interpolation and plotting of digital height models. *Proc. ACSM-ASP Meeting, Washington DC* (reprinted by Wild Heerbrugg), 10 pp.

Strodachs, J.J. and Durrant, A.M. (1986) Surface and solid modelling for civil engineering application—an introduction to 'intelligent ground models' *Proc. CADCAM'86*, Birmingham, 243–248.

Turnbull, W.M., Maver, T.W. and Gourlay, I. Visual impact analysis: a case study of a computer based system. *Autocarto London*, Vol. 1, 197–206.

12 Survey software for digital terrain modelling

P.H. MILNE

12.1 Introduction

The methods used for the recording and processing of field survey data obtained in land surveying have changed dramatically in the last decade. Advances in microprocessor technology combined with miniaturization have now led to the production of compact electro-optical field surveying equipment with the capability of not only presenting the difference in distance and level, but also of calculating the coordinates of the observed survey point. Once data has been stored in digital form, modern computer technology can simplify the previously manual tasks of contouring and volumetric analysis in engineering surveying.

The requirements for using such an automated survey system for digital terrain modelling (DTM) are, firstly, that it can read data already stored on disk from other software, data logging files, etc., and secondly, that it can produce colour graphic displays of contours, surface models and cross-sections on a screen or plotter. An additional advantage of the Survey Software suite of modules (Fig. 12.1) is that data can be exported in a CAD (computer-aided design) format for use in GIS (geographical information systems) etc.

12.2 Field data acquisition

The PC Field Survey module provides a range of methods for field data acquisition, tacheometry, EDM, data logging and digitizing, as described by Kennie (Chapter 2). The data logging facility allows data collected from land, hydrographic, geographical and geological surveys, by a range of dedicated data loggers, to be transferred to Survey Software. The current range, which is always expanding, includes:

- Ellar Databuild
- GeoMEM borehole data
- Geotronics MAP 400
- Kern POST02
- Optimal GRAD
- Wild GRE3/4.

Once a survey has been processed, the results can be saved in one of two formats, depending on the DTM method of interpolation to be used. If a grid model is to be prepared, the data are saved in a '.DAH' format (Milne, 1984). If required in a triangulated irregular network (TIN), then the data are saved in a '.TIN' format. Both

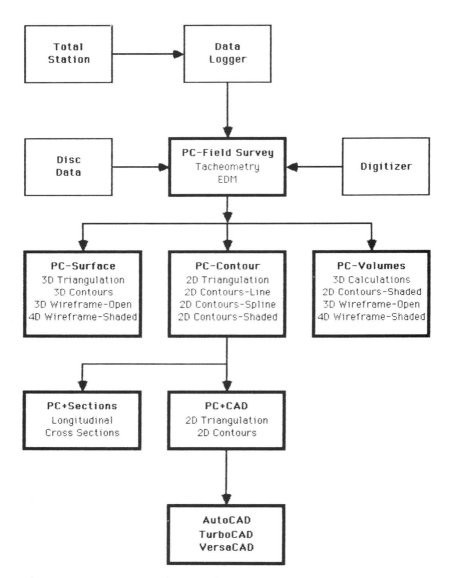

From Field Survey to CAD in Surveying and Mapping

Figure 12.1 Modular concept of survey software from field survey to CAD in surveying and mapping.

types of data files can be viewed on screen or plotted on a range of Hewlett-Packard (or HPGL-compatible) colour plotters at precise scales (Fig. 12.2). Editing facilities are provided to add data points as required, e.g., the fourth corner of a building not seen from the survey station. One advantage of the '.TIN' format is that several days' surveys can be merged into one large data file for processing. Another advantage of the '.TIN' format is that data collected from different types of surveys, e.g., EDM and digitizing, can also be merged or compared easily.

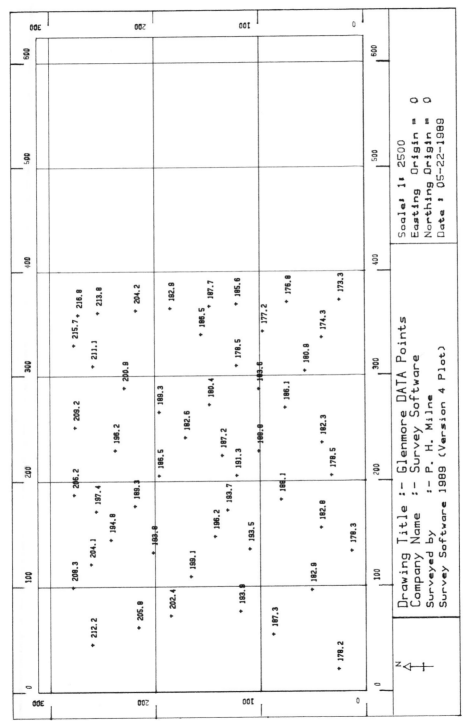

Figure 12.2 DATA points from Glenmore tacheometry survey.

12.3 Grid-based terrain modelling

Gridded models represent the terrain by interpolating from the input data on to a fixed grid. It is generally recognized that the technique used in PC-Grid has limitations, but it is an easy method to implement and store on a microcomputer. The main disadvantages are that the grid size must be fixed in advance and must be sufficiently small to accurately represent irregular surfaces, which naturally leads to excess data in smooth areas. If contouring from irregular data, a random-to grid interpolation is required by PC-Random. This interpolates known data points on to a grid node and means that one is not dealing with actual input data points, but a mean for that area. This has the tendency to smooth out surface irregularities, which may not be a limitation in general topography. It should be noted that a gridded model cannot accurately represent irregular features, such as breaklines at the top and bottom of ridges or embankments. Conflicting opinions exist as to the 'best' method of interpolating random spaced data, as discussed by Petrie (Chapter 8). The program PC-Random provided by Survey Software uses a pointwise method where the nearest five neighbours to the grid node are identified and then the level calculated using an inverse-square distance-weighted interpolation (Milne, 1987).

12.4 Triangulation-based terrain modelling

This method, often referred to as TIN, is being used to an ever-increasing extent in terrain and surface modelling. The reason is that all measured data points are used and honoured directly, as they form the vertices of the triangles used to model the terrain and construct the contours. Thus triangular models represent the ground surface as a series of non-overlapping contiguous triangles with a data point at each node (Fig. 12.3). The triangulated method therefore overcomes all the above mentioned grid-based difficulties, and provides a much more efficient method of representing surface terrain. With the use of breaklines it is also possible to accurately define irregularities such as sharp ridges, quarry faces, embankments, etc. The triangulation method used by Survey Software is based on the Delaunay triangulation method as described in Chapters 8 and 9.

12.5 Contouring

With the formation of the DTM of 3D data, either as a grid model or a triangular model, contours can now be interpolated through the area.

12.5.1 Contours from grid-based model
The DTM data are held in a large matrix of grid cells, and contours are threaded through each cell using a linear interpolation method. For large cells, the straight-line vectors so generated can be quite long, with obvious abrupt changes in direction at cell boundaries. To overcome this problem, each cell is often subdivided and a subgrid generated within each grid cell. The contours are then interpolated linearly and threaded between the points on the subgrid cell, thus giving short vectors with a better visual appearance. A choice of output, either line or shaded contours, is available on screen. Only line contours are available on a plotter, but can be plotted at precise scales.

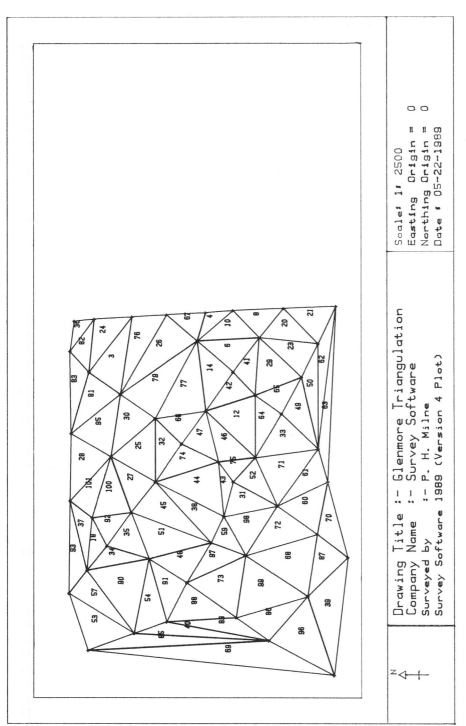

Drawing Title :- Glenmore Triangulation
Company Name :- Survey Software
Surveyed by :- P. H. Milne
Survey Software 1989 (Version 4 Plot)

Scale: 1: 2500
Easting Origin = 0
Northing Origin = 0
Date : 05-22-1989

Figure 12.3 Triangulation network formed from the data points of Fig. 12.2.

12.5.2 Contours from triangulation-based model

As with gridded data, contours are threaded through each triangle using a linear interpolation (Fig. 12.4). For large triangles where straight vectors are not acceptable, the generation of smooth curves using cubic splines is offered as an option to improve the visual presentation of contours (Fig. 12.5). Care has to be taken if using spline curves for close contours in steep areas, as crossing contours may occur if the spline is not well defined. Different values of curve tightness can be selected to avoid this problem.

A major advantage of the TIN-based network of triangles is that it is possible to create windows in the data by entering a closed boundary. For example, if in the Macintosh data file (Fig. 12.6) a boundary is created using the nodes:

$$7, 8, 9, 10, 14, 19, 28, 27, 23, 16, 7$$

then there are two options, firstly, to exclude data outwith the boundary (Fig. 12.7), useful for spoil-heap studies, or, secondly, to exclude data within the boundary (Fig. 12.8), useful for quarries, lagoons, lakes, buildings, etc.

12.6 Surface modelling and volumetric analysis

The same DTM database used for contouring can also be used for surface modelling and volumetric analysis. The ability to display the DTM as a 3D surface model is very useful for gaining insight into area structures, illustrating spatial relationships, deriving relationships or trends and communicating vast amounts of information at a glance. Both square grid fishnet or wireframe models and triangular mesh surface models can be used to view the model from different angles. With the advantage of high-resolution colour graphic screens, the model can be viewed as a 3D open mesh model or a 4D shaded mesh model with contour shading (Fig. 12.1). It is also possible to employ some vertical exaggeration to enhance the perspective using PC-Surface Modelling. This has the advantage that any errors introduced into the DTM at the time of the original survey, and not identified in the contour drawing, can be edited before the final drawings are prepared. The ability to rotate and window into segments of the model gives the user the opportunity to visualize the model from different viewpoints.

Once a DTM has been created, volumetric computations can be readily obtained using PC-Volumes, for example for stockpiles, spoil-heaps, lagoons, quarries, etc. Volumes can be computed either from grid-based or triangulation-based DTMs with full printouts of the computations (Table 12.1). Volumes based on TIN models tend to be more accurate than those based on grid models (Milne, 1988a). Volumetric analysis examples are earthwork volumes computed between two models, or a stockpile removal calculated above a basement level.

The combination of surface modelling and volumetric analysis is extremely useful for visual impact studies, e.g., landfill or waste disposal studies, borrow-pits, reservoir construction, the flooding of low-lying land, or landscaping.

12.7 Ground cross-sections

Once a DTM has been prepared, it is easy to plot a ground or longitudinal section between any two points, either on screen or on a plotter with PC + Sections. This is

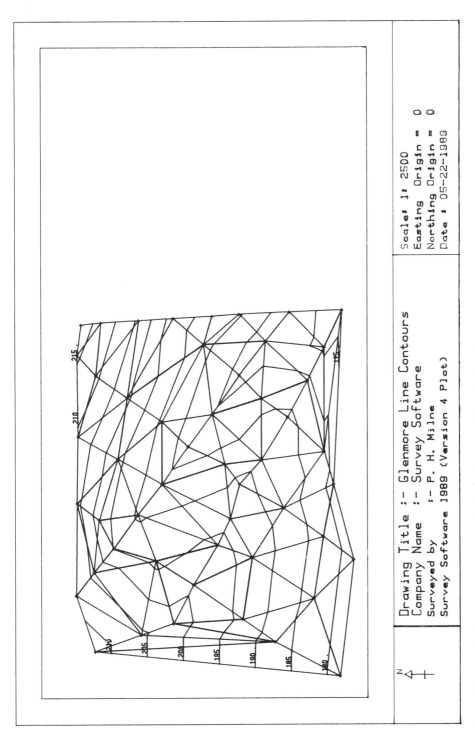

Figure 12.4 Straight-line contours interpolated through Glenmore triangulation network.

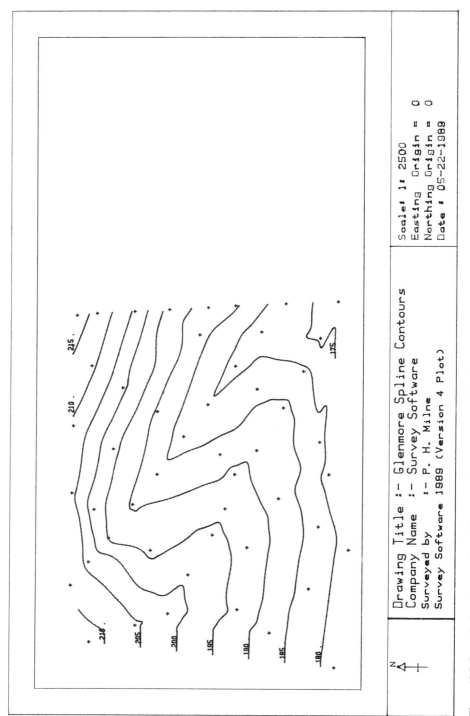

Figure 12.5 Smoothed contours, using spline curves, for comparison with Fig. 12.4.

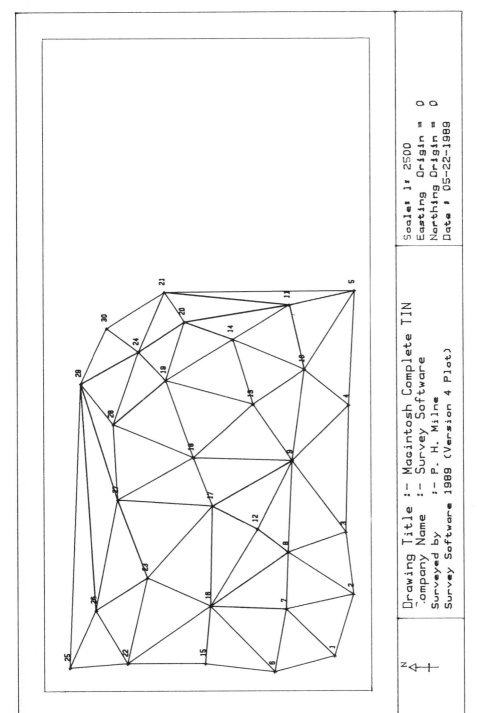

Figure 12.6 Full triangulation network for Macintosh data file.

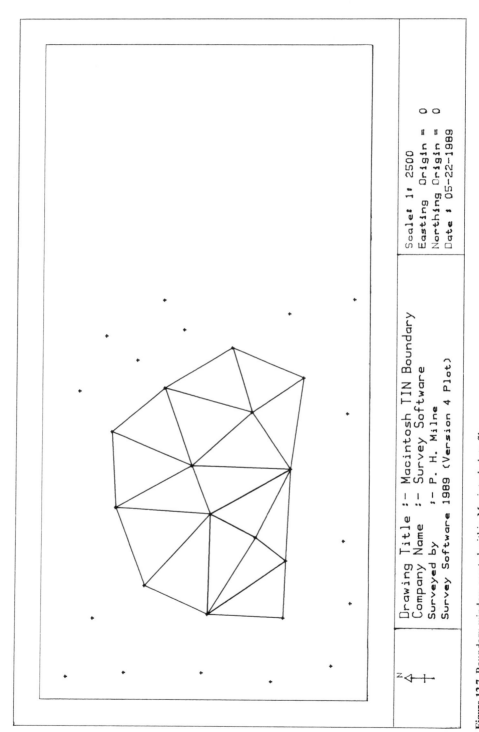

Figure 12.7 Boundary window created within Macintosh data file.

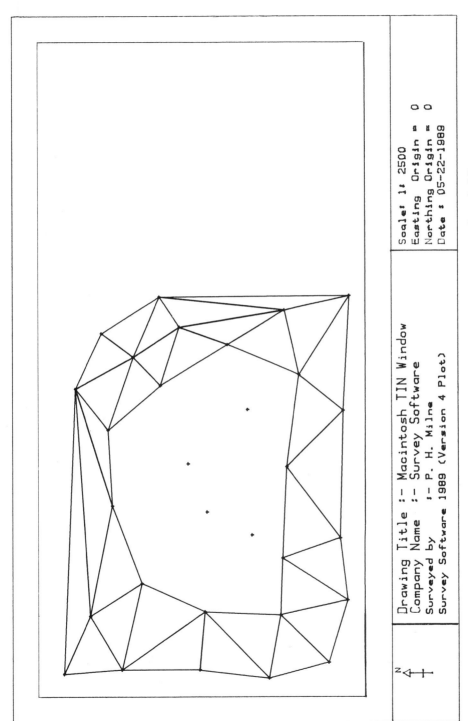

Drawing Title :- Macintosh TIN Window
Company Name :- Survey Software
Surveyed by :- P. H. Milne
Survey Software 1989 (Version 4 Plot)

Scale: 1: 2500
Easting Origin = 0
Northing Origin = 0
Date : 05-22-1989

Figure 12.8 Boundary window omitted from Macintosh data file—useful for lagoons, quarries, reservoirs or buildings.

Table 12.1 Extract from print-out of volume computations for Glenmore data showing the areas and volumes of the first two and last two triangles in the network.

```
DATE                :05-22-1989
DATA FILE NAME      :GLENMORE
NO. OF POINTS     :  50
MINIMUM EASTING   :  21
MAXIMUM EASTING   :  372
MINIMUM NORTHING:   13
MAXIMUM NORTHING:   278
MINIMUM LEVEL     :  173.3
MAXIMUM LEVEL     :  216.8
NO OF TRIANGLES   :  82
```

```
TRIANGLE NUMBER         :  1
TRIANGULATION NUMBER:  3
DATUM LEVEL             :  170
NODE        EASTING          NORTHING          LEVEL

    6       362.000          218.000          204.200
   39       288.000          229.000          200.900
   40       309.000          259.000          211.100
```

```
AREA              = 1225.5
VOLUME            = 43382.7
CUMULATIVE AREA   = 1225.5
CUMULATIVE VOLUME= 43382.7
```

```
TRIANGLE NUMBER         :  2
TRIANGULATION NUMBER:  4
DATUM LEVEL             :  170
NODE        EASTING          NORTHING          LEVEL

    3       368.000          122.000          185.600
    4       366.000          148.000          187.700
   33       339.000          156.000          186.500
```

```
AREA              = 343
VOLUME            = 5693.8
CUMULATIVE AREA   = 1568.5
CUMULATIVE VOLUME= 49076.5
```

```
TRIANGLE NUMBER         :  81
TRIANGULATION NUMBER:  100
DATUM LEVEL             :  170
NODE        EASTING          NORTHING          LEVEL

   14       176.000          219.000          189.300
   46       171.000          256.000          197.400
   50       229.000          238.000          196.200
```

```
AREA              = 1028
VOLUME            = 24980.4
CUMULATIVE AREA   = 82682
CUMULATIVE VOLUME= 1737547
```

```
TRIANGLE NUMBER         :  82
TRIANGULATION NUMBER:  101
DATUM LEVEL             :  170
NODE        EASTING          NORTHING          LEVEL
```

Table 12.1 (*Contd.*)

46	171.000	256.000	197.400
47	187.000	277.000	206.200
50	229.000	238.000	196.200

```
AREA              = 753
VOLUME            = 22539.8
CUMULATIVE AREA   = 83435
CUMULATIVE VOLUME= 1760087
```

extremely useful, for example, in selecting the gradient of a new road, anticipating the level of rockhead for piling, for line-of-sight and intervisibility analysis (Fig. 12.9). The user can select the chainages of the start point, the interpolation interval and the datum for the plot.

After the longitudinal section has been prepared, it is possible to compute cross-sections along the longitudinal section at given chainage intervals on either side of the centre line. The width of the cross-section can also be altered by the user. If any of the cross-section falls outwith the triangulation network, then the minimum default level will be assumed. This Survey Software module can also be used in conjunction with two other modules, PC-Earthworks for road cross-section computation, and PC-Highways for setting out highway alignments and phasing in circular, transition and vertical curves.

12.8 Computer-aided design

The 1980s have seen an increasing interest in CAD programs for the AEC (Architecture—Engineering—Construction) market. To facilitate transfer of data files, most CAD systems are able to import files written in AutoCAD's '.DXF' format (i.e., an ASCII Drawing Exchange format). To allow users of Survey Software to transfer field survey data and contours into CAD programs, interface routines are available in PC + CAD (Fig. 12.1). Data can be stored in different layers, e.g., data points in one layer, triangles in another and contours in yet another. The advantage of storing data in different layers is that the user can then select which layers to display and can also merge the data with design-based data showing new buildings, roads, etc. (Milne, 1988b).The interactive nature of CAD programs allows the user easily to annotate the contour survey with titles, drawing blocks, etc. For users without plotters, scale plots can often be obtained on dot-matrix printers. Numerous CAD programs now use the '.DXF' format, so Survey Software can be exported to programs such as AutoCAD, DAXCAD, TurboCAD, VersaCAD and PC-Dogs.

12.9 Hardware requirements

All of the programs described in Fig. 12.1 will run from a twin floppy machine, either an IBM PC XT, AT or PS/2 or compatible, or an Apple Macintosh, where the master programs are run from the A: drive with a data disk in the B: drive. To save having to change disks from module to module, a hard disk is recommended. Although the software will run on a 512K PC machine with a CGA display, to take advantage of the

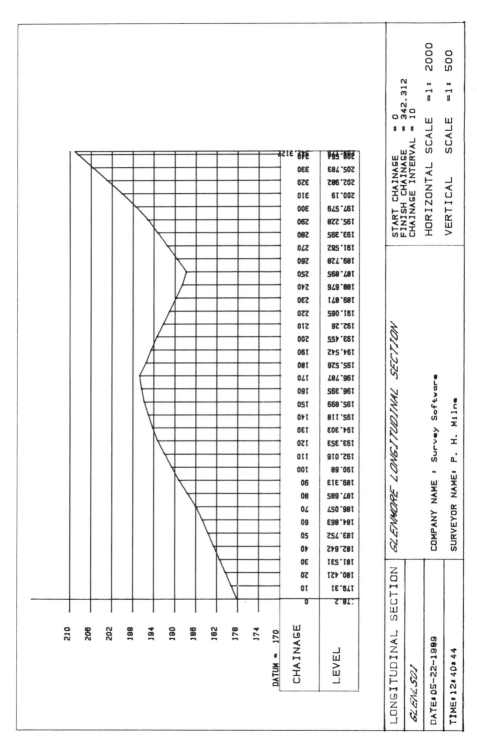

Figure 12.9 Longitudinal section plotted through Glenmore data from lower left to upper right corner.

colour-shading options available, a 640K PC machine with an EGA or VGA colour display is recommended. The Apple Macintosh version requires 1 MB of memory and twin 800K drives, or one drive and a hard disk. A recent software addition has been the ability to save drawings as a plot file, very useful on site, for example, for subsequent plotting in head office, without having to re-enter the survey data.

12.10 Current applications and conclusions

Since Survey Software became available for the educational market in 1982, there has been a considerable expansion of its usage, both by public and commercial organizations in more than 20 countries. By far the largest number of users link the PC-Field Survey module with the PC-Contour and PC-Surface Modelling modules for the preparation of contour drawings varying from A4 up to A1 size plots. Some specialized uses are for archaeological surveys and the plotting of non-destructive testing results, where the colour-shaded screen dumps are especially effective. The ability to read borehole and piling records has contributed to the accurate contouring of substratum layers and rockhead.

Survey Software provides a low-cost solution to the engineering surveyor on site or in head office who requires a quick, efficient and reliable plot of the data on screen or plotter. The modular nature of the software means that, for smaller organizations on a tight budget, the full software suite can be built up module by module to suit the individual application. After each survey, data can be saved to disk for transfer to head office for plotting, etc., or for transfer to a client's CAD system.

Software is updated continuously to take account of modern site survey techniques and advances in computer technology. The interactive nature of the software means that users wishing to customize the graphics screens or hard copy output can easily do so, as the software is not copy-protected. One major advantage is that it is independent of computer and peripherals, giving the surveyor freedom to mix, change or upgrade equipment as required.

References

Milne, P.H. (1984) *BASIC Programs for Land Surveying*, E. & F.N. Spon Ltd., London and New York, 442 pp.

Milne, P.H. (1987) *Computer Graphics for Surveying*, E. & F.N. Spon Ltd., London and New York, 214 pp.

Milne, PH. (1988a) Accuracy of data processing from creation of Digital Terrain Model to contouring and surface modelling. In *Proc. Standards and Accuracies*, Institution of Civil Engineering Surveyors, Reading University, March 1988, Paper 5A, 14 pp.

Milne, P.H. (1988b) Speed + consistent accuracy and more from CAD-generated data. *Civil Engineering Surveyor,* **13**(2) 18–20.

13 The Panterra interactive ground modelling system

STEPHEN GOULD

13.1 Introduction

Surveying in the Falkland Islands, road design in Canada, open-cast mining in Australia, housing estate development in the United Kingdom—these are just a few examples of the international use of the Panterra Interactive Ground Modelling System (Fig. 13.1).

Panterra is a PC-based system comprising a suite of modular programs catering for the needs of land surveyors, cartographers, civil engineers, estate developers and mining and quarrying engineers. Links have been developed with architectural design and detailing software, so that for site development work the system may be used from original site survey through the automatic production of engineering drawings and architectural designs and details.

13.2 System overview

Several important concepts underpin the continuing development of the Panterra system. These are: integration, interactivity, information, user interface and input from the user.

13.2.1 Integration

Panterra is a digital ground modelling system with extensive state-of-the-art facilities for survey processing, digital mapping and civil engineering design. It is a single system accessing a common integrated database comprising raw survey data, a three-dimensional digital terrain model (DTM), engineering designs, and automatically produced drawings. Structuring the database in this way means that, at each stage of the survey processing–modelling–design–drawing production process, the integrity of each area of the database is maintained. Fundamental to the system is the ability to edit each area of this database, and the user always has recourse to an earlier stage of his work if required.

A much greater advantage of integrating software modules around a common database lies in the range of options which the system offers and the way they interact with each other. There are a number of packages available on the market for calculating coordinates and producing contoured survey plots, but Panterra offers a

Note: The names Panterra and Ground Modelling Systems Ltd replace Eclipse and Eclipse Associates, respectively.

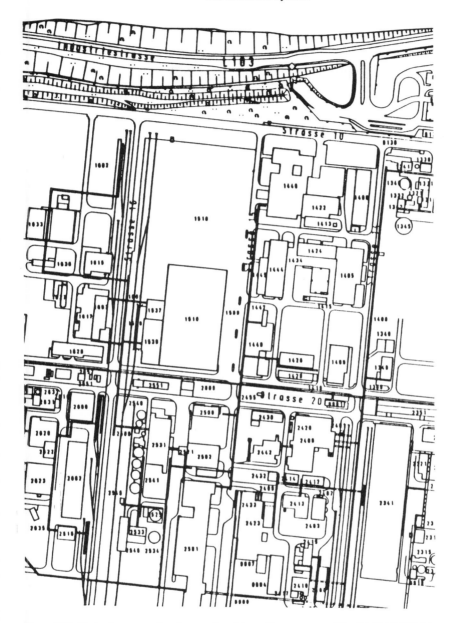

Figure 13.1 Part of a survey drawing produced by a German user of the Panterra system.

wide range of options which build on the concept of computerized survey and site data processing.

For example, from the DTM a volumetric analysis package allows irregular surfaces to be compared automatically for accurate computation of volumes. Long- and cross-sections may be extracted automatically from the DTM to allow ground levels to be included on road or drainage long-section plots.

Another example of the importance of this standard of integration is in the road design package where the vertical alignment may be designed relative to the existing ground level. Similarly, all manhole cover levels in a pipe network may be automatically extracted from the DTM.

Subterranean interfaces may be modelled by entering borehole data to generate multi-strata sections and DTMs of each undergound stratum. Quantities of the various different materials to be excavated can therefore be calculated.

A quantities take-off option in the system allows the user to specify a library of costs relating to features in the surface model. Thus, the length of kerbing or fencing on a design scheme and its total cost can easily be extracted to build up a cost schedule.

A graphical database module permits user-specified records of textual and numeric data to be connected to points, strings or land parcels in the database. Digital mapping on a PC is therefore just one of the options available through combination of software packages in modular form. Some users may need most of the packages available, some may only require a proportion of them. Expanding the use to which the system is put is then always possible, and the investment of the user maximized.

13.2.2 Interactivity

A high level of interactivity is essential for ease of use and productivity in software. Ground Modelling Systems Ltd were at the forefront in the provision of interactive programs, and this concept is inherent in continuing software development.

Interactivity means that user can input data numerically or graphically using the screen cursor or digitizer, and obtain immediate visual and numeric confirmation of each calculation or design stage.

For example, a road designer may indicate that an alignment must pass through points defined visually on the backcloth of the site survey using a graphics cursor. Numerical information on the resultant curve radius, say, will immediately be displayed. Conversely, the geometric parameters of the elements making up the roads may be entered numerically, with a graphical display of the resulting alignment.

Similarly, on some graphics screens, elements of the surface model or building templates may be 'dragged' around and positioned by the user. The new coordinates of these elements are then calculated, displayed and stored. Alternatively, the coordinates of these elements may be entered at the keyboard and they will be redisplayed in their new location on the graphics screen. This kind of immediate visual interface with the surface model or design is the true test of interactivity.

13.2.3 Information

Another important design criterion is the extent of information available to the user, both on-screen and output from the software.

The Panterra system employs twin screens for each workstation (Fig. 13.2). The text screen, a normal PC monitor, displays system menus, help text, and numerical information. Simultaneously, a graphics screen displays a pictorial representation of work in progress and allows input via the graphics cursor. The graphics screen may also be used to display drawings, which have been created automatically, to verify them before sending to the plotter. Many modelling or design facilities require a great deal of numerical data to be reported alongside the graphical representation. A single-screen system inevitably requires a compromise in using the available screen area. While icons, pull-down menus and similar techniques have their applications, the twin-screen approach is more productive when a great deal of data has to be presented to the user.

Figure 13.2 A high performance Panterra workstation, comprising a 32-bit 25 MHz PC with intelligent colour graphics screen.

Output from the system takes a variety of forms. Naturally, high-quality drawing production is an important feature of the system. Despite the emergence of digital data transfer, drawings are still a heavily used form of data interchange.

Panterra automatically produces finished-quality plans and sections incorporating user-specified line styles, symbols and pen colours. Plans may be generated at any grid orientation, plans and sections produced at any size or scales, and a choice of automatic gridding and labelling is also available.

Drawings may be edited using the software, and additional lines, symbols, text and notes added. Automatic drawing production means that some text overwrite may be unavoidable. The drawing editor enables this and similar problems to be easily corrected.

Extensive printed information is available throughout the Panterra suite of programs. This may take the form of a report on data stored, such as coordinates or design parameters, or the provision of additional information during a batch operation. An example of this is when detail survey observations are processed. If the printer is selected, then a listing of both coordinates and errors will be generated. All printout may also be output direct to disk to act as input to other software, for example a word-processing or spreadsheet program.

Other output options include the ability to produce a screen dump from certain graphics screens to a colour copier. This may be useful in providing solid-colour prints for inclusion in a report, as an alternative to a line drawing.

Once the design of a site has been completed, quantities may be taken off, either interactively or in batch. These may be tested against a user-defined library of feature costs to produce a cost schedule. This may then be used as a basis for a bill of quantities. For example, the length of hedging or fencing used on site can be requested and calculated as a batch operation. Alternatively, individual lengths of fencing can be located interactively using the graphics cursor, and the total cost accumulated. Areas of topsoiling or seeding can be calculated and costed in a similar manner. Transfer of information between systems is essential as the numbers of CAD and database programs increase. Users of Panterra must be able to read in maps in digital format in order to update them or use them as a basis for design. Conversely, they may input maps and plans digitally to output to a third-party digital databank. Survey companies may have clients requesting site survey data to be presented to them in a form compatible with their own computerized drafting systems. In this case, flexibility of output options is a must.

The user may also access the database for input or to write his own programs around it. Examples where this facility has been used are in transferring data into the system from aerial surveys via a specified input format, or in writing specialized design programs around the output data. The integrity of the Panterra database is maintained, as imported information is always filtered to ensure compatibility.

13.2.4 The user interface

The user interface with software and hardware is an important feature of any system. No matter how powerful the programs, if the operator is deterred from using the product because it is difficult to access, then his use of the system will diminish. To accommodate all users, the software must be easy to use and consistent in terms of input methods and appearance throughout.

Screen menus have been adopted as the most easily understood method of program access with defined methods of inputting data. The 'tabular input' method is one of these. A screen full of related information is displayed, which the user is free to move around using the cursor movement keys on the keyboard. The current field is highlighted and, if edited, all related data is automatically updated and redisplayed. This method is easy both to learn and use. Once encountered, the user is then familiar with the principle of operation throughout the suite of programs.

Throughout the suite, internal company standards for program design are followed to ensure that the operation of each program is consistent from the user aspect as new software is developed and added to the system. An interactive graphics-based system is much easier to use than a non-graphics-based system. Much more information is available to, and readily assimilated by, the user. Colour graphics are standard, and the facility to interrogate the database by 'picking' items off the graphics screen using the graphics cursor is much faster than searching sequentially through it to find required items.

13.2.5 Input from the user

Input from the user takes two forms; the methods of getting data into the system, and user involvement in software development. The alternative methods of getting data into the system are detailed later in this paper and this section is concerned with the role of the user in defining the direction of Panterra software. The software has been commercially available since 1978. Since then, the number of users has increased dramatically and the range of available hardware has expanded, becoming faster and more economical. As well as benefiting from technological developments in hardware, the software has been modified and improved to accommodate the shift in professional working practices. The need for change is increased as the software is now marketed internationally via a network of subsidiary companies and authorized agents.

The research and development, sales, support and service bureau departments within Ground Modelling Systems Ltd all contain a high proportion of qualified surveyors, engineers and related professionals. Many ideas for future improvements come from within the company. However, as the number of users grows internationally, users are expected to have an input in this respect. Responsiveness to user requirements results in an improved product, kept up to date with changing working methods.

13.3 Hardware requirements

Panterra was orginally developed on Wang 2200 minicomputers. This permitted fast program development and gave the user a state-of-the-art ground modelling system—at the time! In 1985, the decision was taken to launch Panterra on industry-standard 16-bit personal computers operating at 6-MHz clock speed. together with 10-Mb fixed disks. The software was compiled for higher-speed operation and users benefited from mass-produced hardware with a superior cost/performance ratio.

This decision has since been endorsed by developments in the PC marketplace resulting in the current availability of 32-bit personal computers operating at 33-MHz clock speeds together with 300-Mb fixed disks. Numeric coprocessors are fitted in the PC for even faster operation. In a relatively short time, therefore, users have benefited from faster program operation by a factor of several times via hardware developments alone.

A range of graphics screens and cards are supported, ranging from colour PC monitors through to intelligent screens and high-speed cards with multiple graphics coprocessors. Simple colour monitors allow inexpensive access to interactive graphics for low-cost workstations. If high-speed zoom and pan is required for manipulating larger amounts of graphical information, the intelligent screens hold the graphics image within the memory of the screen itself. It may thus be manipulated using the capabilities inherent in the graphics screen. Alternatively, high-speed graphics cards are available for installation in the PC. These use transputers with multiple graphics coprocessors. The graphics image is held in memory on the card for high-performance manipulation.

Most major makes of plotter with RS232 input are supported. These vary from desktop A3 plotters to high-speed A0 eight-pen friction-feed plotters with built-in buffers and plot optimization facilities. The buffer holds the drawing in the memory of the plotter, releasing the PC for further operation. Optimization means the plotter will attempt to plot everything in one area of the drawing sheet before proceeding, rather than simply plotting everything in the order received from the PC.

Likewise, most makes of digitizer with RS232 compatibility are supported. These are available in a range of sizes up to A0. The digitizing software provides extensive facilities for data input at the cursor, and a 16-button alphanumeric keypad and cursor is recommended to take maximum advantage of this. For example, feature information can be input directly via the digitizer keypad to avoid returning to the PC keyboard. Panterra is normally supplied as a turnkey system. Furthermore, on completion of the warranty period, the system is maintained under a single contract for software and equipment. The advantage for the user is that a single contractor takes full responsibility for the operation of the system. This reduces administration costs and provides a single source of assistance, whatever the problem.

13.4 Who uses Panterra?

The workload of surveyors and engineers today is becoming more and more diverse, while many organizations are broadening their scope of work in order to maintain or increase turnover. This has also focused attention on the necessity of reducing unit costs in tenders. At the same time, public sector organizations have been hard-pressed by the need to maintain work schedules despite shrinking budgets and establishments. Against this background, Panterra provides immediate productivity gains and facilitiates expansion into other areas of activity and markets. The following sections of this chapter highlight some of the uses made of the system by several professional groups.

13.4.1 Panterra for the surveyor

Panterra survey packages support a wide range of options for processing of survey observations. Collection methods vary from tape and offset to radial observations, on to the use of total stations for automatic data collection and transmission. Alternatively, existing drawings can be digitized in two- or three-dimensional formats, making extensive use of the digitizer keypad for coding associated data. The digitizing software has been developed to cope with the exacting specification of the Ordnance Survey, the UK national mapping authority.

Field observations can be output from all well-known total stations to the Husky Surveyor field computer (Fig. 13.3). The Husky Surveyor is a portable hand-held microcomputer produced specifically for field use. As well as running the Panterra Field Survey System, it is a fully programmable computer in its own right. It is thus

Figure 13.3 The Husky Surveyor with battery-operated disk drive and printer.

more flexible in use and more economical than dedicated data loggers. The Husky Surveyor's larger screen allows the use of menu-driven programs, and it is shockproof and waterproof against accidental immersion.

The Panterra Survey Data Collection program allows manual or automatic input from a wide range of total and semi-total survey stations. Feature descriptions, string details, dimensions, offsets and remarks are all easily input in the field, minimizing the need for subsequent office-based editing before data is downloaded to the PC.

The survey processing programs in Panterra automatically calculate both control and detail surveys from manual or automatic input. Control options include resection, intersection, spirit levelling and grid transformation, in addition to traverse calculation and adjustment.

To enable the full ground detail to be captured, Panterra supports variable-sized 'point' and variable-width 'string' features. For example, the spread, bole and height of a tree can all be recorded and used in a variety of user-definable ways. As one possibility, the spread and trunk can be plotted to scale with the height written alongside. Features such as ditches and hedges may be positioned by offsets from their centre line, with varying widths noted at each observed point. A rich variety of offset and double observational facilities permits even the most awkward and inaccessible detail to be recorded accurately. String shapes can incorporate any logical mixture of curves, straights, gaps and discontinuities. Intricate shapes such as car parking areas and road junctions can also be accurately modelled with a minimum of observed points. All these facilities are carried through in the automated production of survey drawings.

Survey observations may be edited individually or in batch before processing. During processing, control stations and surface detail are displayed on the graphics screen for verification. This allows gross errors to be detected more easily than by looking at the raw survey data, as the survey is displayed on the graphics screen in exactly the same order in which it was carried out.

Although survey data may be input into the system by any of the methods previously described, once it has been processed all data is stored in a common 2D or 3D database.

A wide variety of options is provided to edit the points and strings in the database interactively prior to the production of finished drawings or output of data to other systems. This allows the visual correction of errors in the survey or the construction of new points and features. Extensive coordinate geometry calculation facilities are included.

The DTM package automatically forms a finite element triangular mesh representing the ground surface. Strings are recognized as breaklines, so that if the surveyor picks up discontinuities in the ground profile (tops and bottoms of banks for example) as strings in the field, these will be recognized and respected when the triangle model is formed. As a consequence, no triangles will pass through the air or under the ground.

DTMs may represent the interface between underground strata or be used in calculating volumes between irregular surfaces. The surveyor's main interest lies in the ability to interpolate and plot contours from them. Survey details and contours may be plotted using a variety of user-defined symbols, line styles and pen colours. Contours may be plotted at any intervals or even be excluded altogether from areas of the plot. Label lines may be defined interactively on the graphics screen, along which intersecting contours will be automatically labelled with their heights.

Survey data may be used as the basis for an engineering design or site development, or to update existing land information. In future, the digital data transfer facilities will be used increasingly for this purpose in conjunction with the other survey input options. In the United Kingdom, the national 1:1 250 and 1:2 500 map series is being digitized and stored in digital format. Although coverage is less than 20% complete, some users are already reading in data for their area directly into Panterra.

13.4.2 Panterra for the civil engineer

The system provides the civil engineer with interactive design facilities for highways, foul and surface water drainage, estate layout and associated work. These have been entirely integrated with the DTM; for example, roads designed in Panterra can easily be incorporated into an 'as-built' surface model for subsequent volumetric calculations.

Road cross-sections may be generated automatically along the line of the road, extracting existing ground or substrata offsets and elevations from stored DTMs. Additionally, the vertical alignment of a road may be designed interactively on the graphics screen relative to the existing ground line, a very useful optimization device. Existing ground levels can of course be included automatically on road and sewer long-section plots, and designs may be undertaken relative to existing site features, to maintain building clearances, for example.

For the highway engineer, the road design packages allow the rapid design and optimization of estate roads or major highways by a choice of methods. Design or editing of an alignment proceeds interactively, with geometric details displayed on the text screen while a pictorial representation is superimposed over the ground plan on the graphics screen. Graphical input allows points through which the alignment must pass to be specified visually using a graphics cursor controlled via a mouse or through the keyboard. Alternative alignments can quickly be assessed, facilitating production of the optimum solution.

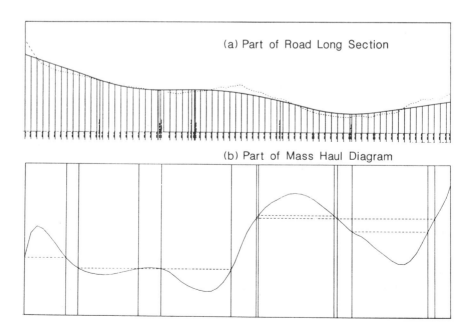

Figure 13.4 Road long-section plot and mass haul diagram.

Vertical alignment design is accomplished in a similarly interactive fashion. Superelevation may be specified as desired, with longitudinal application calculated automatically, either over reverse curves or linearly.

Road cross-sections can automatically be generated from a library of specified templates. Intelligent templates permit the generation of varying design profiles by testing conditions specified in the template. For example, at a particular offset, is the design profile in cut or fill? Varying heights of cut or fill may produce different side slopes, berms or verges to accommodate safety fencing. All of this can be specified in a single template. Cross-sections may contain multiple layers which are used to represent different road construction materials or different ground strata. End areas and volumes of different materials may be calculated from sections, and facilities exist to allow for skew bridge locations, etc. There is a mass haul calculation and plotting option, in addition to multiple cross-section plot formation (Fig. 13.4). Volumes may also be obtained by transferring the road to the 3D ground model and using the mass cut/fill options; these calculate quantities by comparing DTMs or by comparing a DTM to a series of specified planes. Isopachyte contour formation assists the designer to minimize cut and fill. For example, by clearly identifying areas of overlay and planing, this facility is vitally important when overlaying an existing road.

Regardless of the design method employed, plan and section drawings to contract quality are immediately available. Plans may be created to any reasonable size, scale and orientation and can vary in detail from a setting-out drawing to a finished design plan. Print-outs of road geometry, chainages, coordinates, levels or setting-out data are instantly obtained by recalling relevant road alignments and survey control stations. Setting-out data may be in the form of bearings, relative angles or tape and offset measurements.

Panterra highway design options are widely used for both estate roads and major highways. The volumetric analysis capabilities have been employed to calculate cut/fill quantities for all types of project including new motorway constructions. The geometric design and modelling facilities also have obvious applications for drainage channels, railway design, airport layout and other similar developments.

Surface and foul water drainage systems may be designed using different commonly accepted methods of pipe flow analysis. Initially, the pipe network is specified interactively via the keyboard, graphics screen or digitizer. Pipes and junctions may be inserted or deleted, and pipe sizes, gradients and manhole positions changed to optimize layouts. The design options also allow the production of plans, section plots and manhole schedules, which, like all drawings, can be enhanced with notes and text if required. The high standard of automatically produced drawings can be improved still further using the drawing editor, which includes sophisticated facilities such as splitting lines and moving text to eliminate any text overwrite.

Accurate quantities can be taken off interactively from any Panterra model displayed on the graphics screen, or in batch mode direct from the three-dimensional database. A library of unit costs per item, length, area or volume may be stored and used to extract an extended schedule of quantities and rates. These can then be used as the basis of a bill of quantities.

13.4.3 Panterra for the mining engineer

Combining Panterra survey and 3D modelling options allows engineers involved in land reclamation, quarrying, or mining to design their excavation, evaluate useful deposits, experiment with reclamation profiles and record interim excavations. The survey options allow a variety of means of inputting existing surface profiles, as

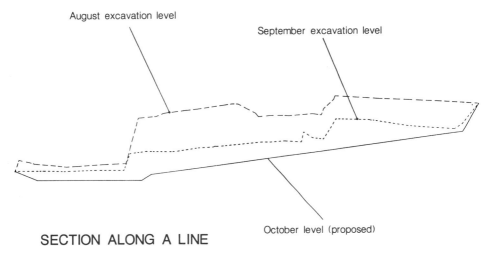

August excavation level

September excavation level

October level (proposed)

SECTION ALONG A LINE

Figure 13.5 Long-section plot through an opencast mining site.

previously described. A separate multistrata modelling option then allows the specification of subterranean models from borehole logs or similar data. Plotting and calculation facilities enable the completion of a digital model which can be used to interpolate borehole information and create multistrata section plots. Ground sections so formed can be viewed on the graphics screen to enhance visualization of the ground strata. Faults, lenses and deposit boundaries are easily modelled to accurately reflect the full detail of the site.

From the survey information and borehole data, a series of models are formed to represent the original ground surface and the interfaces between subterranean strata. Three-dimensional modelling facilities in the suite then allow features such as cut slopes and haul roads to be quickly designed and evaluated. All these design facilities are interactive and graphics-based. For example, excavations can be designed by visually indicating boundary lines and specifying a slope of fixed grade to meet another surface. The boundary of a lower surface will automatically be calculated and displayed on the graphics screen for verification (Fig. 13.5).

The extensive volumetric analysis capabilities of the system permit volumes between irregular surfaces, defined by the original ground level, cut slopes, or strata levels, to be calculated to determine overburden quantities or viable mineral volumes. Repeating the process optimizes the pit design, taking into account any constraints such as site boundaries and excavation slopes.

13.4.4 Panterra for the estate developer

Here the survey, road and drainage design, and the quantities take-off facilities are combined with a sophisticated interactive slab positioning system. This provides the estate developer with all the tools needed to quickly carry out both initial feasibility studies and detailed costed designs. The speed of design facilitates optimization, and, on completion of design work, contract-quality plans and setting-out details may be generated immediately (Fig. 13.6).

Where there is a legal requirement to produce a drawing of a site divided into plots, the Land Packaging option in the software is available. Land parcel boundaries are

Figure 13.6 A housing estate designed using Panterra software.

defined initially by a combination of points and strings, and subdivision design is accomplished by coordinate geometry calculation. Plots of land parcels may be produced which include areas, perimeters, coordinates and boundary bearings and distances.

The slab positioning option allows buildings, or other structures such as paved areas, drives or car parks, to be defined graphically as three-dimensional templates. Additional features such as rooflines and drainage outfall positions may be added. Site layouts are rapidly achieved and optimized by positioning building templates at specified coordinates or relative to other buildings or features. Templates may also be positioned visually against a background of the site survey. Slabs may be 'dragged' and rotated in real time. Individual slabs may also be grouped together and repositioned by offset from a determined construction line, or rotated about any point while maintaining the relative orientation of slabs within the group.

Slab levels can be optimized by generating an 'as-built' surface model and comparing this with a previously generated DTM. Layout plots, building coordinates and setting-out information are available at any time. Three-dimensional features in the database, whether of existing or design detail, can also be viewed on the graphics screen in either isometric or perspective view. They may also be plotted using a perspective plot option in the software. This three-dimensional visualization capability is useful in optimizing the layout of roads and buildings and checking sight lines. A recent example of the use of this option was in investigating the view of a proposed waste transfer station from the bedroom window of a house adjacent to the site, prior to a formal inquiry into the project.

13.4.5 Panterra for digital mapping
The Panterra package is ideally suited to the creation, editing, manipulation and storage of digital maps. The database is compatible with the Ordnance Survey databank in the United Kingdom via an option to transfer data from Panterra to OS format and vice versa. A scaleless map database permits the identification of features in two or three dimensions to 1 mm ground resolution. Drawings consisting of different levels of information can be created at variable scales, orientations and sizes. Drawings and maps can be edited most productively by using the cursor on an intelligent colour graphics screen. High-speed pan and zoom allows easy manipulation of map images that might otherwise appear too dense for accurate editing.

However, graphical mapping accuracy is not limited to the resolution of the screen. New information can be added by specifying accurate 2D or 3D coordinates, or offsets from existing features.

Individual features are identified as points, strings or land parcels. Different libraries may be created, defining how individual features are to be displayed on the graphics screen or plotted on finished drawings. Original ground information may be input from land survey observations, digitized directly into the database or transferred from other systems. Service utilities, street furniture, traffic regulation orders, and details of property ownership are just a very few examples of the type of information which can be mapped two- or three-dimensionally. User-defined textual databases can also be quickly specified and linked to the map. For example, manhole information may be linked to point feature detail. The manhole database may then either be interrogated by traditional methods (sort and list for screen or hard copy), or graphically by using the graphics cursor. This will instantly retrieve and display the manhole record on the 'text' screen. String and area (parcel) information may be similarly linked.

Panterra is compatible with a number of other two-dimensional drafting systems and three-dimensional modelling systems. This flexibility may be important in maximizing the use of facilities such as those described previously. In particular, the ability to read in digital information from the Ordnance Survey databank is an important factor in the future creation of the linked textual and graphical databases outlined previously.

13.5 Summary

Panterra provides an extensive range of computer-aided facilities to surveyors, civil engineers, cartographers, mining engineers, housing estate developers and related professional groups. No special operating environment is required for modern PCs— more than one user has a workstation in a portable office on a mining site! Nor is there any requirement for the user to be a computer expert. The software is structured to be easy to use. Only a minimal knowledge of the computer operating system is required, as file handling, housekeeping and disk maintenance facilities are all menu-driven.

The software licence fee structure allows the user to purchase only those modules which are appropriate to his immediate needs. Further modules may be added later without a cost penalty. A variation allows some software packages to be purchased at a reduced price as an introduction to the system. These may be upgraded later, again without a cost penalty on the licence fees. Thus, the demand of the smaller survey company or for on-site installations is satisfied. At the other end of the scale, the policy seeks to encourage the larger company or public authority to put a PC on the desk of every surveyor or engineer by offering a 'network licence'. The unit cost of each work-station becomes progressively less as the network is extended. In recent years, hardware has increased in power and decreased in price. Software functionality has improved apace. The overall price and performance of PC-based systems like Panterra must surely prompt a rigorous rethink by those who continue to use manual methods.

14 'ProSURVEYOR'—a down-to-earth CADD solution

J. STRODACHS and A. DURRANT

14.1 Introduction

During the mid-1980s, definite trends were beginning to materialize in CAD. The low-cost yet powerful personal computers (PCs) were starting to dominate the CAD hardware market. The adage of 'a CAD system on every professional's desk' was beginning to come true.

The larger CAD vendors, following the hardware trends, began to offer cut down versions of their software. Software design standardized on screen icons (or picturegrams) and pull-down menus. The head up, on-screen and mouse input method of working became common amongst many of the popular CAD packages. ProCAD PC (Professional Computer-Aided Drafting) was one such system.

In addition to its flexible and interactive drawing facilities, the ProCAD system offered useful tools to third-party software developers. These tools were collectively referred to as the 'toolkit'. The 'toolkit' allowed ProCAD's graphics functionality to be shared amongst a wide range of applications. Such applications are individually referred to as Design Modules.

Applications in CADD have taken advantage of ProCAD's 'toolkit', to offer surveyors and engineers a real alternative to the usual fragmented approach to design and computer drafting.

14.2 ProCAD PC

14.2.1 The professional computer-aided drafting system

ProCAD derives its ease of use by promoting the flow and continuity of thought between the user and the database. This user–data interaction is conducted through a ProCAD drawing. Thus the on-screen drawing itself is but a graphical representation of elements within the ProCAD database.

The product is devoid of long and abstract typed commands or distracting bit pad menus. A head up, on-screen working attitude has been adopted to facilitate simultaneous thought and action. Input is carried out via a mouse and on-screen cursor. Drafting facilities are selected by simply directing the cursor to the set of icons, or picturegrams, displayed down the right-hand side of the screen (see Fig. 14.2); and pressing the accept button on the mouse which then activates the selected facility. A pop-up menu will appear on the screen, to the left of the icons, offering associated drafting functions.

Briefly, ProCAD facilities include:

The interactive inclusion and editing of POINTS, LINES, ARCS and TEXT
- The creation and manipulation of user-definable SYMBOLS
- The addition of advanced GEOMETRY to an existing drawing
- ZOOM/PAN/COPY/SCALE/ROTATE/MOVE/DELETE facilities to modify singular or grouped drawing elements
- DIMENSIONING/HATCHING/MEASURING
- Changing the on-screen DISPLAY PARAMETERS so as to switch grids and drawing layers on or off
- RASTER and PEN PLOT hardcopying of on-screen displays.

(i) The ProCAD 'toolkit'. The special feature that makes ProCAD unique when compared to other PC-based CAD systems is its 'toolkit' facility. The 'toolkit' is a library of Pascal functions and procedures that allow *direct* access to ProCAD's drawing database. These can be included in an application program, and run from within ProCAD, thus achieving true integration. The application program is first installed as a design module, and then activated during a ProCAD working session.

(ii) Programming with ProCAD's 'toolkit'. A number of CAD systems claim to provide user programming tools, but in reality these tend to be crude macro-languages of limited capability. Many software houses have adapted existing programs to generate neutral format transfer files, such as AutoCAD's DXF files, which can be read directly into a computer drafting package. The CAD system is then used to produce the final drawings. However, communications rarely take place in both directions. In many cases, only 2D information is transferred, thereby producing a 'dumb' drawing.

The ProCAD organization set out to produce a CAD system with a fully accessible database. This 'window' into the database is called the design module. Applications software was written with embedded calls to the database, and linked together with ProCAD's object code. All programs are written in MicroSoft Pascal.

Graphic elements, such as points, lines, arcs, text and symbols, are stored with a unique integer flag in the drawing database. Each element has ten real, ten integer and five Boolean attributes. A line, for example, has a start and end X, Y, Z coordinate. The layer, line style and pen used to draw the element are also defined. Elements can be 'on' or 'off'; displayed or 'invisible'.

The Pascal functions and procedures used to access the drawing database are collectively called the 'toolkit'. Figure 14.1 illustrates a Pascal program to draw and remove lines on the screen, using the pull-down menus in conjunction with mouse interaction.

There are restrictions on program development. The MicroSoft Disk Operating System (MSDOS) introduces constraints on program size. The overlaying of programs is extensively used to surmount this problem. Pascal has limited array addressing, which can be overcome by using temporary or random access files. Where information cannot be stored in ProCAD's database, control files are used. These are identified by the drawing filename, but with unique extensions. For example; the control file——'COD' contains all the survey feature codes and associated line, pen and symbol types;——'CTL' contains information relating to the drawing origin, text size, grid spacing, contour defaults and the drawing last used;——'MOD' contains all reference to the Digital Terrain Model(s) used;——'TRI' contains the link list of the DTM triangulation.

ProCAD allows up to two separate design modules to be activated at any one time,

```
{******************************************************************************}

PROGRAM Align(INPUT,OUTPUT);

USES      ProDefs,ProcToolKit,PSupp0,PSupp1,PSupp2,PSupp3,PSupp4,PSupp5,PSupp6,
          PSupp7,PSupp8,PSuppA,PSuppB,PSuppE,PSuppX,ProMenu,

                              {Library of ProCAD Toolkit FUNCTONS & PROCEDURES}

{-----------------------------------------------------------------------------}

PROCEDURE LocLine;

BEGIN
 WHILE GetCoordinate('Indicate Start of Line',CPoint,PAbs,FALSE,0,0,x1,y1) DO
                                            {Prompt for start of linked lines}
  BEGIN
  point1.px := x1;
  point1.py := y1;
  point1.pz := 0.0;
  WHILE GetCoordinate('Indicate Next Point',CLine,PAbs,FALSE,x1,y1,x2,y2) DO
                                        {Prompt for line end - Using mouse}
   BEGIN
   point2.px := x2;
   point2.py := y2;
   point2.pz := 0.0;
   lelem := InsertLine(point1,point2);{Enter Line Element in DataBase}
   point1 := point2;
   x1 := x2;
   y1 := y2;
   END;
  END;
END;

{-----------------------------------------------------------------------------}

PROCEDURE DelLine;

BEGIN
 WHILE GetCoordinate('Indicate Line to delete',CPoint,PAbs,FALSE,0,0,x1,y1) DO
                                            {Prompt for line - Using mouse}
  BEGIN
  point1.px := x1;
  point1.py := y1;
  lelem := FindLine(point1);                {Find line element in dataBase}
  IF lelem <> 0 THEN DeleteLine(lelem);     {If line found, then delete it}
  END;
END;

{-----------------------------------------------------------------------------}

BEGIN
  PTKOpen;                                           {Open drawing database}
  submenuend := FALSE;
  WHILE NOT submenuend DO
   BEGIN
   WITH TheMenu DO                                   {Set up the pop-up menu}
    BEGIN
    Header       := '      Line Menu';
    NumCommands  := 4;
    Commands[1]  := '    Locate Line';
    Commands[2]  := '    Delete Line';
    Commands[3]  := '_____';
    Commands[4]  := '  Leave Program';
    END;
   MenuMessage('Straight Options');                      {Toolkit Procedures}
   SelectFromMenu(TheMenu,subchoice);    {Choose from the menu - Using mouse}
   ClearDialogArea;
```

[*Contd.*

```
CASE subchoice OF
   1 : LocLine;                             {User defined procedures}
   2 : DelLine;                             {    "        "        "     }
   3 : ;
   4 : submenuend := TRUE;
   OTHERWISE;
   END;
  END;
 PTKClose;                                  {Close drawing database}
END.
```

{ *** }

Figure 14.1 Sample PASCAL 'Design Module' program using ProCAD's 'toolkit'.

both fully supported by the Bill of Materials option depicted by the 'bomb' icon. The drawing database can contain flagged elements, along with additional user-defined attributes, which are all DBASE-III-compatible. Although no attempt has been made to make use of this feature, it does, however, offer interesting prospects in context to graphically orientated information systems.

14.3 ProSURVEYOR

ProCAD emulates normal survey mapping practice. Having reduced observations into point coordinates, the surveyor chooses the units, scale and paper size for his working survey drawing. A 'clean' drawing sheet is selected from within ProCAD. Each ProCAD sheet can accommodate up to 255 discrete drawing layers, and can include up to 32 000 drawing elements.

14.3.1 Survey mapping
ProSURVEYOR is activated via the design module facility available on the ProCAD main menu. Once selected, the ProSURVEYOR pop-up menu will appear on the screen (Fig. 14.2). Up to 255 individual survey data files can be loaded from the hard disk at any one time, to make up a composite survey job. Point and station files are selected from a pop-up directory menu accessed from within the INPUT/OUTPUT facility. Original data files are not altered. Once loaded on-to the ProCAD drawing sheet, the survey points can be queried, deleted, moved and inserted interactively on the screen. Survey grid lines can be automatically generated, together with coded feature lines and symbols.

Special mapping related features within ProSURVEYOR include BLOCK DATA and PARTITIONS. The former facility allows user-definable data blocks to be interactively manipulated independently of the rest of the survey job. The latter enables the survey jobs to be partitioned on to separate ProCAD drawings, at different sheet sizes and scales (Fig. 14.3(a), (b)). At any stage during mapping, the ProCAD zoom, pan redraw, window, draw lines, symbols and text, delete, etc., facilities can be used. Additionally, the DISPLAY facility can be used to switch the ProCAD drawing sheet layers on or off, according to the degree of screen and plot detail required. A separate pull-down menu is provided for this facility (Fig. 14.4). The mapping capability, and the flexibility described above enables full survey working drawings to be furnished at the required sheet size, scale and orientation, as well as the desired level of detail and annotation.

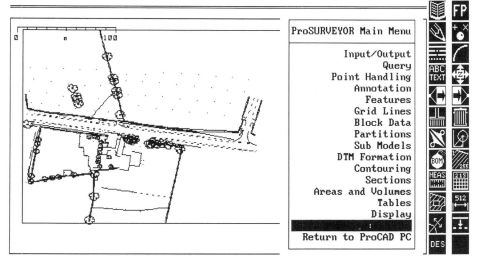

Figure 14.2 ProSURVEYOR main menu.

14.3.2 Digital terrain modelling

Terrain models are created by linking together selected points to form a continuous surface. DTM FORMATION invokes a Delaunay autotriangulation algorithm, which links points with regular-shaped, non-overlapping triangles. Breaklines and boundary strings are recognized, either automatically as coded constraints in the original data file, or by interactive on-screen insertion. EDITING facilities are provided to enable interactive triangle deletion, insertion and switching. Once satisfactorily formed, the terrain model can be used to derive contours, sections, areas and volumes. A special facility enables a survey to be divided up into discrete SUBMODELS (Fig. 14.5). Each submodel can be processed independently or grouped together to form a composite model. A multisurface modelling facility allows the user to compare two DTMs for isopachytes, multistrata sections and interactive design model creation, e.g., overburden/coal seam stripping volumes, etc. For example, Fig. 14.6 shows the contours of a disused quarry considered as a potential wastefill site. The original and proposed contours have been displayed and plotted on the same ProCAD drawing. The fill capacity of the site can be determined by projecting the proposed contours down to the existing quarry floor.

'Quick' and curve-fitting CONTOURING algorithms have been implemented. Additionally, any intermediate modification to the DTM will automatically update 'quick' contours.

Individual SECTIONS can be defined along with user-specified annotation boxes (Fig. 14.7). Multiple sections can be stored, and then downloaded on-to a separate ProCAD sheet, for section presentation purposes. The AREAS facility provides an on-screen 'planimeter'; slope and projected areas can be computed. Both prismoidal and section VOLUMES can be derived from the active model(s), either separately or

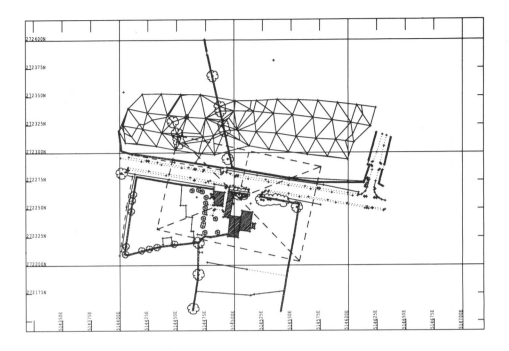

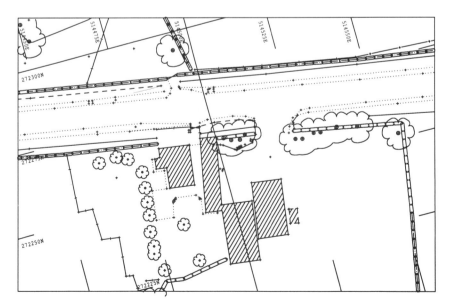

Figure 14.3 (Top) 1:500 scale PARTITION defined on the ProSURVEYOR drawing at 1:1000 scale. (Bottom) 1:500 scale PARTITION displayed on ProSURVEYOR drawing.

ProSURVEYOR (v3.03) Display Options - Make Selection

Current Job : HEAP

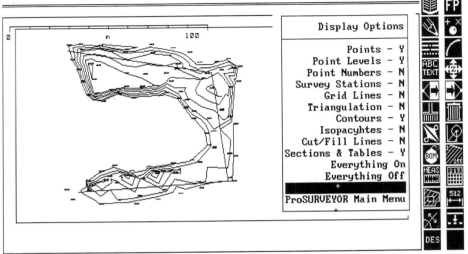

Figure 14.4 ProSURVEYOR's on-screen display menu.

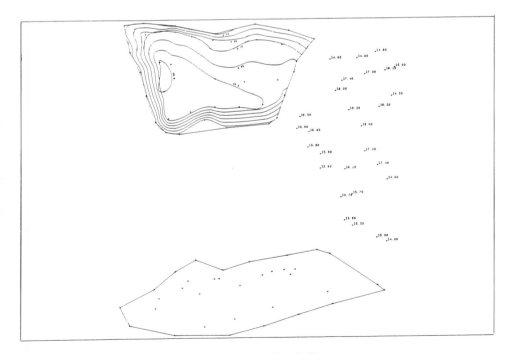

Figure 14.5 Illustration of ProSURVEYOR's submodelling facility.

H

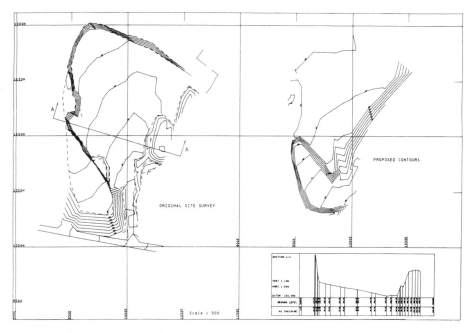

Figure 14.6 Surface projection of proposed contours to original quarry floor.

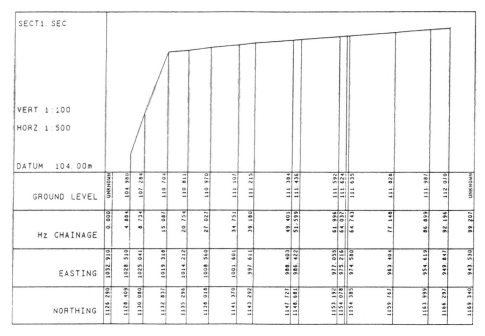

Figure 14.7 Fully annotated section generated using ProSURVEYOR.

with a cut-and-fill balance. As with mapping, the DISPLAY facility provides selective on-screen viewing of the model(s) (triangles) and its attributes (contours and sections).

14.4 ProCAD/ProSURVEYOR software and hardware requirements

Rapid technological developments within the computer industry have dictated that careful consideration be given to both software transportability and ease of upgrading. Thus suitable hardware platforms and software languages must be chosen carefully at the conception stage of software development.

The Pascal software language has been adopted by many application development houses, due to its near English vocabulary, comprehensive error checking, flexible data structure and transportability. The ability to mix languages is a useful aid, since Pascal imposes restrictions on array sizes. ProSURVEYOR has thus been developed to run on the IBM PC, the industry standard, using both MicroSoft and Turbo versions of Pascal.

ProCAD/ProSURVEYOR runs on IBM PCs and compatibles, provided that they conform to a minimum hardware requirement of:

- 640 kb RAM (or more, preferred)
- 20/40 Mb hard disk
- Graphics card (Hercules, EGA, VGA, Olivetti)
- Standard monitor and keyboard
- Maths co-processor (preferred)
- Mouse.

14.4.1 Current applications of ProSURVEYOR
ProCAD/ProSURVEYOR has been used on a variety of interesting surveying and engineering projects. Its users include land and mineral surveyors, quarry contractors, civil engineering consultants and contractors, water authorities and airport authorities. The product has assisted in site development, quarry planning and rehabilitation, road design, and airport construction as well as industrial and domestic estate developments. In fact, ProCAD/ProSURVEYOR is applicable to any project where the emphasis is placed on producing professional working drawings as quickly and as efficiently as possible.

14.4.2 Future software developments
The very basis of ProCAD/ProSURVEYOR leaves it wide open for further development. The concept of totally integrated design modules, 'feeding' off a central database, offers exciting prospects to today's surveying and engineering professionals. A drainage package will soon supplement the ProALIGNMENT design module. Additional developments include a collaborative venture with an existing client, with the objective of producing a practical and easy-to-use reinforced concrete detailing system. This particular venture furthers the company's aim, to 'give clients the tools they need'.

14.5 Conclusions

The ProSURVEYOR product from Applications in CADD is a flexible, easy-to-use application tool, which resides as a totally integrated design module within the

ProCAD computer-aided drafting system. ProCAD/ProSURVEYOR is transportable across the popular IBM—compatible range of computers, including the new laptop generation of portable machines. Applications in CADD is a progressive application development company, whose sole aim is to support the needs of its clients. As such, further software developments will be client-driven. In ProCAD, the company has found the perfect vehicle with which to develop application software, such as ProSURVEYOR.

15 Digital terrain models and their incorporation in a database management system

F. STEIDLER, C. DUPONT, G. FUNCKE, C. VUATTOUX and A. WYATT

15.1 Introduction

In order to make full use of the possibilities of the database concept of SYSTEM 9, Wild developed a new DTM (Digital Terrain Model) package, separate from the existing CIP program (Steidler *et al.*, 1984) which runs in batch mode and uses sequential file structures. The new concept is based on a modified relational database management system, which has been developed by Wild (McLaren and Brunner, 1986; Heggli, 1986). The database management software and the software for data capture and data editing—the basic software of SYSTEM 9—are the tools used for the development of the DTM. The DTM package is written in the programming language C and runs under the UNIX operating system (UNIX™ 4.2 BSD) on the SYSTEM 9 workstations S9-D, S9-E and S9-AP. It makes use of the GKS graphics kernel system.

New characteristics of the DTM package, which are different from those of conventional DTM packages, are mainly:

- Input of database features (nodes, lines, polygons, holes)
- Editing of input data with standard system tools
- Recomputation procedures for smaller windows and their integration into the digital terrain model
- Interactive editing of the contours, the triangle network and the input points
- Graphic output with standard system tools
- Possibility of image injection for quality control (S9–AP only)
- Lines or triangles can be picked within the 3D perspective model and automatically displayed in a plan view, or coordinates can be displayed.

Applications of the DTM are, apart from the interpolation of contours, the computation of regular grids, graphic DTM representations such as perspective views, engineering projects (profiles and volumes), intervisibility problems, slope maps, multisurfaces (oil exploration), differential models and non-topographic uses.

Highly sophisticated algorithms are used for the derivation of the DTM and for its applications. For example, lines defined in the database can be used as breaklines (sharp or round) or structure lines. Breaklines define discontinuities in the slope of the ground surface, where the contours bend; sharply for manmade features (sharp breaklines) or with some curvature (round breaklines). Structure lines are lines which the contours cross at right angles, but with no discontinuity in the slope of the ground surface. The triangulation method is used for the DTM, i.e., all observed points are connected in a set of spatial triangles. Two conditions are applied in the formation of the network of triangles. Firstly, sections of lines or polygons (known in X, Y, Z) always form edges of triangles. Secondly, a triangle is only created if no further point lies within a circle defined by three nodes (secondary condition). Bicubic functions are used for the interpolation of the ground surface within a triangle. Different filter values for mass points, breaklines sharp, breaklines round, structure lines and holes may be used to smooth the surface.

Large data sets may optionally be processed in batch mode, but interactive generation of the DTM, or parts of it, is also possible.

Many editing functions are provided for the graphic output. For example, the positions of annotation text can be specified very easily and the text itself is easily edited.

A sophisticated menu allows an easy input of parameters, many of which have preset default values so the user need only enter a few. Within a menu, the user can jump directly to any other desired menu, so avoiding the disadvantage of paging in menus. The graphics screen with graphic menus can be used, as well as a simple alphanumeric terminal, for conventional menu operation.

15.2 General data flow (Fig. 15.1)

The heart of the computation process is the basic SYSTEM 9 software package, which includes the modified relational database management system, the data capture and editing modules, the graphics handler and, in addition, the plot task. The modified part of the relational database management system we call the object handler, because it deals with the actual objects within the database. It provides the necessary interfaces for the application programmer.

The database management system allows the project manager to define a 'PROJECT'. Within this project he defines working areas, called 'PARTITIONS', to which one operator has access at any one time. Each partition is, in effect, a window with boundaries parallel to the axes of the real-world coordinate system and with specified feature classes, all of which must have been defined in the project before a partition is checked out for operation—for example, for this type of application, input feature classes such as nodes and rivers and output feature classes, such as TIN (Triangulated Irregular Network) and CONTOUR. Within the partition, the operator has fast access to all features.

The DTM package talks to the SYSTEM 9 software via the application interfaces. A DTM administrator, which consists mainly of a parameter file and certain interactive commands, is responsible for the interaction with the database as well as for the data flow and the processing within the package. The parameters are stored within the partition database. The administrator tells the computation parts which feature classes have to be placed in which DTM data group. Each group may include several feature classes. For instance, the feature classes nodes, spot heights and grid points may be placed in the 'mass points' group, while the feature classes rivers, ridges and road edges may be placed in the 'breaklines round' group, etc.. The administrator makes sure that, for example, nodes cannot be placed in a 'breaklines' group, because for this group data of type 'line' are mandatory. After the assignment of feature classes to the different DTM input groups, all necessary parameters are specified via the menus. The parameters are written into a file which is also stored within the database.

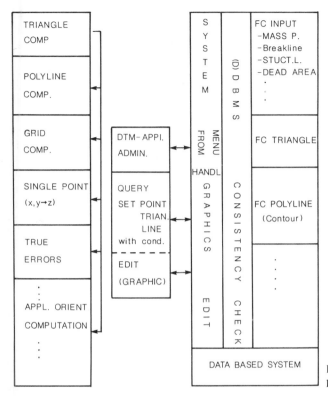

Figure 15.1 Data flow in the DTM process.

The various DTM computation parts now follow. With the aid of the administrative information, the required feature classes are read in from the database and new, DTM-specific feature classes such as TINs, contours or grids are computed and immediately stored in the database. Any data stored within the database may be queried, displayed, edited or plotted at any time, as long as it exists in the working partition. The graphic representation is specified by the user in a 'THEME' which is defined at project level and used within the partition.

When all computation and editing is complete, the output feature classes and the (modified) input feature classes are brought back into the project. During this action, consistency checks are performed and, if an inconsistency is detected, the user is informed and asked to correct it.

15.3 Parameter and data input

A private DTM-parameter file can be generated for each partition. It is also possible to copy the parameter file from another partition or even from another project partition. The DTM menu program checks whether this file exists, and, if not, a default parameter file is created automatically. The input and output feature classes are entered in the menu and the parameter file can be changed. The feature classes are checked for existence and proper use (e.g., for breaklines, feature classes of type 'line' and with defined Z values are mandatory). They are also checked for improper use in two different groups of data, e.g., if an attempt is made to enter a feature class 'river' in the groups 'breakline round' and 'structure line' simultaneously. Such a double use will not be accepted.

The parameters can be input sequentially (in a 'browse' mode) or selectively by typing the line number. The parameters are immediately checked for plausibility (range, syntax) and consistency. HELP functions are provided for the inexperienced user, called from a background menu. For the experienced user, short cuts, which can be user-defined, are provided so that he can jump from any menu page to any other without traversing the menu hierarchy. As a default, it is always possible to step forward to the next menu page or back to the previous page.

Each menu page has a status which tells the user whether his input was correct or not, and, if incorrect, the reason why. Incorrect parameter values are therefore impossible. In the background menu, which is always displayed together with the current menu, a CANCEL function is provided, which allows the user to release the menu at any time with the originally read-in parameters.

The computation parts of DTM may now be started from the menu.

15.4 Input of DTM data

The input data for DTM may be any set of arbitrarily distributed points with known XYZ coordinates. These may be single points, observed by field survey (tacheometry), or in single-point mode in a stereoplotter. Profile or grid measurements can also be used, for example from optimized data-capturing methods such as 'Progressive Sampling'. Breaklines and structure lines (see above) can be entered, as can fault lines, which are lines in which there may be two Z values for each XY position. Fault lines are important in geology and minerals exploration, where different surfaces have to be considered.

Polygons which represent holes (which may be known in XYZ) may also be input. They describe areas where the DTM is not defined and where, for example, the contours must be clipped. Polygons representing dead areas (known only in XY) are not used for the triangulation, only for clipping the contours.

All input data are read directly from the database via feature classes which represent categories of features within a project. Feature classes impose a structure on geographic data through an association of features and their component primitives. Feature-class names are defined in a table within a DTM-specific input group.

15.5 Forming the TIN

The generation of the DTM corresponds to setting up a spatial, triangular irregular network together with information on the surface normals at each node and on connections to neighbouring triangle points. To form the triangles, first all line segments (breaklines round, breaklines sharp, structure lines, fault lines or holes) are stored as edges of triangles. Then mass points (points known in XYZ) are introduced in the triangulation, i.e., the vertices of the triangles will always be measured points. The procedure proposed by Delaunay (explained, for example, in Gottschalk, 1981) is used for the construction of the triangular net. In this procedure a triangle is formed if the circle passing through the three points contains no other point. Lines may be subdivided individually if their segments are longer than a user-defined length (default value is the mean length of the triangle sides).

The surface normals at each node or vertex are then computed. The surface is defined by Zienkiewicz functions (Bazeley *et al.*, 1965) which include all neighbouring measured points, and the normals are computed from tangent planes derived from a least-squares fit. When a segment of a breakline (round or sharp) forms an edge, more than one normal will be computed and stored. Smoothing filters for heights are also introduced; these may be different for each group of input data.

The generated surface is smooth within the triangles and continuous from triangle to triangle except across breaklines. Segments of breaklines are assumed to be straight lines in space with discontinuities in slope perpendicular to them. Segments of structure lines are considered as straight, but the ground surface is continuous across them.

15.6 Editing

15.6.1 Editing input data

The user may delete, add or modify features, input points and lines with the standard tools of the basic SYSTEM 9 software. The consequences of this editing are automatically taken into account in the program. Thus, for example, the triangles affected by such an editing change must be found and deleted from the database and new triangles computed and inserted in the network. New contour segments must also be computed and inserted into the database. The consequences are shown below.

Correction of input data	*Consequences*
Editing of input data with standard system editing tools	Database interactions • Find triangles involved • Delete these triangles
• Delete, Add, Modify points and/or lines	• Compute new triangles • Insert new triangles into TIN • Compute new contour segments • Replace old contour segment by new

15.6.2 Editing the triangle net

The user has the opportunity to optimize the triangulation. Although a very sophisticated triangulation algorithm is used to establish the network, the user may

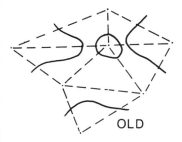
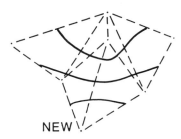

OLD   NEW

Figure 15.2 Editing triangle edges.

EDIT FUNCTIONS

a) Modify contour

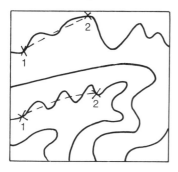

b) Delete contour

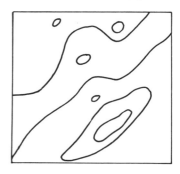

1) Pick point 1 of contour
2) Pick point 2 of contour
3) Highlight contour segment
4) Define new contour segment

1) Pick contour
2) Highlight picked contour
3) Delete contour

Figure 15.3 Contour editing. (a) Modify contour. (b) Delete contour.

wish to modify the triangulation in some areas. He also may wish to change the edge types, for instance to change a feature from breakline round to breakline sharp. The consequences of such changes are also automatically taken into account in the same way as described above. Figure 15.2 shows an example of editing the edges of triangles.

15.6.3 Contour editing
Contour editing provides an improvement in the cartographic quality of the inter-polated contours which represent the mathematical interpretation of the ground surface. Cartographic contours, on the other hand, represent the geomorphologic interpretation of the surface. The computed contours are totally dependent on the input data and they are not necessarily identical with the contours required by the cartographer. Although the interpolation method used in the DTM package comes very close to cartographers' requirements, there sometimes remains the wish to modify small sections of some contours. This may be done by smoothing, generalizing or highlighting shapes.

If the user does edit contours, the appropriate triangles, and possibly some input data must be updated because the next time he interpolates contours he wants to obtain the proper result without further editing. This updating will be an important feature in the DTM package. It will be performed as automatically as possible (Fig. 15.3).

15.7 Applications

(i) Contours. The major application of the DTM package is the generation of a high-quality contour map. In order to reach this high quality, the triangles of the DTM are

subdivided into subtriangles, with a selectable degree of subdivision. The range is from zero (no subdivision) up to 10 divisions of a triangle edge corresponding to 100 subtriangles. The choice obviously depends upon the type of terrain. The XY coordinates of each intersection of the specified contours and the edges of the subtriangles are then computed, leading to a set of polylines defining the contours. Of course, the higher the degree of subdivision, the longer the computation time, and the larger the required storage capacity.

Several options are provided for the graphic representation of the contours. Index contours can be annotated at different vertical intervals and at different distances along the contours. Contours may be marked with arrows to show the downhill slope direction, different colours may be used and contours on steep slopes may be omitted if the slope exceeds a user-defined value, to name only a few of these important practical features.

(ii) Perspective views. An option for the DTM is the generation of perspective views of the ground surface, for example for planning in road design and also for error detection. The perspective view is derived from the TIN. The advantage of this method of derivation is the automatic inclusion of breaklines, structure lines and dead areas, because their segments are always edges of triangles. The perspective centre and the direction of projection can be specified by the user.

The fastest solution is a wire-frame model with optional suppression of hidden lines. A more sophisticated solution is also available in which the triangles are coloured as a function of their slope (so-called flat shading). Smooth shading is also possible, with, as an option, discontinuities of shading at breaklines. The contours may be shown on top of the shaded perspective and all displays can be rotated. Editing within the perspective view will be possible; if a point is chosen within the perspective view the corresponding planimetric view can be displayed and the cursor will show the same point and triangle. The colour graphics screen is used for the displays.

(iii) Regular grid computation. Several applications of a DTM may need a regular grid of points as input data. The Wild DTM package provides the option of interpolating such a grid. The mesh widths in X and Y must be specified, together with the window within which the grid is to be computed.

(iv) Single-point and true-error computation. The essence of this DTM package is the interpolation of heights of points with known planimetric coordinates. Thus, a feature class of single interpolated points can be created simply by entering the appropriate XY coordinates, either by typing them in or by picking them on the graphics screen. A special application of this technique is particularly important—the computation of true errors. The points to be specified for this purpose are selected from the observed points. The interpolated heights are then compared with the measured heights and a mean-square error formed from the differences. This error is then compared with a predefined tolerance and the user is warned if the tolerance is exceeded.

(v) Intervisibility. The intervisibility program shows those areas visible from a specified standpoint by checking each point in the DTM for visibility from the standpoint. By varying the position of the standpoint, the size of the visible area can be maximized.

(vi) Slope maps. A slope map is a representation of the ground surface in which, instead of contour lines, isolines of the maximum slopes are drawn. Thus, slope values (tangents) are introduced instead of heights. Areas between certain isolines may be shaded.

Slope maps are used mainly for land consolidation studies, for example to estimate ground values (value may be greatly affected by slope) or to design road and drainage networks.

(vii) Difference model. A difference model can be derived from two DTMs, representing two different epochs. Instead of normal contour lines, isolines of height differences are computed. A typical example is in road design, where the original surface is compared with the designed surface.

(viii) Profiles and volumes. This program will allow the computation of longitudinal and cross-sections for a specified centreline. The cross-sections are computed from the intersections of the cross-section line and the edges of triangles. The designed cross-sections can be superimposed. Earthwork quantities can be computed, together with the necessary 'cut-and-fill', as an additional feature. All results can be displayed on screen, edited there and then plotted on a table.

Thus, the engineer is able to compute and design interactively all the details he needs for his decision making. The system has produced for him complete longitudinal sections and cross-sections, earthwork quantities, mass-haul diagrams, etc. After studying this information, he may decide to regrade in certain areas and perhaps realign in others. He changes certain parameters and remeasures some points and recomputes. Thus, a road can virtually be restaked at the computer any number of times, at a very low cost.

References

Bazeley, G.P., Cheung, Y.K., Irons, Y.K. and Zienkiewicz, B.M. (1965) Triangular elements in bending-conforming and nonconforming solutions. *Proc. First Conference on Matrix Methods in Structural Mechanics*, Air Force Inst. of Technology, Wright Patterson A.F. Base, Ohio, 547.

Gottschalk, H.J. (1981) Relationen im Raster. *Nachrichten aus dem Karten- und Vermessungswesen* 1, Heft 82.

Heggli, S. (1986) An integrated system for geodetic and photogrammetric data capture and processing, the Wild SYSTEM 9 approach. *Proc. ISPRS-Symposium*, Commission II, Baltimore, USA.

Hofmann-Wellenhof, B. (1981) The use of linear lists for the derivation of contour lines from heights in a regular grid and along irregular breaklines. *Manuscripta Geodetica* 6, 355–374.

Koch, K.-R. (1985) Digitales Gelaendemodell mittels Dreiecksvermaschung. *Vermessungswesen und Raumordnung* 47, Heft 3 and 4, 129–135.

McLaren, R.A. and Brunner, W. (1986) The next generation of manual data capture and editing techniques: The Wild SYSTEM 9 approach. *Proc. ACSM-ASPRS Spring Meeting*, Washington, Vol. 4, 50–59.

Preusser, A. (1984) Bivariate Interpolation ueber Dreieckselementen durch Polynome 5. Ordnung mit C1-Kontinuitaet. *Zeitschrift fuer Vermessungswesen* 109, 292–301.

Steidler, F., Zumofen, G. and Haitzmann, H. (1984) CIP—A program package for interpolation and plotting of digital height models. *ISPRS Congress*, Rio de Janeiro, Commission III, 1984.

Zumofen, G. and Leoni, M. (1977) Neue Programmsysteme zur Berechnung und Darstellung von Isolinien mit Hilfe von Kleincomputern. *Vermessung, Photogrammetrie, Kulturtechnik, Zurich.*

Part D
DTM applications to engineering

16 Application of the MOSS system to civil engineering projects

G.S. CRAINE

16.1 Introduction

The range of projects where the MOSS System has been used is almost endless, covering roads, railways, airports, canals, dams, landscaping, land reclamation and mining. The system has over 200 users worldwide and its major attraction is the unique ability to combine survey, engineering design, analysis and drawing production within a single system utilizing a common database.

Within the survey profession, MOSS is often referred to as an 'engineering system', but this is only a half-truth. The system was originally developed in an engineering environment with the aim of automating both the survey and engineering design functions. The success of the concept is indicated by its worldwide acceptance and its extension into mining, mapping and architectural applications.

16.2 Concepts and system design

The MOSS concept of data storage considers survey and design information as either point or string data representing features which contribute to a surface model. These models can depict the existing ground as recorded by aerial or ground survey methods, geological strata and, most importantly, new works. This allows the creation of new designs consisting of feature lines defining exactly all the detail of the new surface, which is the basis for graphical design.

Previous design methods were based on cross-sections which were restricted in practical application to simple linear schemes. In addition, because they are computerized manual methods their potential for automation and graphic design is limited.

The common approach of creating surface models for both existing and new surfaces in a three-dimensional format within a common database is ideally suited to computer manipulation and the development of graphic design methods. Furthermore, the system has wide application within an organization providing a single system with common standards of input, output and drawing production and with the added benefit of uniting the various departments within organizations working on common projects.

The system was conceived in 1972 when only mainframe computers were generally available. Several principles were adopted that have had a major influence on the way the system has developed and they are just as relevant today. Perhaps the most important decisions were the adoption of Fortran as the programming language and the portability of the code-which resulted in the concept of one program accessing a single project database. Although revolutionary at the time, this has made it possible to mount the system on at least 12 different computers and operating systems with relative ease. Graphical output via plotters or screens is essential to the string concept, and it was a major requirement that all commonly available devices should be supported. It was also decided to adopt 32-bit systems to allow both the storage of national grid coordinates and to ensure the numerical accuracy required for design calculations.

These features have limited MOSS to certain machines, and thus no effort has been made to develop micro-based systems during the intervening time because they would have reduced the function and application of the system. However, with the advent of minicomputers and now workstations such as the Apollo and the MicroVax, it is possible to have mainframe processing in a desktop machine and offer the full MOSS function at a realistic cost.

More recently, with the introduction of interactive graphics, the international GKS standards have been adopted to maintain portability, which also allows advantage to be taken of the special firmware being incorporated in graphics devices.

The modelling concept requires complete freedom of project definition with the storage of infinite surfaces. There is no restriction on the number of models (surface layers) stored in the database, the extent of a surface, and the number of strings or points in a model or string, and the database is structured accordingly.

A further influence was the potential size of projects, and all processing and analysis routines have been developed to handle large volumes of data efficiently. It is not uncommon to have models in excess of 300 000 points, and the user has freedom to use any MOSS function without having to reorganize the data.

16.3 System features

The modules within MOSS provide for survey, engineering design, analysis and visualization, including the production of contract drawings. The range of facilities included in the modules permits the survey and design of schemes in all previously listed application areas. Additional features are being added constantly for specific industry requirements and to improve the general facilities and user-friendliness of the system.

Major extensions have been introduced to make the system easier for the user with no previous computer experience. An additional method of data entry allows the selection of required facilities from menus or lists and 'fill in the blanks' for data entry. When users become familiar with input requirements, they may choose the expert mode of direct free format line entry for faster response. However, it is possible to switch between the two modes of entry at any time during input.

A completely revised automated drawing production system gives greater flexibility and control over drawing layout and content and is the basis for a new set of options which allows the unique three-dimensional design and editing facilities of the MOSS system to be accessed through an interactive graphics screen.

The existing interactive command language and batch operation facilities are retained, allowing the user to run those parts of the system that need immediate response in an interactive mode while computing intensive analysis work can be run in the background.

16.4 Survey

Surface models may be created by aerial survey, digitizing of existing drawings or by ground survey methods. Aerial survey proves attractive for large projects that are planned in advance, and digitizing provides a rapid method of data collection from existing maps. Ground survey techniques have been developed to introduce a high level of automation into the recording of ground and surface detail. The ground control may be established by traversing, and options provide for the calculation and adjustment of traverses by several commonly used techniques. Survey detail may be recorded as points or strings, and the strings may be a series of sequential points following a feature or recorded as separate points having a common label for automatic sorting to produce the required feature.

The linking of points or strings to complete the surface model is a simple task with the interactive editing facilities and the final feature coding may be undertaken in the office. Alternatively, a survey consisting of all point data may be edited to produce the required coded features using the interactive facilities. Usually the site survey will produce a combination of points and strings, but the greater the site stringing, the greater the automation of the survey processing and drawing production.

Data may be recorded using any instrumental techniques including stadia methods, chain and offset, tacheometers, EDM equipment and data collectors. Preprocessing programs are provided for transferring and converting information from all the major manufacturers' data collecting equipment and a fully prompting data collecting system has recently been introduced (Chapter 3).

The new MOSS Site Module has been developed as a special version of the ruggedized Husy Hunter II offering standard 352k and up to 496k RAM and including 64k PROMs to handle special software routines. It is equipped with a multiple-line LCD display, 58-key QWERTY keyboard and a weather- and shockproof case.

The Site Module offers data capture, recording and editing procedures which are fully compatible with new MOSS survey software and which make full use of its advanced features such as flexible survey and recording methods, feature coding systems and choice of a split of activities between work in the field and at the computer graphics screen. In addition, special data-recording techniques have been developed to simplify complex urban survey and cartographic mapping work. Simplification and streamlining of the survey work has been achieved through the use of field verification techniques, screen menus and user prompts which offer guidance through standard and non-standard processes and play an important role in eliminating errors at the data-capture stage along with the need to resurvey.

A key development is provided by a facility for downloading complex three-dimensional design information into the Husky Hunter site computer which allows for greater flexibility, accuracy and accountability to be brought to setting-out work. This special facility allows for the selection of setting-out stations on site along with automatic generation and monitoring of setting-out data.

The MOSS Site Module comprises a suite of programs offering integrated traverse, detail survey, resection/free station and setting-out processing facilities. Additional optional programs are available for specific applications, and these include on-site

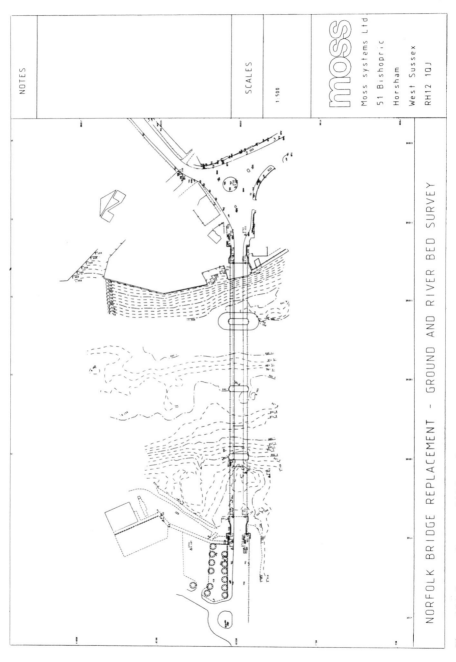

Figure 16.1 Survey of proposed highway improvement.

plotting, on-site levelling, on-site volumes and dimensional control. Compatible with all major electronic instruments, the Site Module can also be interfaced with any HP GL compatible plotter for on-site work.

The survey techniques are very flexible and the use of data collectors guarantees a high level of automation. Figure 16.1 illustrates a MOSS survey and the automation and quality of drawing production using stored feature symbols.

The survey facilities are used by the majority of MOSS users for general land survey, preparation of new designs and surveys, and volume calculations, e.g., for quarries, stockpiles and construction earthworks. A major user is British Rail which uses the survey facilities extensively for both new and minor works, and also permanent way records which are effectively a database of track equipment.

A special feature in MOSS is the contouring facility which automatically produces 'surveyors' contours from the model information. This requires the creation of a triangulation of all the model data, and the algorithms ensure that the strings links are the sides of triangles. The contours produced from the triangulation automatically reflect all the angular features and accurately describe the surface. The contouring technique is automatic, does not require an intermediate triangulation editing process and there is no limit to the number of points to be processed. A typical example of a survey consisting of 2 000 data points produced 17 000 contour points in under 15 minutes on a MicroVax II computer, including the production of the drawing of both the survey and contours.

The process is applicable to any model, and one of the original requirements was to contour design surfaces, such as highway junctions, in order to introduce an automated method of visualizing the surface for checking both design and drainage requirements. Figure 16.2 shows the resulting contours for a merged new design and ground surface model, which highlights the accuracy of the method.

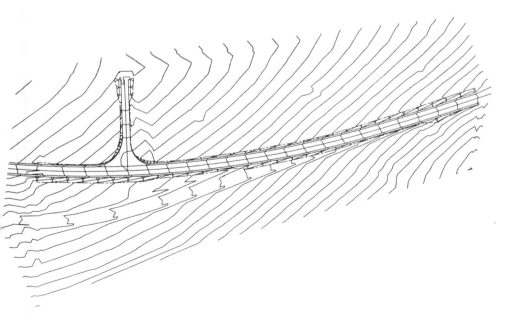

Figure 16.2 Contours generated from merged design and existing surface models.

16.5 Design

The MOSS design facilities are unique in that they allow the design of surfaces in string model form. The concept is widely applicable to engineering and mining and provides an automatic graphical approach to surface design which is universally applicable to all major and minor engineering and architectural works, whatever their complexity.

The technique is to initially identify the geometric skeleton of the design and create combined horizontal and vertical alignment strings to define the major features. Feature design facilities allow the generation of associated features in string form by simple offsetting techniques based on this framework. The model of the new works is completed by generating strings for the batter slopes where the new model merges with the existing surface. Figure 16.3 illustrates a simple highway junction design.

Design with MOSS is a graphical technique. Thus the ideal situation is to have graphics terminals that will allow simple viewing of each operation in the modelling process from the skeletal alignments through to the feature creation. Sections through the model may be extracted at any stage of development and used as a visual check of detail.

Contouring is the perfect method for checking for irregularities and smoothness in designed surfaces. The triangulation methods were developed to ensure perfect surface reproduction from the string information, and the resulting contours may be superimposed on the design detail for visual confirmation.

The comprehensive model generation facilities permit the creation of any model from simple to complex layouts which have been used on such diverse schemes as the M25 London Orbital Motorway and Australian opencast mining projects. Many of the design functions are available in the interactive version of the system which considerably simplifies the generation of complex and detailed designs.

Landscape surfaces are of simple form and usually require the digitizing of a manually developed surface. These techniques may be applied to reclamation of waste material or introducing landscaped areas within a designed project.

16.6 Analysis

A range of functions is provided to permit the analysis of the stored models to produce sections, areas, volumes, setting-out information, triangulation of surfaces and contouring.

The string surfaces are the ideal technique for recording surface features, but the actual surface is best described by the triangulation generated from the strings and points, which accurately records all the surface detail. This triangular surface is used

Figure 16.3 Highway junction design.

for contour production but may also be stored as a model and the sections and volume options can operate on either string or triangular models.

Longitudinal and cross-sections through the ground and new surfaces may be extracted for checking and visualization of the model and to allow the determination of interfaces and for earthwork calculations.

Construction quantities for topsoil strip and soiling may be determined by calculating the area within boundaries created from existing string detail or between pairs of adjacent strings.

A major feature of MOSS is the potential for determining earthwork volumes for both linear projects and in irregular sites such as those encountered in landscaping, mining and highway interchanges. In this situation, the area of interest is identified within a boundary, and volumes determined using either sections or triangular methods. The choice remains with the user and the chosen method will be selected on the basis of both accuracy and efficiency, depending upon the nature and size of the project.

Volumes by section methods are necessary for linear projects to produce running totals of volumes relative to a design line. Section methods can also be used to produce volumes of irregular areas by defining a baseline direction and section interval. The advantage of continuous surface models is that they allow the selection of section direction and intervals depending on the complexity and irregularity of the surfaces. Alternative solutions can be determined, and research has shown that the techniques do produce accurate volumes. Also, they have proved to be very efficient when used on large models.

Volumes can also be determined by triangular methods based on the triangulations of the two surfaces. Unlike many other systems, MOSS produces both the cut-and-fill volumes rather than the balance. This technique is recommended when an exact solution is essential but it requires more processing than the previous technique.

The volume methods have been used in many complex situations in mining, land reclamation and complex highway schemes. The accuracy of the modelling and volumetric analysis has created greater confidence in contract earthwork measurements, and the system has been purchased by Tarmac Construction in the UK for internal use, after they had proven its acceptability on motorway contracts.

The production of contours based on the triangulation of a surface is applicable to survey and design models and is especially useful for checking new designs for geometric criteria, smoothness and drainage. The contouring feature has been extended to produce isopachytes or contours of thickness between two surfaces. Working from triangulated surfaces derived from the original surface models, the isopachytes will accurately reflect the differences between the two surfaces, including angular changes.

This powerful facility has wide application in mining and for pavement resurfacing works where the optimization of overlay thickness and minimization of scarification can lead to significant cost savings.

In hydrology, isopachytes can show the deposition and scour of underwater materials over a period of time and in land reclamation, tip reinstatement and landscaping work, they can be used to give a quantitative assessment of settlement. This technique is very visual and illustrates and emphasizes the use of graphical design and analysis techniques within MOSS.

A major advantage of digital modelling is the confidence with which new work can be set out within the survey control. A string may be set out by using either deflection angles, intersecting rays or distance and offset techniques. The confidence in

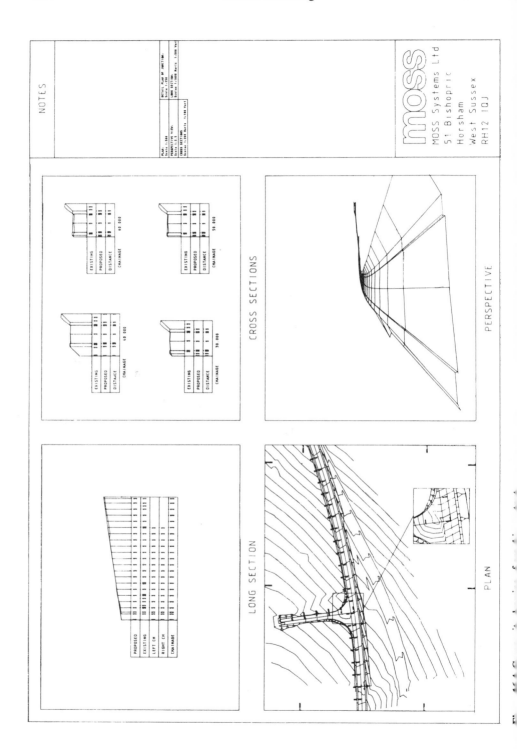

establishing feature position allows schemes to be set out initially from the perimeter to establish topsoil strip limits before setting out design and feature lines.

16.7 Visualization

MOSS relies on graphic output, and to obtain maximum benefit from the system requires a plotter and preferably graphics screens. This output may take the form of plans, sections, contours and perspective views. Drawings and displays are necessary for checking field surveys and the results of editing and contouring. The use of contours is an excellent visual means of checking for survey level errors. Graphical displays are an essential part of the design process and should be used to confirm each of the steps when creating new models.

A major feature and advantage of MOSS is the ease with which drawings and displays are produced. Drawing layouts and feature interpretation can be predefined within the system to allow the production of standard drawings within an organization with minimal input.

The powerful macro-system allows the storage of symbol and line features that can be associated with feature codes when producing drawings. If the surveyor introduces a coding system for identifying line features such as fences and hedges, and point features such as trees, manholes, etc., the plotting macros allow the automatic interpretation and production of finished high-quality drawings. There is no limit to the number of plotting macros that can be stored in the library.

The recently released DRAW facility together with the job macro feature provides a simple and foolproof method for creating standard drawings. The required drawing layouts are defined in advance and the basic commands are stored in a macro file under a unique name, together with default values for all variables. Minimal input is required to generate a drawing, and a plan would be produced by referring to the named plan macro and defining, scale, location and orientation. The output of sheets of sections requires only the specification of horizontal and vertical scales with the appropriate macro.

The macros simplify the production of standard plans and sections within an organization, but where a survey company or consulting engineer is producing drawings for several clients, then individual plan and section macros can be created for each style of drawing presentation.

Additional features allow the paging of plans and section drawings into several sheets and it is possible to produce composite drawings containing several plan and section drawings. The interactive facilities, in addition to providing screen edit and design functions, also provide simple drafting functions for the addition of text and lines to allow the completion of the survey and design drawings produced by MOSS.

Visualization of projects is not complete without perspectives, and both wire-line and hidden-line views can be produced for any model. Perhaps the most impressive output is photomontage where perspective views can be related to photographs, which are invaluable for scheme presentation at public enquiries. Both ground and aerial photomontage views are possible and, in the latter case, the aerial camera position and viewing parameters are determined within the program from measurements of photographic detail. Figure 16.4 is a composite drawing of graphic output.

16.8 Hardware

The most recent release (6.3) of the system, which includes user-friendly menus and graphics preview facilities, is available on a range of machines. However, this is essentially a batch system, and future developments will be concentrated on interactive graphics and graphic-based design facilities. This requires suitable hardware and a commitment to GKS graphics software. The interactive release of MOSS is available on IBM, Vax, Prime and Apollo computers and evaluations are in progress on ICL and Sun equipment. The company continually assesses alternative equipment and is prepared to undertake new versions. The graphics terminals must support GKS, and currently Sigmex, Tektronix and Westward models are supported. There are few problems in supporting plotters, and currently MOSS will drive Benson, Calcomp and HP equipment.

The attraction of present trends is the falling cost of hardware. The latest 32-bit single-user workstations have mainframe performance in a desk-top machine, and the hardware cost is less that the average salary (including overheads) of a surveyor/engineer. The company offers turnkey systems on both Apollo and Vax workstations with additional terminals, plotters and digitizers as required.

17 The use of HASP systems in civil engineering and earthwork design

R.E. HOGAN and M.R. KETTEMAN

17.1 Introduction

The computation of earthwork quantities has traditionally been a non-challenging but time-consuming part of the engineering surveyor's task, whether calculating volumes of materials or evaluating new designs. Often short-cut, non-rigorous estimating methods have been used where great accuracy was not critical in order to shorten the time and effort. HASP has been playing a key role in the automation of survey, photogrammetry and drafting for over 14 years and is committed to improving the sophistication of existing products as well as diversifying into a progressively wider range of applications. The need for improving the efficiency of earthwork computations and providing more productive and new creative tools for civil engineering designers has encouraged HASP to introduce a variety of earthwork-related products.

The main concepts upon which the present earthwork programs are based are interactive graphics design and visualization tools, rigorous TIN (Triangulated Irregular Network)-based surface models and object-linked database management. The system obtains its speed and efficiency by combining an intuitive friendly human interface with the use of highly developed languages and good accessible database structures. This chapter describes the conception, development and operational features of the HASP systems with reference to modelling applications to site grading in estate development. Also discussed is the integration of traditional highway profile design with HASP Automated Superelevated Object Templates. This chapter also describes the visual impact study tools provided by HASP. Much of the early work on HASP volumes determination was associated with mine surveying. Since the determination of mine stockpile volumes has a general applicability, this example is also used to show the general principles of surface modelling.

17.2 The triangulated irregular network

The HASP TIN-based surface modelling programs were originally written to allow contours and volumes to be determined automatically for stockpiles of coal. Traditionally, these tasks have been labour-intensive and time-consuming. The use of the Electronic Total Station for the field work and a computer system for data processing seemed a practical solution. To facilitate accurate and fast data processing, the surface modelling program was conceived with the primary objectives of producing automatic contours and volumetric calculations.

The basic principle of the TIN is the linking of triplets of adjacent data points to create a planar triangular facet. This basic structure is extended over the entire data area, representing the surface as a set of non-overlapping contiguous triangular facets, whose nodes occur at the original data points. This structure uniquely defines the elevation at any point within the data area. The elevation at any point within the TIN may be computed using a method known as area coordinates (Gold *et al.*, 1977). This method first locates the triangular facet enclosing the point, then uses this triangular plane to interpolate the elevation of the point.

The method of generating the TIN has been the subject of numerous technical papers (e.g. Gold *et al.*, 1977; McLain, 1976; McCullagh and Ross, 1980), and has become a popular topic for investigation. The method selected for use with the HASP system was the radial sweep algorithm (Mirante and Weingarten, 1982). This technique forms an initial approximate triangulation through the data points radially. This is then modified iteratively until an optimization sweep through the database produces no further changes (Fig. 17.1). This method was selected because of its high efficiency and the ability to re-optimize the network after post-formation modifications. An important feature of the network formation procedure is the

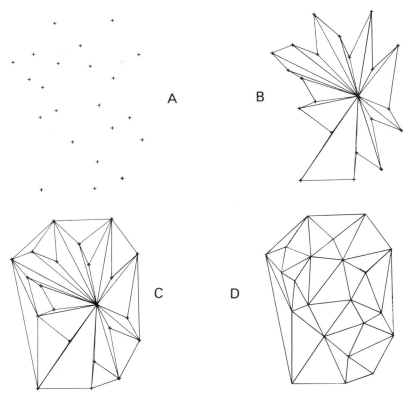

Figure 17.1 The HASP TIN generation process. (A) Raw data. (B) Initial sweep with all points connected radially to the central points and triangles formed. (C) Concavities filled around the perimeter. (D) Triangles reformed to give a better shape using the 'swapping' process.

recognition of breakline information. This technique is used to prevent triangle sides crossing known surface discontinuities. Obvious examples of such discontinuities include the tops and toes of bank and ridges, etc. An equally important yet not so obvious example is the addition of controlling dip lines. Dip lines are the lines that run down the slope, and by defining dip lines, the operator can force the interpolation to take place along these lines. Experience has shown that a surface accurately described by dip lines requires about 1/10 to 1/20 of the number of points required to accurately describe the same surface by contours. The program will align the triangle sides to these features, producing a very accurate representation of the surface.

Most authors when discussing network formation do not consider breakline recognition. However, some information is available (Yoeli, 1977; Elfick, 1979). Many TIN programs have appeared on less sophisticated personal computer applications in the USA in recent years and most of those programs allow the user to edit the triangles to force conformance with the user's interpretation of the surface. This is acceptable for small models, but when the number of triangles is over 100 or so, it is impractical to hand-edit the triangles. HASP models often have over 30 000 triangles and the only practical solution is to have automatic rule-based formation of the model. It was decided not to use any of the common methods but instead to utilize a post-formation technique. This method involves checking each triangle side which a breakline crosses and swapping out the crossings until the two points along the breakline coincide with a triangular side. The 'swapping' consists of locating two adjacent triangles whose common side is being crossed by a breakline and moving the triangle partition within the resultant quadrilateral to the opposite vertices. This 'swapping' process is also utilized during the initial model optimization. The necessity for the recognition and interpretation of breaklines can be seen clearly in the isometric projections with contours superimposed on a model formed from coal stock pile data (Fig. 17.2). As can be seen, omission of the breaklines has resulted in the model forming across the gully between two of the heaps.

Utilization of this technique (including breakline definition) allows the surveyor to pick up data in the field in the same manner as if manual calculations were to be performed. As such, only relevant points of ground change need to be surveyed (e.g., the tops and toes of banks and also dip lines where needed to clarify the shape), with breaklines defined either in the field or interactively back in the office. The system allows the user to interactively define which symbols and lines are to be incorporated in any model at the model formation stage.

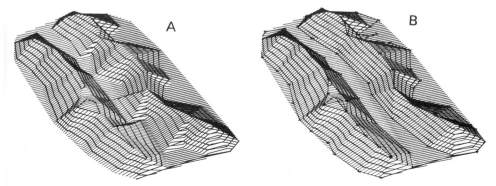

Figure 17.2 The effect of breaklines on a triangular model showing contours superimposed on to a surface projection. (A) Model formed ignoring breaklines. (B) Model formed including breaklines.

The surface model programs were originally developed in 1983 using the HPL language. HPL, originally designed by Hewlett Packard for real-time process control applications, is an interpreted language offering compiled language speed. It features dynamic re-linking through look-up tables, and can even be modified while running. Typical process times for model formation with the original versions of the programs were in the order of 1 to 1.5 hours per 1000 points. During 1984, the package was rewritten in compiled Pascal resulting in dramatic improvements in speed. The model formation on the HP 9816 computer with a Motorola MC68000 0.6 MIPS processor was three minutes, and with fine-tuning and rewriting of time-critical sections of code in assembly language, the time was reduced to one minute per 1000 points. This was the first commercial version offered by HASP. Present HASP computer platforms are the Hewlett Packard 9000 series 300 computers running in the HP-UX (UNIX-based) window-managed, multitasking, multiuser environment. Current processing times for model formation on the model 345 are in the order of 6 seconds per 1000 points. This processing efficiency has allowed a new range of applications beyond simple volumetrics and contouring to be addressed, such as interactive earthwork design, the automatic production of pictorial views and visual impact analysis. Before discussing examples of the uses of the applications of surface modelling, a brief overview of the philosophy behind the applications is provided.

17.3 Pascal system

With the translation of the surface model program to Pascal and the phenomenal performance gains came a commitment to port the other HASP applications to a compiled language. Also about the same time, HASP began serious efforts toward moving to the rapidly evolving industry standard UNIX environment. Hardware platforms for UNIX in 1984 were beyond the budget of most survey offices but our sources at Hewlett Packard assured us that, in two years, the UNIX machines would be available at about the price of the present high end of the systems we were then using. 'C' was the language of choice for UNIX, but a 'C' compiler was not available for the 9816. HASP had a number of very mature applications in HPL and some high-performance applications in Pascal on the 9816. Since Pascal was also supported under UNIX, Pascal was chosen as our primary compiled language and an all-out effort was made to move the HPL programs' capability to Pascal. We did not simply 'port' programs, as we wanted to make them more interactive with the graphics and in keeping with our future UNIX system. The success and size of the HASP HPL system is attested by the fact that it was not until mid-1989 that HASP HPL users were finally willing to give up their HPL systems completely for our UNIX products.

During the translation and porting, it was quickly realized that the most critical area of system efficiency is database manipulation, since almost all operations involve database access. Hence efficiency in this area provides a strong springboard for a very powerful system. Many new data structures became available about that time, some of them being well documented and in the public domain and others being developed internally at HASP.

Many of the data structures are memory-intensive due to the presence of spatial pointers to adjacent areas and to objects in the data. A good example is the storage associated with the triangulated network. For each triangle facet, the following information is stored: pointers to the three coordinated apex points and the three adjacent triangles, Boolean fields stating which sides lie on breaklines, and whether the triangle lies on the boundary of a model or other areas which may be nominated to be excluded from contouring, volumes or other analysis. If the triangle lies on a boundary, then pointers to the adjacent boundary segments are stored. Analogous data structures exist for all non-graphics and graphics database elements.

Obviously, this approach is memory-intensive. The advantage is that the entire database never has to be searched. Instead, a graphic element is retrieved by progressively moving closer to it. Hence only the neighboring element has to be searched for closest proximity to the point required. In this way, the program attempts to mimic the way a human would search for information when locating an object in a room. He does not have to look at each object in turn, but by using his prior knowledge of the room's contents and their spatial relationships, the object is rapidly located by the associative links.

HASP has been a participant in Object Oriented Programming Simulation and Animation conferences and is poised to be among the first to apply and implement pertinent tools deriving from computer animation and advanced true three-dimensional mechanical CAD environments. Window-managed UNIX provides a unique environment for several interdependent processes running concurrently, with updated data being swapped interactively between programs. This will be increasingly important as HASP computer systems are linked to earth-moving equipment equipped with real-time position sensors so that the computer may guide the operator or optionally control blade positioning so that most construction setting-out surveys may be eliminated.

17.4 Volumes of stockpiles

One of the most common tasks of the mining surveyor is the determination of volumes for stockpiles, lagoons, quarries, etc. (Hodges *et al.*, 1984). The HASP system allows volumes to be determined with great speed and a high degree of accuracy. The method of obtaining volumes is dependent on the formation of two models, namely the surface detail model and a base model. Volumes are simply calculated as the sum of the triangular prisms generated from the TIN taken down to a datum level (Fig. 17.3). Hence, inclusive volumes are determined by obtaining the surface model volume down to a set datum and then subtracting the base model down to the same datum. This method has proved to be very flexible, as the user has interactive control over which symbols or lines are included in any model formation. For large areas of

Figure 17.3 The optimized triangular model superimposed on to a surface projection.

ground such as dirt tips, where only a small portion of the site is likely to change annually, the facility is given to merge any altered information into the last data set, thereby removing the need to re-survey the entire area. From this, the annual change can quickly be verified.

As an example of the increased efficiency such a system offers, a direct comparison was arranged with the North Notts area of British Coal in 1985. The test site was a large-area stock pile for which the volume of coal was required. Using conventional techniques, two days were required in the field, and then approximately three weeks of manual processing elapsed before the volume figure was available. Using an Electronic Total Station with automatic field data acquisition, the field survey time was reduced to one day and the subsequent automatic processing of the data on the HASP system provided a fully contoured plot and the volume of material within one hour of returning to the office. The two volumes actually agreed to within a quarter of one percent. This highlights the vast increase in efficiency which such computer systems can offer. At present a number of users from the extractive industry are performing this very task with the HASP system. Some of those users gather the data by aerial photogrammetry with the stereoplotter linked on-line with the HASP system through the HASP EL stereoplotter encoder link interface. It is not uncommon with this arrangement for more comprehensive accurate data to be gathered at rates exceeding 1 000 points per hour!

17.5 Earthwork design

The process described for volumetric calculations allows accurate results to be produced quickly, but certain limitations exist which limit its range of practical applications. The method could not easily be applied to engineering earthwork calculations because it gives no indication of the spatial distribution of the quantities or the split between cut and fill. It was therefore necessary to develop more advanced techniques based on the same concepts, but allowing a much more rigorous analysis of the data.

The earthwork programs are included in two software packages, one oriented towards road and highway design with linear haulage planning and balancing, and the other aimed primarily at complex site-grading design. The highway design program initially developed for use in the USA incorporates design criteria variables specified by the American Association of State Highway and Transportation Officials (AASHTO) design manual. By superimposing the alignment on the surface model, cross-sectional and profile data are generated automatically. Section points are generated at every intersected crossing of the section line with the triangle boundaries creating a set of section data of the general form of station, with the left side (minus) progressing to the centre, and with the right side having plus values. Profiles are then generated as offsets from the described centreline with the ability to have up to nine profiles simultaneously. Usually profiles are generated along the centreline but may be at any offset included within the limits of the data. The operator designs the vertical alignment consisting of VPIs and vertical curves (both symmetrical and asymmetrical). Road projects often include the addition of drain lines, water lines, sewer lines, manholes and catch basins. Facilities are included for the design of these features as well.

The HASP Highway design program introduces the concept of Automated Superelevated Object Templates (ASOT). The AASHTO manual has a library of

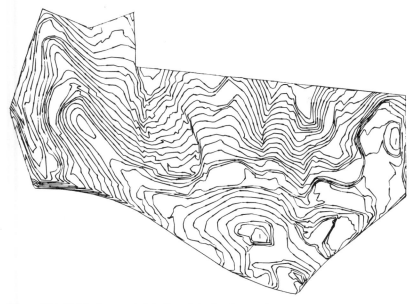

Figure 17.4 Original ground data for estate development.

standard form factors for pavements, curbs, walks and side finishes. The HASP
ASOTs consist of three classes of objects, the first being *wearing surface*. The wearing
surface object may have up to nine subsurface objects which summarize volumes of
each contiguous object. The second class of object is the *curb* object. The curb objects
remain vertical during superelevation rotations. The third class of object is *side finish
objects*. Side finish objects may have any number of slope and or right-of-way
constraints, so the catchment is automatic, but under the constraints placed by the
operator. The design and catchments are automatically stored with the design data so
cross-sections may be transferred on command to a standard graphics file for editing,
additions and plotting. The concept is the creation of a graphics 'snapshot' with user-
defined automatic labelling, placement on sheet, etc., in order to minimize the work
that the operator has to do to complete the necessary graphics exhibits. The graphics
files may also be output to any other graphics system as well. The program
automatically creates the plan view in real-world coordinates. The plan view may
then be merged into the original ground model for pictorial visualization and more
precise volume determination by the HASP Intersurface model. Traditional average
end area sections and reports may be generated, along with mass haul diagrams,
summaries and tables.

HASP has developed a method of extracting a rigorously defined model between
two surfaces. The intersurface data set contains all of the surface attributes of both the
upper and lower surface so that the result is a very accurate isopach or cut-and-fill
model. To illustrate the principles involved as well as different types of surface
description data, we have chosen the original ground data set (Fig. 17.4) with contour
line descriptions and the design data (Fig. 17.5) with practically all edge, top and flow
line breaklines. Original ground is often described using field surveyed spot elevations
and breaklines or contours. Practically all design surface data is of the form of
breaklines and a few spot elevations. The original ground data level was input into the

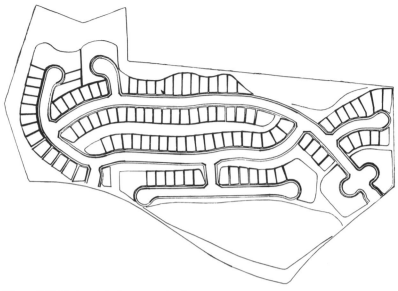

Figure 17.5 Design data for estate development.

system in about one hour by stream digitizing contour lines from an existing contour map. The design data level was input into the system in about 1.5 h by digitizing linework with individual elevations. The two data levels were modelled with the HASP surface modelling program. The HASP SITE DESIGN I program was then used to extract the intersurface data creating a new data called the intersurface. The program compares the two models and develops the difference between the two surfaces. All discontinuities of the ground surface are already represented in the model of the ground surface and each one of the known points in the ground model is compared with a corresponding elevation in the design surface, the difference being the elevation of a corresponding data point in the intersurface data level. Additionally, the breaklines of the design surface are traversed and, at each crossing of a triangle boundary in the ground model, a point is created with the difference also being determined and stored in the intersurface data level. This new intersurface data level when modelled using the breaklines of the design surface rigorously represents both the upper and lower surface.

The resultant intersurface thickness model may be contoured to yield equal depth contours, the zero contour representing the daylight (no cut or fill) line. The elevation at any point in the intersurface model will represent the depth of cut or fill at the point. The SITE DESIGN II program contains a number of the features of the surface model program in that it can produce contours, slope analysis maps, sections, profiles and projections. It has a more comprehensive group of tools to analyse and manipulate the data by altering the elevation of individual points or groups of points selected by drawing a polygon around them. The cut-and-fill volumes may be calculated from the triangular columns, the program shading the areas of cut-and-fill in different colours. The volume within a specific area may also be determined. The user defines a polygon, and the program determines the volume of the triangular columns within the polygon. Those columns which the polygons sides cross are split into subcolumns for inclusion in the calculations. These volumes determined from the original data points are therefore inherently more accurate than is possible with those using cross-sectional or grid methods. The method requires very little operator input and only the original data from the ground and design models are used. Therefore the data processing is very fast, appearing instantly even on the slowest of the HP UNIX computers which HASP offers. The times required for the various operations using an HP 370 computer are as follows:

(1) To form the natural ground surface model (4850 points, 9692 triangles): 29 seconds.
(2) To form the design surface model (3055 points, 6108 triangles): 27 seconds.
(3) To extract the intersurface data: 144 seconds.
(4) To form the intersurface model (21852 points, 37446 triangles): 180 seconds.
(5) To calculate the volumes of cut and fill: 6 seconds.

The complexity and rigorous description of the surfaces can be seen by referring to Figs. 17.6, 17.7 and 17.8. 'What if' scenarios are easily formulated with immediate real-

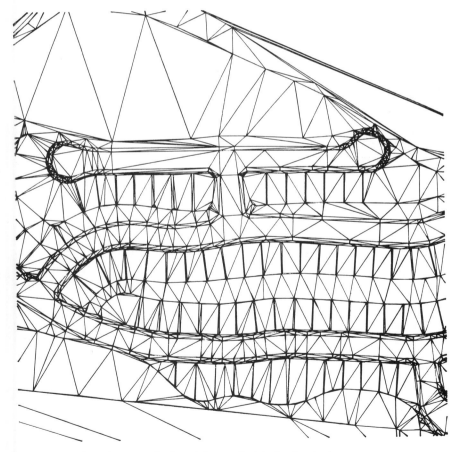

Figure 17.6 Triangular network of partial area of estate development.

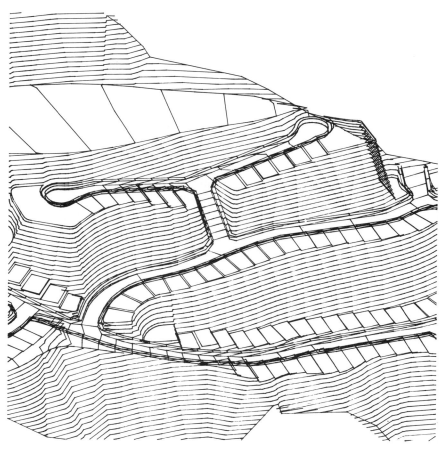

Figure 17.7 Isometric plot of surface on partial area of estate development showing projection of design data and finish grade contours.

time results in the site design program. The editing of the design is accomplished using a combination of the graphics editor and the site-design programs. Major changes may be made in a cut-and-paste type of mode where the operator draws a polygon around the elements to be affected. He may turn triangles on and off, raise and lower elevations, and extract groups of graphics elements inside or outside the polygon. Minor modifications may be made by editing the elevations only and not the X, Y coordinates. The obvious advantage of this technique is that the surface models do not have to be regenerated. This is due to the fact that the surface model only controls the X, Y distribution of data points. The quantity changes may therefore be computed instantaneously as editing proceeds. This gives the designer a very powerful and efficient tool to optimize the quantities distribution. Groups of points may be elevated by defining a polygon around them using a similar method to that used for area deletions. This enables contours, slopes or building pads to be raised or lowered to optimize grading operations.

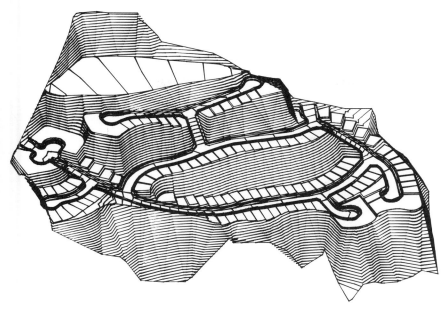

Figure 17.8 Isometric plot of surface of estate development showing projection of design data and finish grade contours.

17.6 Visual impact analysis

Preparing information for planning applications for proposed new mine sites, solids waste disposal sites or other developments or extensions to existing ones, is both time-consuming and expensive. One of the main considerations is the effect of the proposed development on the environment. An important aspect of this environmental research is the visual impact of the proposal on the surrounding area. Traditionally a visual impact impression is obtained by manual sectioning techniques or by simply utilizing the skill and experience of the landscape architect. In the early stages of planning, numerous sites may be under consideration and so it is desirable to have a method of quickly establishing the viability of each.

Having generated a ground model of the proposed area, this visual impact analysis can be performed automatically once the position of the proposed site is defined (Fig. 17.9). The method for calculating intervisibility to all points on the design (whether above or below ground) is based on the use of a user-defined grid interval placed over the surrounding land. A section is computed through the model from the current grid point to each point defined on the feature. The criterion used for intervisibility is that the section line should not cut the line of sight at any point. Different parameters may be input into the program and the results analysed in various ways. For example, it is possible to specify any eye height above the ground surface from which the feature can be viewed. The resultant plan highlights the area of intervisibility over the whole area. Specific areas can then be addressed and analysed in more detail by selecting a window that is to be viewed or by defining a single point. Information such as the number of points on the feature which are visible and the degree of their visibility is also presented.

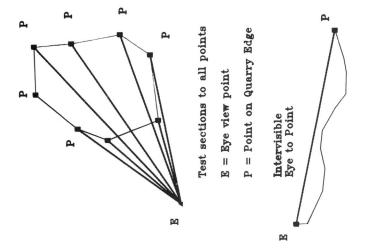

Test sections to all points

E = Eye view point

P = Point on Quarry Edge

Intervisible
Eye to Point

Not Intervisible
Eye to Point

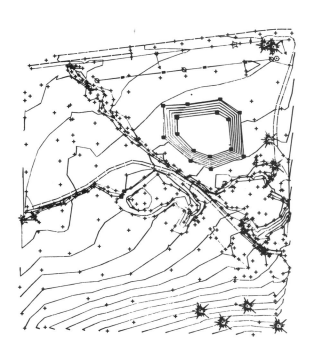

Figure 17.9 Intervisibility analysis. Top, raw data used for model generation. Upper right, analysis of intervisibility from a point. Centre right, section showing intervisible point. Lower right, section showing non-intervisible point.

Using this technique, numerous proposals can be rapidly assessed for their viability, and once a site has been chosen, the impact associated over the whole life of the development can also be assessed. For example, the proposed 5-, 10- and 15-year development plan for a quarry can be entered and analysed. Because of the speed of the operation, the effects of design features such as surrounding banking can also be included. In this way, the landscape architect now has the capability of accurately determining the visual impact over the whole life of any proposal with confidence in the results obtained.

HASP also supports the Hewlett Packard SRX solids rendering hardware for more realistic light rendered presentations of the surface model. Once the design is in the HASP database, it can be written out to a format recognized by the SRX system. With the minimum SRX system, the operator may select up to 16 simultaneous light sources with a choice of up to 256 colours. With an interactive 9-knob control box, either the position and intensity of the light sources or the position of the viewer can be moved and the vertical scale can be manipulated. The new images are recreated and repainted in from 1 to 5 seconds. The Turbo SRX with additional memory buffers and graphics accelerators achieve real-time flicker-free fly-by effects.

17.7 Conclusions and future directions

The main advantage of automated data processing is one of increased efficiency which is dependent upon the power of the computer system used and system's ease of use. By allowing interaction through a friendly user interface, the user, whether calculating volumes, determining single sections or optimizing the earthwork design for a new site, is allowed full control over the environment in which he is working. This leads to confidence in the results which are produced. The designer can consider more alternatives in his investigations and produce more soundly based solutions and more cost-effective projects.

HASP development personnel actively participate in a number of professional groups and also follow closely the new capabilities presented by the Hewlett Packard hardware on which our systems operate. The aerospace industry has invested billions of dollars in simulation and animation systems. The movie industry has also developed significant new tools for computer animation as well. We are beginning to see some exciting new tools coming from research by Alan H. Barr of the computer graphics group at California Institute of Technology. Dr Barr and his associates have introduced a new approach to modelling, called 'teleological modelling'. Dr Barr describes this approach as follows:

> Teleological modelling techniques provide an extension of the current notions about how to make mathematical models of objects; the approach utilizes forces, torques, internal stress, energy, and other physically derivable quantities which allow us to simulate many of the fundamental properties of the shapes, combining-operations, and constraints which govern the formation and motion of objects.... Intuitively, a teleological model is 'goal-oriented' modelling. It is a mathematical representation which calculates the object's behavior from what the object is 'supposed' to do. Each modelling element within this context is called a 'teleological' modelling element, because its character is directed towards an end or shaped by a purpose (from the Greek work *teleos*, meaning end, or goal).

Animation videotapes demonstrating the techniques applied to the automated assembly of mechanical parts are truly exciting and impressive. It is clear that similar techniques applied to earthwork modelling and civil engineering works could save countless hours of design and analysis work. The engineer would be able to accomplish more work in less time and have more consistent results.

References

Elfick, M.H. (1979) Contouring by use of a triangular mesh. *The Cartographic Journal* **16**.

Gold, C.M., Charters, T.D. and Ramsden, J. (1977) Automated contour mapping using triangular element data structures and an interpolant over irregular triangular domain. *Computer Graphics* **11**, 170–175.

Hodges, D.J., Nicholson, T.J., and Ketteman, M.R. (1984) Surface and underground automated mine surveys. *The Mining Engineer* **143**, 271.

McCullagh, M.J., and Ross, C.G. (1980) Delaunay triangulation of a random data set for isarithmic mapping. *The Cartographic Journal* **17**, 93–99.

McLain, D.H. (1976) Two dimensional interpolation from random data. *The Computer Journal* **19**.

Mirante, A., and Weingarten, N.H. (1982) The radial sweep algorithm for constructing triangulated irregular networks. *IEEE Computer Graphics and Applications* **14**, 11–21.

Yoeli, P. (1977) Computer executed interpolation of contours into arrays of randomly distributed height points. *The Cartographic Journal* **14**, 103–108.

18 The use of a Digital Terrain Model within a geographical information system for simulating overland hydraulic mine waste disposal

ADRIAN M. DURRANT

18.1 Introduction

The concept of 'intelligent terrain models' (ITMs) was first introduced by Strodachs and Durrant (1986), at CADCAM '86. Further research work has since been carried out, some of which is presented in this chapter. An ITM is defined as 'the logical structuring and management of terrain geometric and attribute data, for representing terrain topology and supporting the dynamic requirements of terrain-related simulation applications'.

In late 1985, the Department of Civil Engineering at Loughborough University of Technology, UK, conducted the Beta test on the Medusa Terrain Modeller (MTM). The MTM was developed by Cambridge Interactive Systems (CIS), authors of the Medusa range of engineering software. The software generated a terrain model from a random data set, based on the Delaunay triangulation algorithm. In addition to terrain representation, the integration of 'free form design' solid models was available, through the use of Boolean operation facilities.

Recommendations for further software enhancements were made, subsequent to which a Geographical Information System (GIS) was developed. The Medusa GIS was built around a relational Database Management System (DBMS). Medusa GIS can be described as a system to manage, analyse and display spatial and attribute data in a geographically continuous manner. The GIS was designed for large data applications, such as mapping, utilities management and terrain modelling. Its powerful applications to terrain-related simulation activities are discussed in this chapter.

18.2 Concepts of a relational DBMS-based terrain model

It is essential that the user understand not only the application of this data, but, more fundamentally, its very nature, before assessing some ideal structure for it. In some cases, this involves taking one step back, and looking at it afresh, as will be briefly attempted in the context of terrain modelling.

Terrain has an explicitly defined shape, or topology, which is represented digitally by some form of connectivity. This connectivity is defined between the base components, or entities, the data points. These points may be ordered, semi-ordered or random in their orientation to one another. It is a number of such connections between existing points, or approximations, which represent a terrain surface. For example, in the case of a Triangulated Irregular Network (TIN) type of terrain model, there exists a linear connectivity between randomly ordered data points, in the form of a set of non-overlapping triangular polygons. This connectivity may be defined by a Delaunay triangulation, for example, of the data set. Therefore terrain modelling naturally lends itself to a topological GIS data structure, where geometric relationships are explicitly defined.

As with a multitude of other engineering-related applications, terrain modelling, requires that additional salient information be associated either spatially or implicitly with points or areas of the terrain. Taking the simple example of the data point, in addition to its spatial X, Y, Z coordinates and its connectivity information, the user may wish to know the date and method of capture, and the date on which it was last modified. With reference to a surface polygon, a further example may include the presence of existing surface features, such as soil type and vegetation, or existing structures such as the alignment of pipelines, roads, etc. All components, or entities, also have graphical associations, such as the line between two vertices, which represents the connectivity between two points on the terrain surface, or the colour of a particular soil type. Interactive computer graphics enables the user to display and manipulate dynamically all this information in a visual, or symbolic, form.

The greatest importance should be placed on the actual model which represents the data. A data model is defined as 'a set of well-defined primitives, or concepts, used to represent the static and dynamic properties of data to support the desired transactions'. The relational model, for example, is based on one such concept, the relation. Spatially related data model primitives can be grouped into two categories; geometry, and spatially related non-geometric attributes.

As terrain models become larger, greater attention must be focused on data management. Database Management Systems (DBMS) can store all information salient to a project within a single database. Large data volumes can be managed and analysed according to user-specific rules or theories. Multiuser access to the same data is also possible, such as would be the case with most multidisciplinary projects. Version controls and privilege levels ensure that the data is not corrupted, either by design staff working on the same area of the database, or by unauthorized personnel gaining access to confidential project information. Furthermore, data security facilities are also offered, whereby the database contents are safeguarded against computer 'crashes'.

18.3 The Medusa geographical information system

The Medusa GIS arose from a need to maintain data continuity and structural topology, as well as to support large data applications. It was primarily geared to mapping and utilities management. As the system as already been comprehensively described by Bundock (1987), this chapter will concentrate on the implications of Medusa GIS for terrain modelling.

The data model used in Medusa GIS for representing cartographic data is shown in Fig. 18.1. Briefly, the SOLID entity represents a volume in space that is bounded by a

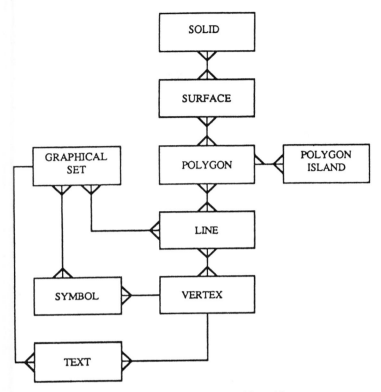

Figure 18.1 Data model of Medusa GIS cartographic entities.

set of surfaces. A many-to-many relationship exists between solids and surfaces, since a single surface may also be common to more than one solid, an example being the complex geometry that may exist between overburden and an underlying faulted coal seam.

The SURFACE entity represents a single continuous surface formed by a set of planar polygons, each polygon of which may be shared by multiple surfaces. Hence a many-to-many relationship exists between surfaces and polygons, an example being the modelling of borehole data as a series of surfaces.

POLYGON entities represent closed areas formed by a set of lines. Any number of lines may be used to form a closed polygon, for example, three in a TIN and four in a grid-based terrain model. Each line may form part of the boundary of more than one polygon (i.e., adjacent polygons). Hence a many-to-many relationship is said to exist between polygons and lines.

LINE entities represent any type of line segment, be it a straight line, arc, smooth curve or a string consisting of many vertices. The VERTEX is the basic component, or entity, that describes a unique spatial location. As a line is defined by a set of vertices, each vertex of which may be shared by more than one line, a many-to-many relationship exists. Therefore, from any vertex, adjacent vertices can be identified, via their explicit line connectivity. Hence their implied geometric attributes, such as height difference, distance, slope, etc., can be derived.

The previous paragraphs describe the structural, or geometric, nature of the GIS data model, to which non-geometric attribute information may be assigned. This may be numerical, such as a point number assigned to a vertex; textual, such as a description; or symbolic, such as a cross representing the vertex location. A graphical set entity relates to a set of entities, which by necessity need to be referenced as a single entity, examples being feature strings and breaklines.

The vertex is the only entity to contain coordinate information directly, and each vertex, and all lines referencing it, are only stored once, thus ensuring that no data duplication occurs. When modifying a vertex, all endpoints of lines referencing it are amended accordingly, thereby maintaining data consistency. Likewise, with adjacent polygonal facets, no duplicate lines are required, since each line is shared by all polygons referencing it. The advantage of such a data model is apparent when interactively modifying the terrain, as data integrity is maintained. Further advantages of shared, non-duplicated and spatially referenced data become obvious in large-scale terrain applications, such as river sedimentation and catchment studies, as response degradation is not linearly proportional to data volume. In addition, land masses are modelled according to their natural phenomenon, that is, spatial continuity, as no data segmentation occurs, and terrain model size is limited only by physical disk capacity. Furthermore, data access times do not suffer the linear degradation of traditional systems. For example, the searching time for a record in a 20-megabyte terrain model is 100 milliseconds, while that for 20 gigabytes is only 500 milliseconds! This is achieved by access to database records via a key, or index, which, for geometric data, is created as a function of its spatial location. Hence an ordering of two- and three-dimensional space can be provided. Prior to model implementation in the relational database, a procedure termed 'normalization' is invoked, essentially to remove data redundancies in order to provide a 'clean' terrain structure with which to work.

The Medusa GIS provides a data dictionary facility, which can be regarded as a system for managing the data in the database. In addition to acting as a repository for

Table 18.1 Data definition of the DAILYVALVE database table

	Table DAILYVALVE contains daily valve records			
Table	Dailyvalve		Channel 5	—Daily valve records
Field	valvenum	1	int	—Valve number
				Type int 1
Field	startdate	1	int	—Start date
				Type date 1
Field	starttime	1	int	—Start time
				Type int 1
Field	stopdate	1	int	—Stop date
				Type date 1
Field	stoptime	1	int	—Stop time
				Type int 1
Field	oftonnes	1	rea	—Overflow tonnage
				Type reas 1
Field	uftonnes	1	rea	—Underflow tonnage
				Type reas 1
Field	valvehours	1	rea	—Valve hours
				Type reas 1
Key	startdate	starttime	valvenum	
Index	valvenum	startdate		

descriptions of the data, the names of the data elements, verbal descriptions of their meanings, and descriptions of their characteristics, it describes the logical and physical structure of the database. This includes the relationships between data elements, the grouping of data elements into tables, or records, the assignment of tables to physical database files, and also the protection of the data from unwanted viewing, creation or modification. It effectively provides the mechanism by which the user can specify and design a database best suited to his particular application. Table 18.1 shows the database definition of one such table, which is pertinent to tailings dams, as will be discussed later. A database query language (QL) allows interactive access to, and modification of, the database. In addition, a high-level language called BaCIS is available for user applications programming.

18.4 An application of Medusa GIS to overland hydraulic mine waste disposal

Tailings dams are structures built from mine waste and are a special category of an hydraulic-fill embankment. They are unique in that the dam is built to dispose of the wastes from a particular mine, and hence there is no choice of materials. These structures are continuously constructed throughout their service lives, and therefore require ongoing technical management. Tailings dams have become topical since the tragic failure of an NCB coal tip in 1966, at Aberfan, Wales, and more recently the disaster in Stava, Italy, in mid-1985.

Novel techniques have been developed and implemented over the last 35 years to dispose of mine waste. Once such method involves the cycloning, or centrifugal separation, of the tailings material into a coarse, more stable, wall-building fraction, and a finer fraction which is discharged into the storage area. The author had access to the tailings dam which services the Wheal Jane tin mine near Baldhu, Cornwall, UK, where the cycloning technique is used. The spoil which is extracted from the underground workings is processed in a mill, the non-ore-bearing material being finely ground and mixed into a slurry, prior to transportation to the dam.

Two surveys were carried in April and July 1986 out to assess the geometry of the subaerial, or beach, area of the dam along with its salient feature points, such as the cyclone valve locations. Most of the ground survey operations were conducted using a Zeiss Elta Total Station, linked to a Husky Hunter data recorder. In addition, a photographic analysis of the beach build-up, as well as material sampling, was carried out. The survey data files were reduced on a BBC microcomputer, prior to their transfer to Prime and Vax minicomputers, and consequently into the original Medusa Terrain Modeller and the Medusa GIS, respectively.

Figure 18.2 shows the data model that was used to describe the Wheal Jane tailings dam project. Tailings production figures were provided by the mine for each month, and were held as a series of database records. On each day of every month, the mine operators collated site records. There was therefore a one-to-many relationship between tailings production and daily site records. Likewise, for every day, each individual cyclone valve, of which there were 26 on the dam embankment, may be either opened or closed, between times logged by the operators. Each individual cyclone has associated with it an underflow, for wall-building, and an overflow, for beach-building.

The survey data, as well as the digitized embankment geometry, was loaded into the Wheal Jane project database along with the salient attribute tables. These included

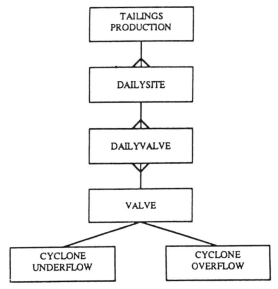

Figure 18.2 Data model for hydraulic tailings disposal at Wheal Jane.

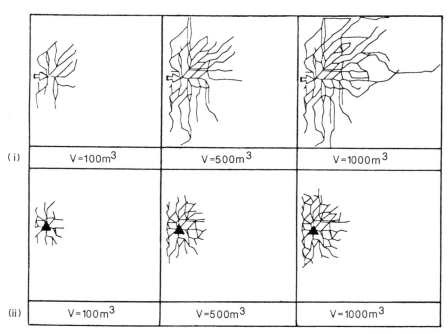

Figure 18.3 Channel patterns derived for slope and inertia Tailfan alignment criteria. (i) [$f_{slope} = 1$; $f_{inertia} = 1$]. (ii) For slope-only criteria [$f_{slope} = 1$; $f_{inertia} = 0$]. Both with stream branching [$P_b = 1$].

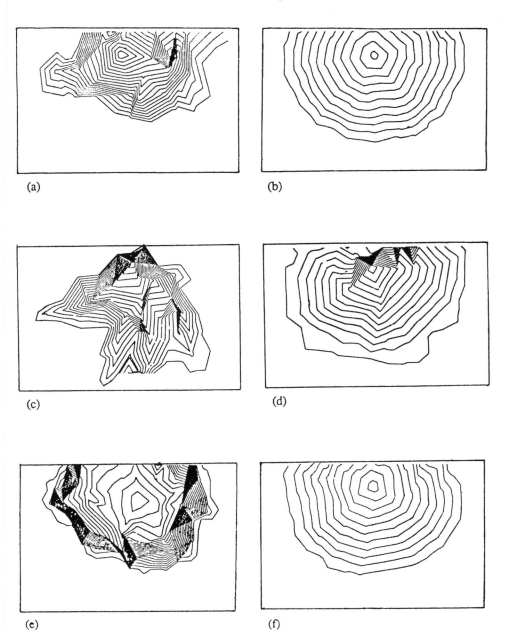

(a)

(b)

(c)

(d)

(e)

(f)

Figure 18.4 Contour plots of Tailfan simulated tailings deposits, with and without stream branching ($P_b = 1$ and 0). (a) Slope dependency without branching ($f_{slope} = 1$; $f_{inertia} = 0$; $P_b = 0$). (b) Slope dependency with branching ($f_{slope} = 1$; $f_{inertia} = 0$; $P_b = 1$). (c) Inertia dependency without branching ($f_{slope} = 0$; $f_{inertia} = 1$; $P_b = 0$). (d) Inertia dependency with branching ($f_{slope} = 0$; $f_{inertia} = 1$; $P_b = 1$). (e) Slope/inertia dependency (unbiased without branching ($f_{slope} = 1$; $f_{inertia} = 1$; $P_b = 0$). (f) Slope/inertia depending (unbiased) with branching ($f_{slope} = 1$; $f_{inertia} = 1$; $P_b = 1$).

the tailings production figures, the daily site records and the individual cyclone valve records. In addition, graphical symbols were created to represent the valves, cyclone underflows and overflows, and other project attributes. In the case of the overflow symbol (an arrow) its graphics entity represented not only its spatial location on the beach, but also the initial direction of its discharge. These, along with any other database entities, could be interactively interrogated and moved according to user requirements.

The problem of describing, monitoring and analysing the dynamic nature of the tailings discharge and deposition was ideally suited to a solution using the GIS. For any given, date the DAILYVALVE table (Table 18.1) was scanned, and the corresponding operational cyclones were sorted into an active list, according to their start-up times and numeric identifiers. The location of each active cyclone overflow was then determined, together with the total tonnage emitted from it for that day. Observations at Wheal Jane showed that tailings deposition occurred as lateral stream washing, adjacent to the stream channels which meandered across the sub-aerial sections of the cycloned beach. The simulated stream channel alignments were empirically determined, according to weighted values (f_{slope} and $f_{inertia}$) of local terrain slope and inertia of flow. It was possible to initiate simulated stream branching, in the relatively flat, low-flow-energy areas of the beach, by specifying a branching probability value (P_b). Tailings deposition was modelled according to a surveyed equilibrium beach profile, which was a function of the tailings' characteristics and the environmental conditions. Deposition took place according to an algorithm which 'matched' the tailings' equilibrium slope profile to the local beach geometry. Vertices which qualified for deposition were duly uplifted, and the resultant volume of deposited material was calculated. Each cyclone in the active list was deactivated, or shut down, once its recorded volume for that day had been discharged. The above described simulation model was called 'Tailfan'.

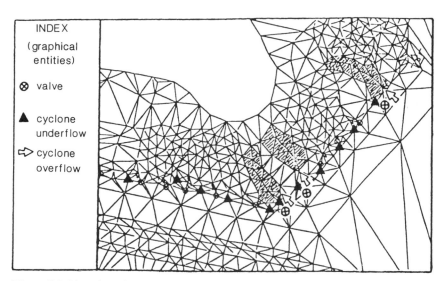

Figure 18.5 Plan view of eastern embankment at Wheal Jane, showing cyclone overflows operation.

Controlled filling basins, or digital 'swimming pools', were created, to study the effects of the Tailfan simulation criteria, and the terrain data distribution and density used, upon both the patterns of the simulated stream channels and the geometry of the resulting simulated tailings deposits. Figure 18.3 shows the effects of the terrain slope, inertia of flow and stream branching, upon the patterns of simulated stream channels. Figure 18.4 shows the geometry of simulated tailings deposits, resulting from a variety of stream alignment criteria. It was found that the best approximation to a radially issuing tailings fan was derived from slope-dependent alignment criteria, with included stream branching (see Fig. 18.4b).

The findings of the above 'pool' tests greatly assisted in calibrating the Tailfan simulation model for the Wheal Jane project. Figure 18.5 shows cyclone overflows in operation on the eastern embankment at Wheal Jane, as recorded by the mine operators for 15 May 1986. The interactive computer graphics facilities, integrated within the Medusa GIS, were used to highlight active stream channels, with adjacent channels appearing in different colours. In addition to the graphical interaction, useful project reports could also be compiled by interrogating the database tables, according to user-specified conditions, using the query language (QL). For example, the DAILYVALVE table could be searched, to provide such details as the total individual valve operating hours, and the total tonnage emitted between specified dates and filling sequences, tailings output rates and service schedules could thereby be derived.

18.5 Conclusion

The application of new philosophies, such as those which have been discussed in this chapter, offers exciting prospects to all potential users of Digital Terrain Modelling techniques. Whereas the traditional area of DTM use was highway engineering, there is now a greater appreciation of its applicability to other spheres of engineering and the natural earth sciences.

This chapter has described one methodology by which terrain modelling techniques can be applied in a manner, beyond that of representing static terrain geometry. The term Intelligent Terrain Model (ITM) refers not to an expert sysem, but, more fundamentally, to a logical description of terrain-related information used for dynamic simulation purposes. The Medusa GIS offers a simple and flexible data model that not only represents terrain topology, but also supports a dynamic platform for simulation applications, tailings dams being one such example. Furthermore, the Database Management System (DBMS) approach directly addresses the data management requirements of large-scale engineering projects.

It is possible, that in the not too distant future, river hydrologists, sedimentation specialists and pipeline design engineers will greatly benefit from the use of such simulation tools.

Acknowledgements

The author wishes to thank the Science and Engineering Research Council and WLPU Consultants, Consulting Civil Engineers, for jointly sponsoring his work, part of which is presented in this chapter. In addition, gratitude is extended to Carnon Consolidated Ltd for access to the Wheal Jane site, and to Mike Bundock, formerly Mapping Products Manager, Cambridge Interactive Systems (CIS), Harston, and the CIS organization, for their generous assistance and use of the Medusa software.

References

Bundock, M.S. (1987) An integrated DBMS approach to Geographical Information Systems. *Proceedings, Eighth Symposium on Automation in Cartography*, Baltimore, USA.

Strodachs, J.J. and Durrant, A.M. (1986) Surface and solid modelling for civil engineering applications—an introduction to 'intelligent ground models'. *Proceedings, CADCAM 86 Conference*, N.E.C, Birmingham, 243–248.

Part E
Visualization—especially landscape and military applications of DTMs

19 Visualization for planning and design

T.J.M. KENNIE and R.A. McLAREN

19.1 Introduction

Increasing emphasis has been directed recently towards the development of computer-based display techniques which produce a high degree of image realism. Where such techniques are used to describe the three-dimensional shape of the Earth's surface and man-made or other 'cultural' information, the process is referred to as digital terrain and landscape visualization. The input data for such a process will normally consist of a digital terrain model (DTM) to define the geometric shape of the Earth's surface, together with further digital data to describe landscape features. For applications at small scales, this landscape information may be polygonal land use data, while at larger scales it may include explicit geometric descriptions of individual features or blocks of features. Although the photogrammetric acquisition of DTM data is well established and has been reviewed by Petrie (Chapters 4 and 5), to date much less attention has been directed towards the photogrammetric acquisition of geometric and visual descriptions of significant objects on the terrain. Although the x, y, z coordinates of features such as buildings may be observed, in many cases the Z coordinate values are deemed superfluous and do not form part of the recorded dataset.

While there is no shortage of techniques for general visualization purposes, the characteristics of the Earth's surface can significantly limit the applicability of many of the more general-purpose techniques. Some of the more important differences between terrain and landscape visualization and other forms of visualization, especially those found in the advertising and entertainment industries, are that:

(a) natural phenomena are inherently more complex than man-made objects and thus more difficult to model;
(b) the Earth's surface is not geometrical in character and cannot be modelled effectively by using higher-order primitives such as those used in solid modelling CAD/CAM applications;
(c) the terrain and landscape model dataset sizes are considerably larger than the datasets used in many other forms of visualization;
(d) there are generally higher constraints on geometric accuracy than in many other applications;
(e) the scenes are not spatially compact and therefore the modelling may involve multiple levels of detail based on the object–viewpoint relationship;
(f) the optical model is more complex due to the effects of atmospheric refraction and Earth curvature which are encountered in extensive datasets.

19.2 Techniques for displaying digital terrain and landscape visualizations

The degree of realism which can be achieved in the visualization process depends upon several factors, including the nature of the application, the objective of the visualization, the capabilities of the available software and hardware and the amount of detail recorded in the model of the scene. At one end of the realism spectrum are relatively simple, highly abstract, static, monochrome, wireframe models, while at the opposite end are sophisticated, highly realistic, dynamic full-colour images. The former approach is currently more common for site-specific design purposes in fields such as landscape architecture and civil engineering, while the latter tend to be more common in regional applications such as military and flight simulation.

19.2.1 Display of 2D terrain models

A number of relatively simple techniques have been developed for displaying DTMs on a computer graphics display. All are essentially extensions of the traditional cartographic process of representing relief on a 2D surface. Probably the most common method used to indicate height variations on a 2D surface is the contour. Colour graphics display devices may be used to portray contour lines, although colour variations are more commonly used to give an appreciation of the height variations rather than the traditional variable line widths. An extension of the colour coding of contour lines is to colour-code the areas between contour lines as an aid to improving the appreciation of the topographic variations. Faintich (1984) has discussed the use of a number of such techniques, particularly for the detection of gross errors in terrain data.

In addition to the display of contours, information such as slope and aspect can be derived from a DTM. By combining both attributes it is possible to replicate the technique of relief or hill shading. Scholz *et al.* (1987) discussed the procedures adopted to automate the production of relief shading on 1:250 000 scale aeronautical charts. These techniques are capable of meeting the display needs of many users. They do, however, suffer from a number of deficiencies, including the production of a highly abstract view of the terrain from a fixed viewing position with little opportunity for including landscape features in the displayed image.

19.3 Methods of rendering 3D terrain and landscape models

A number of alternative strategies exist for the transformation and display, or 'rendering', of terrain and landscape models on to a 2D raster scan display. The first, and computationally the simplest, involves the use of a video digitizer to 'frame grab' a photographic image. The position of new features may then be added to the model. The second, and computationally the most complex, approach is to mathematically define all features within the model. The basic elements of this approach are discussed below. In some cases a hybrid approach may be adopted where photographic images are used to model complex natural phenomena and are combined with terrain and landscape features which are modelled by computer graphics. An example of this hybrid approach is the 'photomontage' (section 19.4.1).

19.3.1 Review of the rendering process

Having assembled the model of the terrain and associated landscape features contained within the required scene, the scene may then be rendered on to a suitable

SCENE DEFINITION

Digital Terrain Model	Landscape Feature Model
Geometry	
Visual Characteristics	

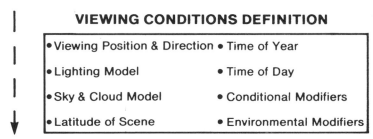

VIEWING CONDITIONS DEFINITION

- Viewing Position & Direction
- Time of Year
- Lighting Model
- Time of Day
- Sky & Cloud Model
- Conditional Modifiers
- Latitude of Scene
- Environmental Modifiers

RENDERING PROCESS

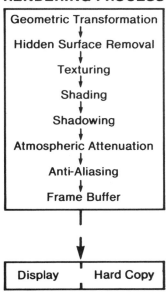

Geometric Transformation
↓
Hidden Surface Removal
↓
Texturing
↓
Shading
↓
Shadowing
↓
Atmospheric Attenuation
↓
Anti-Aliasing
↓
Frame Buffer

Display	Hard Copy

Figure 19.1 Procedures involved in the rendering process.

display system. The rendering process, however, requires a number of other parameters to be defined (Fig. 19.1), including:

(a) the viewing position and direction of view of the observer;
(b) a lighting model to describe the illumination conditions;
(c) a series of 'conditional modifiers'—parameters which describe the viewing condition of the landscape objects (under wet conditions, for example, the surface characteristics of objects are quite different to those shown under dry conditions);

(d) a set of 'environmental modifiers'—parameters which describe atmospheric conditions and may model effects such as haze; and

(e) a sky and cloud model representing the prevailing conditions.

All or part of the information may then be used in the scene-rendering process to generate a two-dimensional array of intensities or pixel values that will be displayed on the raster display device. The complexity of the rendering process is directly dependent upon the degree of image realism required by the user.

(i) Geometric transformations. The 3D terrain and landscape information is normally mapped into 2D space by a perspective projection, where the size of an object in the image is scaled inversely as its distance from the viewer. The projection in effect models a 'pinhole' camera. For site-specific applications where the geometric fidelity of the rendered scene is of vital importance, for example the creation of a photomontage product in visual impact assessment, it may also be necessary to incorporate both Earth curvature and atmospheric refraction corrections into the viewing model.

(ii) Depth cueing. When a 3D scene is rendered into 2D space with any level of abstraction, an ambiguous image may often be portrayed. To compensate for this loss of inherent 3D information, a number of techniques have been developed to increase the 3D interpretability of the scene using depth cueing techniques that attempt to match the perceived computer-generated image to our 'natural' visual cue models.

Firstly, depth cues are inherent in the perspective projection used to create the 2D image. When a projected object is known to contain parallel lines, then this type of projection is excellent at depicting depth, since parallel lines seem to converge at their vanishing points. This very effective visual cueing is highlighted when comparing the projection of a triangular DTM with its equivalent square grid derivative, in their wireframe forms (Fig. 19.2). The depth cueing is pronounced in the square grid form due to the combined, convergent effect of parallel lines and the diminishing size of the uniform squares. The triangular form, with its lack of uniformity and randomness of triangular size, can present a very ambiguous and confusing image that requires further cueing techniques to allow adequate interpretation.

Although images formed by wireframe models with hidden edge removal are primitive images with no pretence of being realistic, they still portray form and geometric fidelity and provide an inexpensive technique for visualization. Major

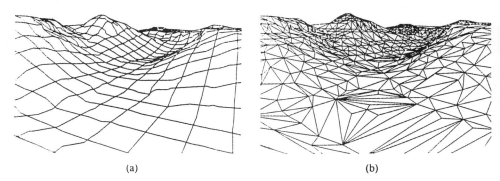

(a)   (b)

Figure 19.2 Depth cueing using (a) gridded structure and (b) triangular structure. (Cambridge Interactive Systems Ltd.)

advantages are the low overheads in their production and their ability to be output on standard vector plotting devices.

A second depth cue corresponds to the relative lighting intensities of objects. Through atmospheric attenuation, more distant objects lose contrast and appear dimmer than closer ones. If the sky is modelled in the scene, then this can also add to the depth cues by ensuring that the sky appears lighter overhead than at the horizon and thus conforms to reality.

Object simplification is a further form of depth cueing. As an object recedes from a viewer, it diffuses through several forms of apparent simplification until it eventually disappears from sight. This is due to the limited resolution of the eye's optical system and is influenced by atmospheric distortion and the shape and texture of the object. Computer graphics can also be used to generate stereoscopic images that provide a very powerful depth cue. Whiteside *et al.* (1987) discuss the production of simulated stereoscopic perspective views of the terrain using digitized vertical aerial photography and digital terrain elevation data (DTED) produced by the Defense Mapping Agency.

(*iii*) *Hidden surface removal.* Hidden surface removal techniques are employed to remove the edges and surfaces that are obscured by other visible surfaces. The implementation of the technique of hidden surface removal is computationally expensive, especially for complex landscape scenes, where the rendering process can involve hundreds or thousands of surfaces. Therefore the challenge has encouraged a wide variation of algorithms. These can be categorized into two fundamental approaches: image space and object space algorithms.

One of the most popular algorithms used is the Z-buffer or refresh buffer image space algorithm. This approach assumes that the display device has a Z-buffer, in which Z-values for each of the pixels can be stored, as well as the frame buffer in which the intensity levels of the pixels are stored. Each polygon is scan-converted into the frame buffer where its depth is compared with the Z-value of the current pixel in the Z-buffer. If the polygon's depth is less than the Z-buffer value then the pixel's frame buffer and Z-buffer values are updated. This is repeated for each polygon, without the need to pre-sort the polygons as in other algorithms. Hence, objects appear on the screen in the order they are processed.

(*iv*) *Anti-aliasing.* Many computer graphics images displayed on raster display devices exhibit disturbing image defects such as jagging of straight lines, distortion of very small or distant objects and the creation of inconsistencies in areas of complicated detail. These distortions are caused by improper sampling of the original image and are called aliasing artefacts. Techniques known as anti-aliasing, which have their roots in sampling theory, have been developed to reduce their influence (Crow, 1977).

(*v*) *Shading.* The next step towards the goal of realism is the shading of visible surfaces within the scene. The appearance of a surface is dependent upon the type of light source(s) illuminating the object, the condition of the intervening atmosphere, the surface properties including colour, reflectance and texture, and the position and orientation of the surface relative to the light sources, other surfaces and the viewer. The objective of the shading stage is to evaluate the illumination of the surfaces within the scene from the viewer's position.

The effectiveness of the shading algorithm is related to the complexity of the model of the light sources. Natural lighting models, in the case of landscapes under daylight conditions, are normally simplified by assuming that there is only a single parallel light source, the Sun. Refinements to this light model have been developed by Nishita and Nakamae (1986), in which the lighting model is considered to be a hemisphere with a large radius that acts as a source of diffuse light with non-uniform intensity, thus simulating the varying intensity of sky lighting.

Two types of light sources are apparent in the environment: ambient and direct. Ambient light is light reaching a surface from multiple reflections from other surfaces and the sky, and is normally approximated by a constant illumination on all surfaces regardless of their orientation. This simplified model produces the least realistic results, since all surfaces of an object are shaded the same. However, more complex modelling of the ambient or indirect component of lighting has produced enhanced realism using ray tracing techniques (Whitted, 1980) to model the contribution from specular interreflections and transmitted rays, and radiosity techniques (Goral *et al.*, 1984) to account for complex diffuse interreflections.

Direct light is light striking a surface directly from its source without any intermediate reflection or refraction. This can be broken into two components, diffuse and specular, depending on the surface characteristics. The diffuse component is re-radiated equally in all directions while the specular component is re-radiated in the reflected direction. Both reflected components must be modelled to create realistic images. The diffuse component is simple to compute since it is independent of the viewing position. The specular component is, however, more complex to model since real objects are not ideal reflectors, a portion of the light being reflected off the axis of the ideal light direction. Phong (1975) developed an empirical approximation which was subsequently refined by Blinn (1977) and Cook and Torrance (1982).

Having modelled the intensity of colour at a point, this must now be expanded to encompass a surface, normally defined by a polygonal mesh. In Gouraud's approach (Gouraud, 1971), the true surface normals at the vertices of each of the polygons are calculated. When the polygon is coverted into pixels, the correct intensities are computed at each vertex and these values are used to interpolate values linearly across the polygonal surface. An alternative, but computationally more expensive approach, was developed by Phong (1975).

(*vi*) *Shadows.* Shadows are an essential scene component in conveying reality in a computer graphics image. A scene that appears 'flat' suddenly comes to life when shadows are included in the scene, allowing the comprehension of spatial relationships amongst objects. Shadows provide the viewer with one of the strongest visual cues needed to interpret the scene.

The fundamental approach is equivalent to the hidden surface algorithm approach. The shadow algorithm determines what surfaces can be 'seen' from the light source. Those surfaces that are visible from both the viewpoint and the light source are not in shadow. Those that are visible from the viewpoint, but not from the light source, are in shadow. A variety of shadow algorithms have been developed and can be categorized into five groups (Max, 1986): Z-buffer, area subdivision, shadow volumes, pre-processing and ray tracing.

(*vii*) *Surface texture detail.* Natural landscape scenes are characterized by features with a wide variety of complex textures. Computer graphics visualizations of landscapes can only achieve an acceptable level of realism if they can simulate these intricate textures. The 'flat' shading algorithms, described in the previous section, do not meet this requirement directly since they produce very smooth and uniform surfaces when applied to planar or bicubic surfaces. Therefore the shading approach must be supplemented by other techniques to either directly model or approximate the natural textures. The explicit modelling approach involves creating a more detailed polygonal and colour model of the landscape and surface features to enable a higher level of detail and texture to be visualized. For landscape visualization, explicit modelling has so far proved impractical due to the size and intricacy of the model that would have to be created to reflect the required level of detail.

Texture mapping provides the illusion of texture complexity at a reasonable computational cost. The approach refined by Blinn and Newell (1976) is essentially a method of 'wallpapering' existing polygons with a user-defined texture map. This texture map can, for example, represent frame-grabbed images of natural textures. A modification of the texture mapping technique is to utilize satellite remote sensing imagery to 'clothe' the terrain model with natural textures. A further approach for texturing terrain models involves the use of fractal surfaces. Fractals (Mandelbrot, 1982) are a class of irregular shapes that are defined according to the laws of probability and can be used to model the natural terrain. Simple models of the terrain are defined using quadrilaterals or triangles, which are subsequently subdivided recursively to produce more detailed terrain models. Some of the most successful examples of the use of fractals have been in the field of animation.

(viii) Atmospheric attenuation. Due to atmospheric moisture content, objects undergo an exponential decay of contrast with respect to distance from the viewpoint. The decay converges to the sky luminance at infinity. This reduction rate is dependent upon the season, weather conditions, level of air pollution and time. The result is a hazing effect.

19.4 Applications

The use of visualization techniques for both military and civilian applications is an area of significant growth at present, and some of the factors which account for this have been outlined previously. Generally these applications are either for small-scale, regional visualization, or conversely for large-scale, site-specific projects. The former group are oriented primarily towards military applications, whereas the latter tend to be more biased towards engineering, mining and landscape planning.

For the majority of small-scale regional applications, the accuracy requirements of the height data are relatively low and may therefore be acquired from existing map sources. Because of the need for wide coverage, it is also common practice for such applications to use one or more of the regional terrain databases which have been developed in recent years. The most commonly available sources are the Digital Terrain Elevation Database (DTED) of the US Defense Mapping Agency, the Digital Elevation Models (DEM) of the US Geological Survey (USGS) or the Digital Terrain Models being produced in the UK for the Ordnance Survey (OS).

19.4.1 Landscape planning—visual impact analysis
Growing public awareness of environmental issues has been recently strengthened by the European Economic Community's Directive 'Assessment of the Effects of Certain Public and Private Projects on the Environment'. This Directive will force proposed changes to the landscape to be publicly assessed for environmental impact. A component of this environmental audit is a statement on the visual intrusion of proposed landscape changes. Consequently, projects such as road construction, transmission line routing and opencast mining, as well as more dynamic phenomena such as forestry, will need to be judged visually.

Traditionally, landscape visualization techniques have involved the building of physical models or the creation of artist's impressions. However, these are time-consuming to create, and are inherently inaccurate and inflexible once created. In order to quantify the level of visual intrusion more accurately, computer graphic modelling and visualization techniques are being used increasingly in the planning and design of landscape projects. These new approaches allow more accurate visualizations and more analytical assessments of visual intrusion to be determined. Due to the flexibility of the approach, many more proposed designs can be evaluated, resulting in a more refined design solution. Turnbull *et al.* (Chapter 20) pioneered the development of a Computer Aided Visual Impact Analysis system (CAVIA) that has been used, for example, to provide evidence at public inquiries related to electricity transmission line routing through environmentally sensitive landscapes (Turnbull *et al.*, 1986). Projects are typically performed at the subregional level with areas up to 40×40 km being analysed. The approach uses DTMs, landscape features and proposed design objects to produce an estimate of the visual intrusion. This visual intrusion toolkit includes intervisibility analysis to produce levels of visual impact, dead-ground analysis, identification of the portions of the landscape forming a backcloth for the design object, situations where the design object appears above the landscape horizon, and the identification of optimal locations for vegetation screen placement.

One of the visualization techniques used by CAVIA and other landscape planning systems is photomontaging. A photomontage is a physical or image composite of photographs of the existing landscape with a registered computer-generated image of the proposed design object(s). In this approach only the proposed design objects have to be rendered, avoiding the rendering of the intricate terrain and landscape detail. However, to achieve total image fidelity, the computer-rendered portion of the image must effectively 'merge' with the photographic image. Therefore, the atmospheric and distancing effects apparent in the photographic image must be inherited by the computer-generated image. Nakamae (1986) has developed techniques for merging image components to compensate for fog and aliasing effects. However, problems still occur when the design object's interaction with the existing landscape is complex, such as in road earthworks, and where foreground features, such as trees, partially obscure the design objects. Future solutions to this problem will use a hybrid approach to rendering a photomontage image, incorporating ray tracing, frame-grabbed images and textures, image processing and pixel painting techniques.

19.4.2 Road and traffic engineering

Visualization has found a number of interesting applications in the field of road design. Many road engineering design systems are now offering visualization capabilities. These form an integral part of the design process and allow the design to be subjectively assessed and refined for safety and visual intrusion in the context of its environment. The Transport and Road Research Laboratory, UK, has developed (Cobb, 1985) a system to model and visualize road designs. Applications of the system include the following.

(i) Road lighting scheme design. In complex road designs or under environmental constraints, the design of an efficient road lighting scheme can be enigmatic. This is the perfect design environment for the application of visualization techniques, where the designer can examine directly the results of his design under a variety of atmospheric conditions.

(ii) Road safety. Visualization tends to imply aesthetic appearance. However, in this application, visualization is concerned with the perceptual problems encountered by road users. Factors contributing to potential perceptual problems could be line-of-sight difficulties, incorrectly positioned street furniture or poor lane markings. In accident-prone sites, the system can be used to identify possible contributing factors and to evaluate solutions.

(iii) Street furniture design and placement. New designs for road signs and street furniture and their optimal positioning can be evaluated. This application has been taken one step further by the West German car manufacturer, Daimler-Benz, who have added dynamics and created a car simulator. The driver can experience driving a range of cars under a variety of driving conditions.

19.4.3 Architecture and urban design
In recent years urban renewal has become an activity increasingly exposed to and controlled by public and Royal opinion. Architects, in an attempt to alleviate public fears of a continuation of the 'Kleenex box' era, have turned to computer-generated images to convince the public of the merits of their proposed building designs. Computer-generated visualizations have become a fashionable marketing tool. Although the architectural industry was one of the first application areas where computer-aided design (CAD) techniques were applied, it is only recently that tools for creating high-quality visualizations of the resulting building designs have been made available. This capability is a natural extension of the CAD process, and many CAD system vendors are now supplying it as an integral part of their system or as an interface to foreign visualization packages.

Typically, architectural visualizations are not just isolated previews of the proposed building, but also include the contextual surroundings to allow appraisal of its applicability to the existing character of its urban environment. This usually involves the creation of a three-dimensional model of the terrain, streets, street furniture and buildings in the immediate vicinity of the site. This approach was recently followed by Arup Associates in their submission for the development of Paternoster Square in London. The computer model of the development was created on a McDonnel-Douglas IGS system and supplemented with the surrounding urban details through photogrammetric measurement and direct input from the Ordnance Survey's digital map series.

Since government-supplied large-scale map series of urban areas normally lack the essential height information of the buildings, users are either forced into the expense of an original survey, or are inhibited in the use of these techniques. In an attempt to encourage the use of these techniques, the City of Glasgow in Scotland sponsored the creation of a digital model of the downtown core of the city (Herbert, 1987). This was performed by the Universities of Strathclyde and Glasgow, and the model included the terrain, all major buildings, streets and rivers. Potential users are now being encouraged to use the 'city database' to site communications equipment and derive tourist information as well as create visualizations for planning purposes.

19.5 Conclusions

Computer-generated visualizations of digital terrain and landscape scenes are now widely accepted in many application areas as efficient technical analysis, design and marketing tools. Visualization techniques have released the world from its traditional two-dimensional approach to display, and in so doing have highlighted the three-dimensional deficiencies in our sources of data in terms of availability and accuracy.

Many large-scale maps have minimal terrain elevation data, and the majority ignore the three-dimensional aspects of surface features. Although the photogrammetric acquisition of DTM data is well established, to date much less attention has been directed towards the photogrammetric acquisition of three-dimensional descriptions of significant objects on the terrain. In many cases the three-dimensional coordinates of features such as buildings may be observed, although often the Z coordinates values are deemed superfluous and do not form part of the recorded dataset. Photogrammetrists therefore have an important role to play in the acquisition of this form of data, particularly in urban environments and at close range. Indeed, the lack of data currently inhibits the wider application of many of these techniques.

In the GIS environment, visualization techniques are recognized as an invaluable system component, aiding in the interpretation of spatially related phenomena and complex data analyses that takes the GIS a step beyond two-dimensional polygonal overlay analyses. Many GIS vendors are including this capability in their systems to help cope in our understanding of the 'fire hose' of data being produced by contemporary sources such as satellites. Over the past decade, substantial advances in the realism of computer-graphic-generated visualizations have been achieved. The techniques of ray tracing and radiosity in modelling global illumination currently form the leading edge. Despite this progress, the visualization of natural phenomena inherent in landscape scenes is still simplistic, forming 'realistic abstractions' or 'abstract realism'. Our visual system is specialized and highly skilled at recognizing them in all their subtle forms, making the objective even more elusive.

At present, apart from flight simulation, the production of realistic scenes using the visualization techniques discussed in this paper have been largely restricted to static applications. Similarly at present, dynamic visualizations are primarily restricted to 'abstractions'. The work of Muller *et al.* (1988), provides a good example of the current developments in the field of video animation using small-scale DTMs. For the future, however, the ability to perform dynamic visualization with high levels of scene realism will become an increasingly attainable goal, although the explicit modelling of many natural phenomena will continue to be a significant problem. For example, many of the current landscape visualization products use photographs, frame-grabbed images or texturing techniques of scene components to avoid the computer-graphic synthesis of natural phenomena. Some creators of the most realistic landscape images synthesized to date have used their artistic licence by overtly approximating the simulation of natural phenomena and 'faking' low-interest scene components with stochastic or texture map processes. Although not creating a scientifically correct image, they have achieved surprisingly effective results using what could be termed artistic techniques. This emphasizes the need for the scientific and artistic communities to work together towards this goal. Certainly more research is needed in perceptual comparison of real scenes and their synthetic image counterparts, with particular emphasis on the synthesis of natural colours and textures.

Inevitably, the successful approach will resort to true simulation of the phenomena based on the laws of physics. This is a computationally expensive option that in most cases is currently prohibitive, but will become feasible with the guaranteed increase in processing power. This is already happening with the recent release of parallel processing architectures, minisupercomputer-based workstations, high bandwidth links between supercomputers and workstations and the availability of customized VLSI for specific applications. The present techniques of approximating or faking (Frenkel, 1988) will be displaced by progressive refinements of the simulation model. This approach has been endorsed by McCormick *et al.*'s (1987) initiative on visualization in scientific computing. Despite realism being a distant target, it acts as a convenient measure of our techniques and understanding and will continue to be relentlessly pursued to our continuing benefit.

References

Blinn, J.F. and Newell, M.E. (1976). Texture and reflection in computer generated images. *Communications of the ACM*, October, **19** (10) 542–547.

Blinn, J.F. (1977) Models of light reflection for computer synthesised pictures. *ACM Computer Graphics*, July **11** (2) 192–198.

Cobb, J. (1985). A new route for graphics. *Computer Graphics '85*. Online Publications, Pinner, UK, 173–188.

Cook, R.L. and Torrance, K.L. (1982). A reflectance model for computer graphics. *ACM Transaction on Graphics*, January **1** (1) 7–24.

Crow, F.C. (1977). The aliasing problem in computer generated shaded images. *Communications of the ACM*, November, **20** (11) 799–805.

Faintich, M.B. (1984). State of the art and future needs for development of terrain models. *International Archives of Photogrammetry and Remote Sensing*, **XXV** (3A) 180–196.

Frenkel, K.A. (1988). The art and science of visualising data. *Communications of the ACM*, **31**(2), 111–121.

Goral, C.M., Torrance, K.E., Greenberg, D.P. and Battaile, B. (1984). Modelling the interaction of light between diffuse surfaces. *ACM Computer Graphics*, July, **18**(3), 213–222.

Gouraud, H. (1971). Continuous shading of curved surfaces. *IEEE Transactions on Computers*, June, C-**20**(6), 623–629.

Herbert, M. (1987). A walk on the Clydeside. *CADCAM International*, **6**(4), 41–42.

Kennie, T.J.M. and McLaren, R.A. (1988). Modelling for digital terrain and landscape visualisation. *Photogrammetric Record*, **12**(72), 711–745.

Mandelbrot, B. (1982). *The Fractal Geometry of Nature*. W.H. Freeman, San Francisco.

Max, N.L. (1986). Atmospheric illumination and shadows. *ACM Computer Graphics*, July, **20**(4), 117–124.

McCormick, B.H., Defanti, T.A. and Brown, M.D. (1987). Visualisation in scientific computing—a synopsis. *IEEE Computer Graphics and Applications*, **7**(7), 61–70.

Muller, J.-P., Day T., Kolbusz, J.P., Dalton, M., Richards, S., and Pearson, J.C. (1988). Visualisation of topographic data using video animation. In *Digital Image Processing in Remote Sensing*, J.-P. Muller (Ed.), Taylor and Francis, London, Chapter 2.

Nakamae, E., Harada, K., Ishizaki, T. and Nishita, T., (1986). A montage method: the overlaying of the computer generated images onto a background photograph. *ACM Computer Graphics* **20**(4), 207–214.

Nishita, T. and Nakamae, E. (1986). Continuous tone representation of three-dimensional objects illuminated by sky light. *ACM Computer Graphics*, July, **20**(4), 125–132.

Phong, B.T. (1975). Illumination for computer generated pictures. *Communications of the ACM*, June, **18**(6), 311–317.

Scholz, D.K., Doescher, S.W. and Hoover, R.A. (1987). Automated generation of shaded relief in aeronautical charts. *ASPRS Annual Meeting*, Baltimore, **4**, 14 pp.

Turnbull, W.M., Maver, T.W. and Gourlay, I. (1986). Visual impact analysis: a case study of a computer based system. *Proc. Auto Carto London* **1**, 197–206.

Whiteside, A., Ellis, M., and Haskell, B. (1987). Digital generation of stereoscopic perspective scenes. *SPIE Conference on True Three Dimensional Imaging Techniques and Display Technologies*, **761**, 7 pp.

Whitted, T. (1980). An improved illumination model for shaded display. *Communications of the ACM*, June, **23**(6), 343–349.

This chapter is based upon Digital Terrain and Landscape Visualisation (Kennie, T.J.M. and McLaren, R.A.) published in *The Photogrammetric Record*, 1988, **12**(72), 711–745.

20 The role of terrain modelling in computer-aided landscape design

M. TURNBULL, I. McAULAY and R.A. McLAREN

General introduction

Under the pressure of growing public awareness of environmental issues, planning applications for the creation or modification of structures in the landscape are now subject to more stringent environmental impact analyses. This was reinforced in 1988 when a new EEC directive governing environmental impact became effective. The directive recognizes landscape and, by implication, visual impact, as an issue within environmental impact analysis. Traditional visual impact analysis techniques have employed artists' impressions and physical scale models. However, this approach has tended to be inaccurate and subjective. In an attempt to improve the techniques, a more analytical approach involving computer-aided analysis of digital landscape models has been developed.

The chapter describes the computer-aided approach to landscape design from the applications or designer's perspective. The new design tools and their use are explained in the context of a recent case study. Areas of research leading to the enhancement of these tools are detailed. Secondly the supportive role of the DTM component of the design package is analysed. The workflow associated with the design process is outlined and the capture, generation and use of the DTM explained within this framework.

PART 1: VISUAL IMPACT ANALYSIS

20.1 Introduction

Over a number of years, the Turnbull Jeffrey Partnership, a firm of planners, architects and landscape architects, has developed a system for computer-aided landscape design. With the support of the Forestry Commission this system has been extended to include forestry planting, felling and replanting design. In collaboration with ABACUS (University of Strathclyde), and supported by the Science and Engineering Research Council (SERC), the Central Electricity Board (CEGB) and the South of Scotland Electricity Board (SSEB), Computer-Aided Visual Impact Analysis (CAVIA) is being validated and a further system is being developed. This work builds on visual impact analysis commissions undertaken over a number of years. In particular it has been based on projects concerned with transmission line routing, since data exists for a number of routes from planning and design through to construction, forming an excellent basis for the validation of the computer techniques used for visual assessment and for presentation at public enquiries.

The software is written in Fortran 77 and uses the GINO-F package for the generation of graphic images. It runs on the University of Strathclyde's VAX cluster under VMS, where the choice of computer ranges from a VAX 11/750 to an 8600 machine. The system supports a number of graphics terminals, digitizers, and plotters. Data are transferred from the VAX cluster to an IRIS 2400 workstation for animation. Frame grabbing and colour work is performed with an IBM PC equipped with a Pluto 2 imaging system.

20.2 CAVIA—data and tools

The system for computer-aided visual impact analysis (CAVIA) can be considered as having four main subsystems.

Landscape objects. Integration of digital terrain models with land-use, topographic features and satellite ground-cover data to generate plans, cross-sections and perspective views of the landscape.

Design objects. Modelling, in plan, section and perspective and, if required in colour, of any man-made objects such as electricity pylons, buildings, bridges, or natural objects such as trees, shrubs, etc.

Visibility. Accurate measurement of the degree of visibility of design objects or topographic features in the landscape and the complementary measurement of 'dead ground'.

Visualization. Generation of images of terrain and design objects, in line or colour form, suitable as visualizations on their own, as montages with black and white or colour photographs or as colour video images by mixing with 'frame-grabbed' or live images.

20.2.1 Techniques used for visibility and visualization

Six principal programs are involved in the process of determining the visibility of design objects or topographic features in the landscape. In addition, many other programs are needed in support for the purposes of data input, checking, modification, manipulation, analysis and display. All programs operate on the same landscape and design object data. The principle of intervisibility is fundamental to the program methodology in identifying the locations in the landscape where a person can or would see a design object or feature.

Based on this principle, the program VIEW1 considers only the topography as defined by a DTM. The visibility of a design object established by the application of this program results in the worst possible case, since no allowance is made for any screening which may result from trees, buildings or other such features in the landscape. In some landscapes, these features may reduce visibility considerably. The height of the design object may be set as required. If this is, for example, the top of a transmission tower, the visibility of the whole tower can be investigated. All the simulations take into account curvature of the Earth and refraction of light which can be significant when viewing distances over 1 km are involved. VIEW1 also allows the investigation of invisible areas, that is, 'dead ground'. This is important in certain siting or routing considerations since the visibility or invisibility of a design object placed in an area of dead ground can be determined immediately.

The VIEW2 program can be used, for example, to analyse views of a transmission tower route from specified viewpoints. This program can modify a DTM to allow for the screening effect of features such as trees or buildings, identify those towers visible from a specified viewpoint and detail the percentage visibility of each tower. It can also determine for each tower whether it is seen against a backcloth, in which case, the area acting as backcloth is identified, or if seen above the horizon, the percentage of the tower thus seen. This program can take into account whether a viewpoint, such as a road, is in a cutting or on an embankment. Additional programs VIEW3 and VIEW4 can identify the significance of features such as trees or hedges in controlling the visibility and screening of towers, and can identify the areas from which the towers would be seen if existing screening (e.g. trees) were removed, thus identifying areas critical to achieving screening. The results of the VIEW1, VIEW2, VIEW3 and VIEW4 programs are represented in map form, either as computer print-outs, computer line drawings or as tables and diagrams. By using additional programs, VIEWER and LANDVU, it is possible to produce an accurate perspective drawing of the visible towers in the landscape from a given viewpoint. This is important in demonstrating the visibility of the towers and their scale in the landscape.

The VIEWER program is a three-point perspective drawing procedure providing 'wireline' drawings with hidden lines removed of design objects for photomontaging. VIEWER processes a three-dimensional geometrical description of a design object, for example the structural steelwork of a transmission tower and/or a DTM, and outputs a three-point perspective view of it. This view is fully corrected and adjusted to accurately depict the exact design object as viewed by an observer from any viewpoint location in the study area. The LANDVU program produces a computer-generated three-point perspective of the DTM from a specified viewpoint in the form of a 'wireline' drawing with hidden lines removed.

Using this data and the appropriate viewing parameters recorded during the site photography, perspective views can be drawn by the computer replicating the photographs exactly. To prepare a photomontage, these perspectives are printed on to transparent sheets and overlaid on the enlarged colour site photographs. An exact registration between a photograph and the computer drawing is achieved by means of control points. These points are known and visible features in the photographs, and are modelled in the computer views to ensure an accurate match in the montage.

20.3 Case study: Torness transmission lines

The CAVIA system has been put to use on various projects in its development and is, to some extent, project-led. The Torness Transmission Lines Study is one of a considerable number of projects undertaken, but was the first in which most of the system was been used. Transmission lines transmit electrical power from the source to load centres, and carry power on conductors suspended at regular intervals from towers of lattice steel construction some 45–50 m high. The Torness Transmission Lines Study involved the detailed visual assessment of two alternative transmission line routes in the south-east of Scotland in the Borders Region, centred on the Lammermuir Hills which rise to 500 m and separate the fertile and populous plains of East Lothian in the north from the Merse and Upper Tweed Valleys in the south. The Lammermuirs have an open treeless appearance, and regionally the hills are subject to environmental policies contained within the structure plans of Lothian and Borders Regional Councils, designed to protect the ecology and the scenic nature of the area.

20.3.1 Visibility: general considerations
In conducting a study of the potential visibility of a transmission line, it is first of all necessary to make some judgement about the area from which the transmission towers will be visible. Distance is a very important factor when viewing a transmission line in the landscape. In general, apparent height in the landscape varies inversely with distance. In many instances, the overall visibility of a transmission line will be limited by surrounding topographic features. Where this is not so, the most significant views are likely to be experienced within a distance of 5 km from the transmission towers. Longer-distance views can be significant, particularly where a transmission line is viewed above the horizon, on the skyline. Accordingly, a study area extending beyond 5 km from the routes was defined to accommodate such longer views. This larger area was selected by examining the topography surrounding the immediate study area.

Use of a computer in visibility studies has distinct advantages over field studies:

(a) to given levels of accuracy, the visibility of design objects in the landscape can be simulated and recorded by analytical techniques which could be impossibly long to undertake manually;
(b) after data acquisition, different alternatives can be evaluated quickly and easily.

20.3.2 Data collection
A 30 km × 30 km area of the Lammermuirs mainly within Borders Region was identified for analysis. Although a square study area was chosen, this is not a requirement of the techniques. The DTM was prepared from Ordnance Survey 1:25 000 series maps by Laser-Scan Laboratories Limited of Cambridge, using specially developed equipment and computer programs. The process involved following the contours with a laser and recording in digital form the location of all points where the contour lines changed direction. The data was then transformed to National Grid coordinates, interpolated to the specific grid of 25 m and verified using contoured drawings and three-dimensional representations of the data. Before the DTM of the study area was prepared, four trial areas, typifying the different topographic features of the study area, were tested to determine the most appropriate grid interval for the DTM and the levels of accuracy required. Roads and woodlands were also digitized and the information converted for use at a 25-m grid.

20.3.3 Application and interpretation
Initially the entire area of 30 km by 30 km was studied using a DTM with a 200-m grid generalized from the 25-m grid to assess the overall visibility of alternative routes. The VIEW1 program produced visibility maps for each of 40 transmission towers and amalgamated these to produce 'contours of visibility' for each route in total (Fig. 20.1). This level of analysis was used to identify broad patterns of visibility and to highlight areas for further detailed study.

The results showed that both of the proposed routes exhibited a similar pattern of visibility. Both crossed a high upland section of the Lammermuir Hills at a height of approximately 400 m. In the area, the local population resides in villages and farm steadings which for the most part lie in lowland areas along valley floors. With the exception of one stretch of trunk road, the same conclusion applied to the major roads

K

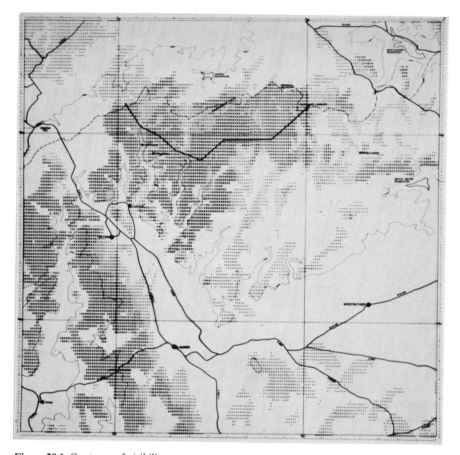

Figure 20.1 Contours of visibility.

in the area. The results demonstrated that the resident and road-using population were entirely screened by intervening topography from either of the transmission routes considered in the study.

Subsequently, a smaller area of 22 km by 22 km was analysed using a 100-m grid to give greater accuracy. The area chosen was based on the 'contours of visibility' map produced using the 200-m grid. The analysis of information generated by the VIEW1 program, when run using the smaller data set, validated the general conclusion and permitted certain more detailed conclusions to be drawn about the similarities and differences in visibility for each of the proposed lines.

Once the main areas of visibility had been defined, a more accurate visual assessment was undertaken from selected viewpoints within these areas. Clearly not every potential viewpoint was of equal importance, and a selected list of viewpoints was identified by field observation and from the results of the VIEW1 program. The visibility from individual viewpoints was assessed using the VIEW2 program. This analysis generated a range of detailed information for each viewpoint. In this way, valuable data was obtained to assist with the assessment of the visibility, particularly in respect of the extent of visibility of particular towers above the skyline.

Whenever possible, the results of the computer studies were correlated with a parallel exercise of field observation recording the nature of the view, type of landscape and vegetation, important buildings, and recreational activities. The height of a transmission tower can be estimated roughly in the field by employing a yardstick device, and estimates of visibility can be recorded by a simple method of classification. The computer results can then be combined to construct an analysis of visibility in the context of landscape setting and type. The VIEW3 program was used to define 'critical areas' where the existence of a feature, such as woodland or a building, of a stated size would make the difference between visibility and invisibility. It was then possible to examine maps and aerial photographs to establish if trees or buildings occurred at these critical locations.

20.3.4 Visualization
The visualization stage of the study was undertaken using VIEWER and LANDVU computer perspective programs. It was originally intended to prepare a photomontage from each of the viewpoints with a view of the alternative lines (see Fig. 20.2). However, bad weather precluded the possibility of obtaining photographs from all but one of the viewpoints, and a perspective view of the landscape from these viewpoints was generated using the LANDVU program. The results of this program were correlated with the results of the VIEW2 program and from the photomontage process obtained for the one possible montage permitted by the weather.

20.3.5 Outcome of case study
A public inquiry proved to be necessary to resolve the issues associated with the transmission routes. The computer-assisted techniques proved to have certain key advantages:

(a) The quantification of the extent of visibility associated with the alternative routes and the exploration of the detailed pattern of change in the visibility was demonstrated by reference to maps with overlays. This directed debate away from speculation about what might or might not be seen, and instead focused it on interpretation of patterns of visibility and the implications of these for the landscape of the area.
(b) The production of accurate photomontage and perspective views of the landscape

Figure 20.2 Photomontage of proposed transmission towers.

from selected viewpoints showed the landscape before and after the transmission towers had been erected. In some cases, these proved particularly interesting, since the towers were screened entirely from view where field observation had indicated that some parts of some towers might be visible. Once again, debate was focused on the interpretation of visual images rather than on their accuracy.

20.3.6 Access tracks

Although the public inquiry found in favour of the SSEB's proposed route, the method of construction and the access tracks for construction and maintenance of the lines were subject to further planning submissions to ensure that their impact on the environment was minimized. It was therefore decided to prepare detailed maps identifying ground cover habitats along 12 km of the most sensitive terrain on the route and to advise on the most appropriate construction techniques and ground reinstatement.

The practical objective was to provide good-quality ground cover data to ensure accurate and prompt drawings for planning permission outlining proposals for full landscape and ecological reinstatement and for subsequent contract documents. In addition to the planning submission, this information was used in design work and as an outline guide in construction planning, and in discussion with relevant conservation bodies to assure them of adequate landscape reinstatement. Finally, the maps with a distribution of ground-cover habitats were linked to prescribed components of landscape advice, detailed techniques of construction, and reinstatement practice to be employed in association with particular habitats.

The distribution of ground-cover habitats was prepared by remote-sensing techniques based on LANDSAT satellite imagery. In this case, the use of LANDSAT data was selected due to the short timescale before construction work began. Traditional methods using published maps of soil and land use classifications linked to a time-consuming site survey would have limited the scope and extent of the study, particularly in the ability to examine the distribution of ground-cover habitats at a variety of scales. The satellite interpretation work was carried out by the Environmental Remote Sensing Applications Centre (ERSAC) Limited using a GEMS image-processing computer system. The interpretation was validated by on-site checking of selected areas. The ground cover data available at the end of this process was held in the form of a classification at 50-m grid intervals for an area of 22 km × 22 km.

After the interpretation and field-checking process, a final ground-cover interpretation was prepared as follows: woodland, water, heather dominant, heather/grass/burnt mix, upland grass, improved grass, bracken dominant, cropland. The ground-cover data derived from the satellite imagery was analysed and then combined with elevation data at a 25 m-grid and data on the location of woodlands and roads. Additional data was derived from the DTM: slope at specified gradients; drainage direction; aspect; elevation at specified contour intervals. This derived data was then combined with the ground-cover data to produce maps combining ground cover distributions such as heather, etc., with physical characteristics such as slope, elevation, etc. An interesting correlation was observed between certain classes of slopes and heather moorland on deep peat.

20.4 Current and future developments

Since the completion of the transmission line and access tracks, extensive validation of the techniques has been undertaken.

Validation of visualizations. The visualizations produced before the line was constructed have been compared with the actual line, and any differences investigated. The performance of software has been evaluated and the programs used for the study assessed against their current versions.

Transmission towers. Various transmission towers have been modelled in detail. Modelling of conductors and wires has been completed. Methods of dealing with distance have been investigated both in terms of line drawing output and colour (Fig. 20.3).

Photographs for montaging. Until now it has been necessary to use professional surveyors to locate photographic viewpoints and control points to the accuracy required for montaging computer images with photographs. Considerable effort is going into accurately calculating and checking viewpoint data from identifiable features in the photographs. In addition, work is being undertaken to identify the most effective photographic and montaging techniques.

Image mixing: the effects of climate. Climatic conditions affect the visibility of middle-ground and distant objects via varying lighting conditions and atmospheric pollutants. In order to mix computer-generated images with video images of a natural landscape, it will be necessary to model in software the effects of distance under varying weather conditions. The approach adopted will be to photograph, for example, receding pylon lines under extremes of visibility and by image processing to establish the degree of atmospheric visual attenuation to apply to the computer-generated images.

Figure 20.3 Perspective of towers using pen plotter to give distance effect.

Image mixing: separation of foreground and background. Realistic visualization of a building often requires viewing from a position which locates the building between a background, for example, hills, and a foreground, for example, trees. Conventional 'cut-and-paste' techniques of mixing the images, although feasible, are laborious and seldom adequate. It is intended to automate the mixing process using the frame-grabbing and buffering capabilities of the PLUTO II imaging system, so that separate parts of the landscape and building images can be successively superimposed to produce the required composite image.

Output display: video. The main issue to be investigated is the cost-effectiveness of producing dynamic images of design and planning proposals. It is already possible to generate a 'walk-through' of buildings and of neighbourhoods, but only by the expensive and time-consuming process of frame-by-frame filming. The IRIS 2400 workstation offers the possibility of real-time dynamic viewing, video recording and manipulation of a model, in either wireline or fully coloured modelling. The remaining problem is then to ensure colour constancy from the designer's materials to a screen view and to the video record.

PART 2: THE ROLE OF TERRAIN MODELLING WITHIN CAVIA

20.5 Introduction

This part of the chapter describes the underlying DTM component of the system described, explaining its role within this design environment. The chapter has been structured to reflect the workflow associated with CAVIA design procedures. The high-level workflow is shown in Fig. 20.4.

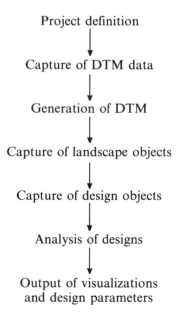

Figure 20.4 CAVIA workflow.

The system's use and the designer's interaction with the DTM at each of the phases are explained, and possible areas of improvement and system extensions are discussed.

The preliminary stage of a project is to set up a series of 'soft' system parameters to customize the CAVIA system for the particular project characteristics. The parameters associated with the DTM are Feature Class to DTM Data Type Relationships.

The user can create a data model for the project through the facility of feature class definition. Having defined the data model, the user can then associate feature classes with DTM data input types to control the generation of the DTM. For example, ridgelines, rivers and roads could be allocated to breaklines.

20.6 Capture of DTM data

20.6.1 Elevation data types to be captured

The DTM package is designed to accept generalized forms of elevation data, including:

- arbitrarily distributed spot heights
- profile or grid measurements
- contour lines
- breaklines.

The most common and readily available form of elevation data used in projects have been contours and spot heights from topographic maps. Although their accuracy is relatively low, it has proved sufficient for projects where evaluating alternative locations of objects in the landscape or evaluating different size of objects in the landscape is the primary objective. However, once the general location plan has been approved, an original survey is carried out to refine the design and to derive engineering construction parameters. This approach minimizes the cost of surveying, particularly in projects involving route selection, as it limits surveying to the corridor of the selected route.

One of the key landscape design analysis tools of the CAVIA system is the accurate measurement of the visibility of objects. This involves line-of-sight calculations using the DTM. The results of this analysis are extremely sensitive to the accuracy of modelling ridge features within the DTM. To meet this requirement, the DTM allows these features to be defined and integrated into the DTM through the mechanism of breaklines.

20.6.2 Methods of capture

The spatial extent of project areas where this technique of landscape design is applied varies enormously from planning subregions, 30 × 30 km, to street sites 100 m across. Therefore, the methods used to capture the terrain information reflect the quantity and quality of data to be collected. For projects involving route selection over large areas, the source of elevations is usually restricted to existing topographic maps. However, for projects whose final product is an engineering design, an original survey using photogrammetry or field surveying is performed.

For a project of large spatial extent, digitizing of the contour information has been contracted out to service bureaus, such as Laser-Scan Laboratories Limited of Cambridge, to digitize automatically on their FASTRAK system. Once the digital contours have been loaded into the system, they are then supplemented by manually digitized spot heights and breaklines. For smaller project areas, all the elevation information has been manually digitized. The contour digitizing procedure involves a user-definable filter based on an automatic, incremental, distance release that substantially reduces the data volume without jeopardizing the resulting DTM fidelity. For large-scale urban projects, an interface to the UK Ordnance Survey digital data has been contemplated. However, apart from the current sparse coverage, the 1:1 250 series has insufficient elevation information to support this application.

20.6.3 Forming the triangular irregular network (TIN)
The generation of the DTM involves setting up a spatial, triangular irregular network including connection information about neighbouring triangles throughout the network. Prior to the formation of the TIN, the elevation input data is filtered so that resulting segment lengths conform to a user-defined constraint. This controls the length of triangle sides. The construction of the triangles uses a constrained version of the procedure proposed by Delaunay (explained in Gottschalk, 1981, for example). As with the Delaunay approach, the triangulation is unique, but the following constraints are imposed during the generation:

- triangles must not cross breaklines (including contours, since they are treated as breaklines in this application)
- triangles must not be composed of more than two vertices from any one breakline.

The result of this process is a continuous triangular mesh that is stored in the associated project database. This is the definitive version of the terrain definition from which all other terrain characteristics are derived.

20.6.4 Capture of landscape objects
As landscape objects such as roads are being digitized from map material, the system automatically forces the features to conform to the DTM. In the case where landscape objects have no explicitly associated elevation information, the system automatically calculates and allocates the elevation to the X, Y points defining the features. This is achieved by floating the landscape object on to the DTM.

Instead of landscape objects being constrained to conform to the existing DTM, some may force modifications to the DTM during placement. This occurs when the landscape object either has explicit elevation information associated with it, or is a 3D object with inherent geometry that does not conform to the existing DTM. In both cases, the landscape object is providing extra elevation information and the DTM is locally modified to accommodate it.

20.6.5 Capture of design objects
The DTM is involved in two aspects of design object capture: the constraint that design objects must conform to the existing terrain and the integration of derived DTM parameters in user-defined placement rules.

(i) Design object—DTM conformity. When the landscape designer is placing design objects into the model of the landscape, the system ensures that there is a consistent fit between the existing DTM and the design object. The interaction of the design object with the DTM is treated in the same way as the landscape objects previously described.

(ii) DTM constraints on design object entry. During the initialization of the project, the user defines the type of design objects to be integrated and analysed in the landscape design, e.g., electricity tower types and tree types. The involves the user defining generic descriptions of the objects in design object libraries. The definition has two components:

(a) *Geometric definitions* are scale-independent definitions describing the geometry of the design objects. These are subsequently instanced through transformations when the design objects are positioned in the landscape model.

(b) *Placement rule definitions* are user-defined rules imposing constraints on the placement of design objects. The rules are normally based on a combination of DTM-derived and other environmental parameters. An example that does not use terrain information is the constraint to place plants within specific soil types. However, terrain parameters can be used to influence the position of the design object. In forestry design, rules governing the placement of tree species include restrictions based on slope and aspect criteria derived from the DTM. In more complex applications, the placement rule is based upon the geometry of the design object in combination with the terrain. This is the situation found in the planning of transmission lines. Apart from cost considerations, the major constraint on the detailed placement of transmission towers is the clearance of the associated catenary formed by the suspended conductor between the towers. Safety regulations dictate the minimum clearances allowable for a variety of land-use types.

During the placement of design objects, the system can display the existing data in perspective as well as the default orthogonal transformation simultaneously. This provides the designer with the possibility of positioning design objects directly on the perspective display that in many cases is more natural to a designer.

20.7 Selection of project area subset for analysis

All data associated with a project, i.e., DTM, landscape objects and design objects, are stored in a single definitive version called the project database. When a designer wishes to work on a project, the required portion (logical and physical subset) is extracted from the project database and copied into a working subset called a partition database. On completion of the design work in the partition database, any changes to the data are integrated back into the project database in a controlled environment to ensure data integrity. This mechanism avoids problems associated with data corruption in a multiuser environment.

20.7.1 Formation of a regular grid DTM
Historically, the CAVIA package used a regular grid DTM. Although this has been superseded by the TIN-based DTM, several of the application analyses still use a regular grid DTM. During many of the preliminary analyses in the design process, the designer is more concerned with broad patterns than with the detailed accuracy of the result. This can be achieved more easily in an interactive environment using a regular grid rather than a triangular DTM, due to the higher speed of processing. Therefore, the system derives a regular grid DTM from the TIN and carries them in tandem for the duration of the analysis. Once the analyses are complete, only the TIN is permanently retained for future use. The TIN is the definitive state of the DTM and all other elevation information is derived from this source.

20.8 Design analysis

Once both the landscape and design objects have been captured, the designer can then analyse and optimize the proposed landscape design using the created computer simulation. Two sets of tools are available: visibility and visualization analysis.

20.8.1 Visibility analysis
Based on the principle of intervisibility, this set of tools allows the designer to:

- determine the degree of visibility of design objects in the landscape
- quantify what portion of the visible design objects protrudes above the horizon
- analyse the degree of sensitivity of visibility to the placement of screens such as hedges and trees
- determine the optimal position of screens.

This result of these analyses is a design that 'hides' or minimizes the visual impact of the design objects in the landscape. In all the visibility tools, the geometry of the terrain is provided by the DTM, and ray analysis takes into account Earth curvature and atmospheric refraction and variable viewing heights. Due to the sensitivity of the degree of visual impact on the influence of landscape objects such as buildings and vegetation, in most cases, it would be false to analyse a bald landscape. Therefore, the system integrates user-specified categories of landscape objects into the visibility analysis. Individual landscape features can be interactively included or excluded from the analysis to determine their influence on visibility. In effect, the landscape objects are integrated into the DTM for the duration of the analyses.

To enable the effect of vegetation screens on visibility to be judged with time, their heights can be varied interactively to simulate growth between analyses. The results of visibility analysis are normally maps illustrating visibility polls for the area of interest.

20.8.2 Visualization analysis
These tools allow the designer to view the proposed design on the graphics screen at varying degrees of realism, depending on the stage of the design process. During the early design phase when a high degree of interaction is required to support the inevitable number of design iterations, the designer can display perspective views of the design proposal on the graphics screen. Due to performance constraints on the current hardware configuration, only approximate wireframe models of the landscape and design objects are portrayed. Hidden line removal is an option.

It has been found that the display of a wireframe, regular-grid DTM is more easily interpreted than the equivalent triangular mesh. Landforms such as ridges can be easily identified, there is less possibility of misinterpretation, and relative distancing more quickly quantified. However, it is felt that a triangular mesh is superior in visual interpretation over a grid when shading is incorporated into the displayed image.

When the final design proposal has to be presented to the client or at a public enquiry, a high-quality visualization of the design is created. At present, this is achieved by combining a field photograph panorama with a corresponding computer-generated vector plot of the design objects to create a photomontage (Fig. 20.4). The interior and exterior orientation parameters of the camera are simulated during the generation of the computer plot to guarantee registration of the component images. The DTM is not normally included in the photomontage, except in the case where the terrain itself will be changed by the design. Since the DTM is generally used to position the design objects, any discrepancies between the plotted positions and the photographed terrain clearly highlight any inadequacies of the DTM's fidelity.

Current research is investigating procedures to create a photomontage directly from the graphics screen where a frame-grabbed image of the photograph is directly combined with an enhanced image of the design objects. The technique applies lighting and distancing effects, matching that of the photograph (Nakemae, 1986), to produce a high degree of realism. This also opens up the possibility of using remotely-sensed data to increase realism. Although Thematic Mapper data has been used to identify vegetation types and aid in the routing through environmentally sensitive areas, the data can also be used to control texturing of perspective views.

20.8.3 Other landscape design tools utilizing the DTM

Apart from the use of the DTM in the visibility and visualization tools of the CAVIA component, it is used extensively within other facets of the landscape design package. This includes:

- slope analysis
- drainage analysis
- cross-sectioning
- cut-and-fill calculations
- 'bill of material' quantity surveying tools.

The DTM's role within the landscape design package ranges from simple cross-sectioning, through intricate visibility analysis to sophisticated imagery textured visualizations. The DTM is fundamental to the computerized approach described and will continue to find new applications within landscape design. However, the widespread use of these techniques is presently inhibited by the current lack of publicly available digital landscape data. The current 1:1 250 digital data has sparse coverage, is only applicable to large scale urban projects and lacks the essential third dimension. This approach and other related ones would benefit if the Ordnance Survey provided full 3D landscape data at large and medium scales.

20.9 General conclusion

The natural and man-made environment is under increasing stress. We are entering a phase when the exploitation of energy resources is likely to cause a dramatic acceleration in our rate of impact on the natural environment; in particular, there is cause for serious concern regarding the damaging visual impact of energy-related developments such as oil terminals, dams, power stations, transmission lines or opencast mining, on remaining areas of relatively unspoilt rural landscape. At the same time, the need to renew our inner cities places enormous responsibilities on architects and planners who seek to integrate elegantly and economically the new with the old.

The political will exists to address the problem; a recent EEC directive recognizes landscape and, by implication, visual impact, as an issue within environmental impact analysis. What is lacking is the means to appraise and compare the visual impact of alternative proposals objectively, economically and, above all, in a manner which is understandable to the range of interests involved in the design and planning process. CAVIA is a computer-aided system which attempts to meet these requirements. Computer techniques such as CAVIA are now begining to be recognized as having an increasingly important role in aiding in the resolution of planning and design problems. The challenge is to integrate a diverse range of techniques from several disciplines into one system.

Acknowledgements

The authors are indebted to the transmission engineers of the South of Scotland Electricity Board and the other members of the Torness Transmission Lines Study Team: William Gillespie and Partners and Dr J Benson, Consultant Ecologist; also Laser-Scan Laboratories Limited, Environmental Remote Sensing Applications Centre (ERSAC) Limited, and Professor G. Petrie of the University of Glasgow.

References

Gottschalk, H.J. (1981) Relation im Raster. *Karten- und Vermessungswesen* **1**, 82.

Nakamae, E. *et al.* (1986) A montage method: the overlaying of the computer generated images onto a background photograph. *ACM SIGGRAPH 1986 Proceedings* **20**(4), 207–214.

Petrie, G., and Kennie, T.J.M. (1986) Terrain modelling in surveying and civil engineering. *Presented paper at the British Computer Society Displays Group's Conference on 'State of the Art in Stereo and Terrain Modelling'*, 32 pp.

Rogers, G. (1980) Dynamic 3D modelling for architectural design. *Computer Aided Design* **12**(1) 13–20.

Turnbull, W.M. *et al.* (1986) Visual impact analysis: a case study of a computer based system. *Proc. Auto Carto London Conference* **1**, 197–206.

21 Military applications of digital terrain models

M.W. GRIFFIN

21.1 Introduction

Over the last decade the increase in the power, capability and availability of computer hardware and software has coincided with the willingness of the analyst and military end-user to exploit computer technologies. At EASAMS Limited, work began in the 1970s on relatively straightforward applications of digital terrain models (DTMs) for use in intervisibility studies for the Army. Initially, these activities merely automated processes previously carried out by means of cartographic exercises and field trials. In replacing the laborious elements of such essentially manual methods, it was realised that DTMs became not just part of the operational analysts' toolkit, but were established as important military system elements in their own right. Subsequent development led to the inclusion of DTMs in Command, Control, Communications and Information (C3I) Systems, Geographical Information Systems (GIS), automated cartography, advanced display systems, simulations, navigation systems and robotics.

The following sections describe three 'case-study' military systems and discuss some of the problems encountered. An attempt is made to demonstrate the ways in which the functionality and power of such systems have been expanded since the 1970s and have reflected the changing pace of technology.

21.2 Military applications

21.2.1 Intervisibility studies (Case Study 1)

(i) Background. During the late 1970s, DTMs and related software were produced to assess the effects of observer-to-target intervisibility on the development of anti-tank guided weapons (ATGWs). The data sets used were limited to a 12 km × 12 km area of gridded datum points at a 50 m resolution, containing spot heights and cultural information. The latter was restricted to buildings, trees and a vector-based representation of road, tracks and powerlines.

(ii) Intervisibility analysis. A set of software routines, written in FORTRAN and APL, was produced in order to manipulate and process the clients' 'raw' data. Running in batch mode on ICL 1900 series and CDC Cyber mainframes, the software permitted the rapid production of intervisibility plots on a map overlay. The area plots or 'sweeps' radiated up to 360° from the observer positions and allowed for user-specified observer-to-target ranges. These sweeps plotted the 'dead' ground for one or

a combination of topography, culture (trees and/or buildings and/or pylons) and atmospheric obscuration. Such results were most usefully presented as radial 'spokes'—lines originating at an observer position—or area infill superimposed on the map background. Multiple, overlapping sweeps helped identify optimum engagement zones or were used to plot dead-ground approaches for route planning and weapon siting. Figure 21.1 shows the relationships, using the DTM, between an observer position and a column of 14 vehicles moving over the terrain. Four exposure windows are plotted against the map background, with target intervisibility indicated for each of three one-second 'snapshots'.

Sectional views of the intervisibility from point A to point B were also produced. Such sections, utilizing vertical scale exaggeration for clarity, were plotted against a map background. Subsequent development addressed the problems of modelling the movement of changing target arrays over the terrain and their intervisibility or detectability in relation to various types of culture. Detection, both visual and by missile seekers in conjunction with various search patterns, was studied and a six-degrees-of-freedom missile model used to represent the flight of various missile types (command to line-of-sight and lock-after-launch) over the terrain. Work was also performed on the modelling of engagement scenarios where missile and target travel over, and interact with, the terrain.

(iii) Perspective viewing. Eventually, perspective plotting routines were developed from first principles. Such techniques are now widely used throughout the military and civil world. Most viewing techniques are concerned with the form of the ground; more sophisticated techniques add cultural, atmospheric and tabular information to the topographic view. Normally, each DTM point (in X, Y, Z space) is transformed with respect to an observer point, with a resultant azimuth and elevation (angular value) mapped to a pixel address on the display device. Enhancements include hidden line/area removal, field-of-view (FOV) restrictions, illumination effects (texture, shadows, haze) and automatic scaling. Unfortunately, processing penalties are incurred by high-resolution perspective viewing algorithms, and even now few systems produce detailed views in real time (e.g., frame rates > 20 Hz). Currently, the need for geographic data standardization is being recognized at EASAMS with the creation of perspective viewing software to handle any common data format.

The solftware also allowed a user-defined choice of algorithms for both fully detailed perspective views and less detailed views for rapid appraisal situations by using lower resolutions, 'mathematical' instead of 'true' perspective algorithms and FOV restrictions. Such perspective views were produced initially in order to synthesize a 'rasterized' view of the target from a simulated missile position. The perspective views were subsequently used for general perusal of the digital landscape (Fig. 21.2). The techniques were also used for non-military work (Griffin, 1981). An important shift in emphasis here was from batch processing to interactive processing, with the adoption of faster, multiuser mainframes in the late 1970s. Perspective views were supported by field surveys, aerial photographs and cartography, mainly as a means of validating the results being produced. For example, photographs of the area of interest were compared with perspective plots using the same grid references, fields of view and viewing direction.

Other interesting results were obtained. Earth curvature, for instance, was calculated and allowed for, but was found to be insignificant for the small engagement ranges (up to 5 km) involved. Darkness and atmospheric obscuration (e.g., smoke, haze, precipitation), on the other hand, were not modelled but were found to affect target intervisibility dramatically during field work.

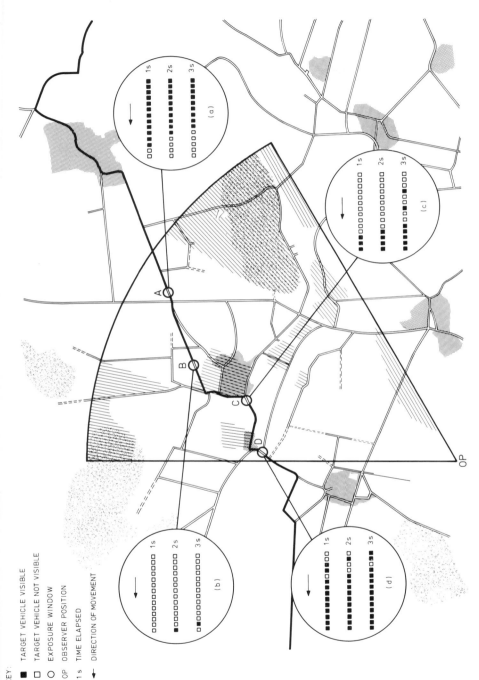

Figure 21.1 Multiple-target intervisibility sampling.

KEY:

■ TARGET VEHICLE VISIBLE

□ TARGET VEHICLE NOT VISIBLE

○ EXPOSURE WINDOW

OP OBSERVER POSITION

1s TIME ELAPSED

↓ DIRECTION OF MOVEMENT

21.2.2 Unmanned vehicles (Case Study 2)

(i) Background. In the early 1980s, EASAMS became involved with the MAID (Mobile Autonomous Intelligent Device) research program, which was initiated by MOD to assess the feasibility of using unmanned vehicles (UMVs) for the Army (Smith, 1986). UMVs generally have potential for a variety of military roles, including surveillance, sentry duty, mine laying or clearing, rescue operations, tactical and terrain surveys or offensive action (Ogorkiewicz, 1986). The UMVs are envisaged as operating in three modes: remotely controlled, semi-autonomous and fully autonomous (robotic). Clearly such UMVs must interact with, and navigate over, the terrain.

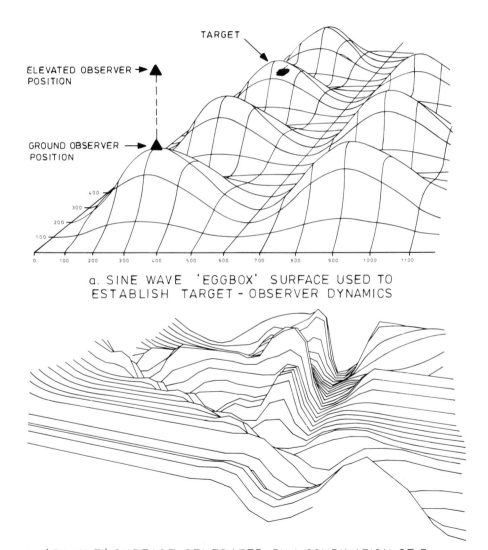

a. SINE WAVE 'EGGBOX' SURFACE USED TO
ESTABLISH TARGET - OBSERVER DYNAMICS

b. 'RAVINE' SURFACE GENERATED BY A COMBINATION OF 5
MATHEMATICAL AND STATISTICAL FUNCTIONS

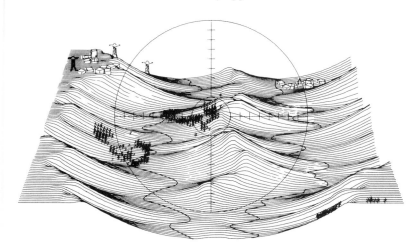

c. SYNTHESIS BY MEANS OF LANDSCAPE EVOLUTION MODEL

Figure 21.2 Synthesized digital landscapes.

What was identified, therefore, was a large investment in software in order to simulate and assess terrain-related UMV subsystems prior to actual construction and testing of the various UMVs.

(ii) Software environment. The first part of this work was devoted to developing new DTMs and refining existing software. Software design was carried out and new 'structured' languages, such as Pascal, adopted. Host computers included DEC PDP 11/44s with Floating Point Systems (FPS) 120 Array Processors, Digital Equipment Corporation (DEC) VAX and MicroVAX minicomputers, and a Control Data Corporation (CDC) Cyber mainframe.

(iii) Data sources. Unlike most C3I or aerospace DTM applications, the UMV study required terrain data at a very high resolution in order to simulate detailed aspects of vehicle motion, since ground features such as ditches and embankments would need to be taken into account. In addition, the vehicle needs to use its sensors (scanning laser range finder, TV cameras, Doppler radar and so on) to process information about the terrain to avoid obstacles such as individual trees, ditches and fences. Less detailed DTMs were also required, as the UMV needs to carry an on-board digital terrain system for position-fixing and dead reckoning. It was the proposed navigation of such an autonomous vehicle over unprepared terrain that produced some of the most profound problems of the research work.

Four data collection methods were investigated and assessed in order to supply high-resolution altitude and culture data.

(a) *Data collection by survey.* Accurate three-dimensional coordinates may be collected by field survey methods, existing map sources and photogrammetry (Petrie and Kennie, 1987 and part A, this volume) depending on the scale and the extent of the area involved. Such survey techniques are now well documented, and were examined by the project team only for their cost-effectiveness and relevance to the particular requirements.

(b) *Existing data.* The requirements of UMV navigation and tactical planning could have made use of existing data, such as the Digital Landmass System (DLMS) levels 1 and 2, and various related data sets. Digital data sets associated with remote sensing applications were also considered (House of Lords Select Committee, 1983). Other digital data sets at a 1:10 000 scale were available under licence from the UK Ordnance Survey (OS). Ordnance Survey digital data sets were selected as the most convenient cost-effective source for the UMV project. Such data sets are provided as vectors or strings, since this is the most convenient way of automating the map facsimiles that show linear features such as road or streams. However, such data sets sometimes require conversion to a grid-based system for more efficient and cost-effective processing as arrays or matrices.

Conversion programs were written in Pascal to provide an 'assembly line' approach to reformatting Ordnance Survey digital data to gridded data sets. This first involved a process of interpolation (using second- or third-order polynomials) to obtain the elevation data from the sparsely-distributed spot heights. The accuracy of fit was verified by field work and aerial photographs, and the level of error was felt to be acceptable for the requirements of the UMV simulation. Culture features were then encoded to the nearest grid point (using a 1-m resolution data set). Interestingly, the same software allowed users to add 'bogus' features to the DTM, thus catering for updates and for experimentation.

Such activities allowed any OS digital data to be encoded to a high-resolution gridded set. Moreover, aerial surveys of selected areas of interest could be commissioned by the client and produced to the OS specification. The processed data represented $1 \, km^2$ and was stored as a set of rapid-access data files in a hierarchical database. Finally, other specialized, high-resolution data sets were encoded by digitizing large-scale maps. This was done for example, for soil survey maps in order to provide the artificial intelligence (AI) software component of the UMV with a 'going' or mobility database.

(c) *Data synthesis by mathematical functions.* The third method of data collection involved the synthesis of topographic and other information by means of computer simulation. This was to provide vehicle simulation and display software with a 'meaningful' data input while other data sources were still in preparation. Initially, a set of mathematical functions was derived and encoded, and, by experimentation, several synthesized terrain 'surfaces' were produced. All functions were based on the 'right-handed' Cartesian axis system where the x-axis points north, the y-axis points east, and the z-axis points down. There is a minor disadvantage here in that all heights above the origin are negative. However, this convention accorded with existing matrix manipulation software and allowed the use of familiar and well-documented mathematical notation (Blakelock, 1965; Britting, 1974).

The mathematical surfaces varied in complexity. A simple sine-wave 'eggbox topography' (Fig. 21.2a) was employed with successful results, using an equation of the form:

$$z = -a(\sin^2 bx)(\sin^2 by)$$

where a is the relief amplitude (e.g., 200 m) and b is a scaling factor (e.g., 0.3).

More complicated (but no less valid) pseudo-terrain surfaces were constructed using a combination of mathematical and statistical functions. This approach involved combining two 2D curves to produce a 3D surface, which was then combined with other curves in an additive process. An example composed of five surfaces is shown in Figure 21.2b. Although some interesting results were obtained, the methods provided only limited inputs to the vehicle simulation.

(d) *Data synthesis by geomorphological simulation.* A technique was devised whereby users could obtain 'geomorphologically-sensible' 3D digital terrain data both cheaply and rapidly. This involved a specially-constructed computer model of

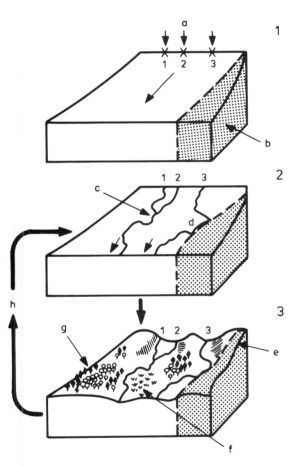

Figure 21.3 Theoretical sequence of synthesized terrain evolution.

Key:
Phase 1: Primary surface and other parameters initialized
      a—Stream entry positions established on highest surface edge.
      b—Band of resistant rock defined.
Phase 2: Stream patterns established by random-walk process.
      c—Stream 1 becomes tributary of Stream 2.
      d—Stream 3 avoids resistant lithology.
Phase 3: Slope profiles established.
      e—Additional slope retreat on steep ground to simulate instability.
      f—Evaluation of terrain features and attributes.
      g—Wooded areas established on 'game-of-life' basis.
      h—Feedback loop to Phase 2.

landscape evolution (Griffin, 1987). The approach adopted a computer program that simulated the establishment of a stream network on a tilted surface, with accompanying fluvial downcutting and slope adjustment. This was achieved by an interactive mechanism that combined deterministic and stochastic processes with geomorphological theory. The effects of differentially resistant lithology and wooded areas were also simulated (Fig. 21.2c). Although more detailed process-response models have been implemented (Anhert, 1977, 1987; Armstrong, 1976, 1980; Craig, 1980), the rapid method (called GEOMIX) allowed data sets to be tailored to specific UMV tasks by altering initial parameters prior to program execution. Moreover, the results—a matrix of high-resolution altitude and mobility data—could be produced in a matter of seconds, and plotted either as a relief (contour) map or as a 3D view (Fig. 21.2c).

The GEOMIX algorithm was evolved in preference to alternative computer-generated artificial landscapes, such as the 'fractal' computations of Mandelbrot (1982) and others (Fournier and Fussell, 1982), since these were founded on purely stochastic principles regardless of the visually impressive and sometimes unrealistically spectacular results obtained.

(iv) Software. By 1987, a suite of software routines designed to manipulate the DTM had been designed, implemented, tested, installed and accepted at the users' site. These routines allowed the user to create, update and maintain the digital terrain models required for the UMV project, and included the software packages shown in Fig. 21.4, together with extensive documentation. Some functions, such as the contour plotting software, were originated in the early 1970s before any reliable commercially-available equivalent was felt to be widely available.

(v) Artificial intelligence. The DTM was intended for use with on-board vehicle artificial intelligence (AI) software. Such AI software, written in the languages PROLOG and LISP, could utilize the DTM for route planning and as part of the intended integrated navigation system. The route planning software was based on various heuristic algorithms which could find routes by means of roads and over 'unprepared' terrain.

21.2.3 EAMACS (Case Study 3)

In parallel with other terrain work, EASAMS produced the EAMACS (EASAMS Architecture for Management and Control Systems) graphics system in the early 1980s. This was a response to the relative lack of processing power and colour displays associated with early DTMs and Geographic Information Systems. The EAMACS System utilized a dedicated PDP 11/73 computer with customized graphics processing boards. The display device employed a touch screen MMI (man-machine interface) for a single display or combination of rasterized, vector or video pictures. The EAMACS system incorporated many of the algorithms for digital map display and terrain analysis described above. The main application and market area of this system was in the C3I field, where the potential for rapid processing and display of tactical information on a map background has long been realized (Neel *et al.*, 1982; Ogier-Collin, 1988). Advanced functionality such as 'panning' and 'zooming', and digital removal or enhancement of cartographic features, were found to be useful in operational environments.

Current developments are making use of relational database management systems, and proposed industry standards such as the X-windows system and secure-UNIX operating environments.

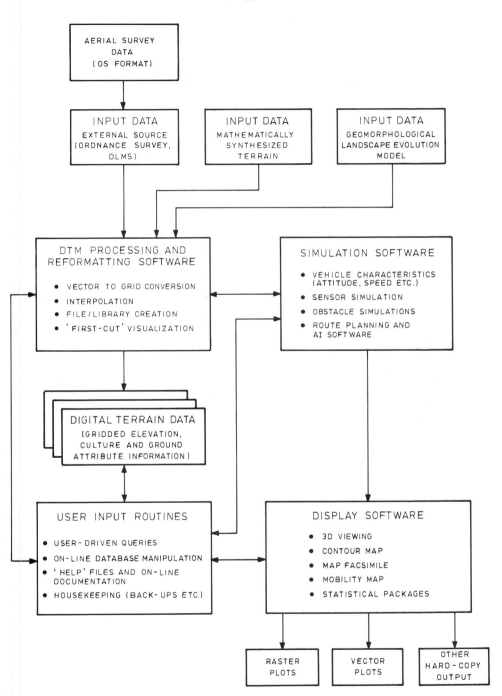

Figure 21.4 Structure of DTM and utility software for the UMV project (simplified).

21.3 Discussion

21.3.1 Overview

This section discusses some of the problems and constraints encountered by terrain analysts at EASAMS with reference to the application 'case study' examples described above. Although many aspects will be recognized by workers in the civil field, there are felt to be some special problems encountered in the military sphere. Systems incorporating digital terrain models and other terrain-related software may be used as tools in operational analysis, or they may be implemented as in-service systems. In the case of the latter, implementations vary between robust, high-performance systems for use in aircraft (e.g., for terrain-referenced navigation) and battlefield C3I systems (Kennie and McLaren, 1988). There may be constraints imposed by the processing power, size or weight of the hardware, which itself may be 'ruggedized' or may conform to the appropriate military specifications for shock, temperature, humidity, electromagnetic (EM) environment and so on.

Military software requirements for security (encryption, access control and operating system limitations), interoperability, robustness and portability, add to the complexity and cost of such systems. Similarly, system designs which cater for high-integrity error-checking and other defensive software, so-called user-friendliness and redundancy will also incur high development and maintenance costs.

21.3.2 Performance aspects

(i) Volume of data. The storage, processing and display of large volumes of terrain data poses one of the most severe and common problems to the software engineer. Consider, for example, the fairly modest requirement of 25-m gridded points over an area 200 km × 200 km. This gives 6.041×10^6 datum points, which (at 2 bytes per point) yields a storage overhead of about 13 megabytes, with associated data transmission and processing penalties. Current trends point to the use of optical storage devices for large volumes of data. Since, at present, optical disks (for example) tend to be read-only devices (Bouwhuis, 1985), such technology is being employed initially for the static components of military terrain information systems, such as elevation data, while more dynamic data components, such as threat or tactical information, may reside on more conventional magnetic media.

(ii) Processor speed. Another limiting factor is the power, performance and configuration of the Central Processing Unit (CPU) and display devices being used. Machines capable of running at more than 50 MIPS (million instructions per second) are now widely available, but they may be prohibitively expensive. Moreover, benchmarks of processing speeds may provide misleading results which are sometimes inappropriate for particular DTM applications.

For flight simulation applications involving terrain models, the processing requirements may be formidable where realistic colour video images are concerned (Yan, 1985).

(iii) Parallel processing. Concurrent or parallel processing concepts are generally implemented as important and useful methods for solving time-critical and large-volume data processing problems. Essentially, concurrency is achieved, both in software and hardware terms, by avoiding serial dependency and instead 'pipelining' or partitioning tasks whenever possible (Spriet and Vansteenkiste, 1982).

Work on the UMV project reflected the general requirement for more processing power, where at first FPS AP-100 series array processors were utilized to carry out limited arithmetic operation in parallel. This gave way to Meiko computing surfaces ('transputers') and an exposure to concurrent programming languages, such as Occam (May and Taylor, 1984; Fisher, 1986). It is worth noting that the large-array or matrix operations that tend to be carried out during DTM processing were found to be conveniently incorporated in concurrent designs. Other languages, such as Ada, offer concurrency (or 'tasking') as a built-in feature (Barnes, 1984), although operating systems, design tools (Brubaker and Case, 1986) and debugging tools must improve before significant gains can be made.

(iv) Display devices. Processing problems are encountered in the area of display devices, where the compromise between colour/resolution and processing speed is well known. The combination of vector images with a raster-scan map background display is a useful technique for presenting high-resolution linear detail (e.g., roads, contours, symbols and feature outlines) while optimizing processing and storage resources. Raster-vector integration and other display processes (such as windows) may be executed by dedicated processors on board the display device. Typically, high-performance graphics workstations are in general use for military terrain modelling and map display applications.

(v) Real-time processing. Real-time processing is characterized by the presence of deadlines, where failure to meet a deadline is considered a system fault. In effect, the use of such a term is an admission that digital computers are sometimes unable to cope with the demands of a 'real' world. Such a time-critical interaction with real events is demonstrated, for example, by a real-time digital terrain navigation and display system on board high-performance military aircraft. Some of the UMV software was targeted at a real-time environment. In these cases, great care was taken at the design and implementation stage to ensure that no overall system degradation would take place. Hopefully, real-time DTM processing problems will be eased by the adoption of more powerful hardware, more efficient software and the use, where appropriate, of concurrent methods.

21.3.3 Standardization

Standardization is clearly of importance to any system distributed between collaborating users, whether they are divisions in an Army Corps or the countries in the NATO alliance. For military terrain-related systems, particularly those intended for future field service, efforts are currently being made to adopt certain hardware and software standards. Some of these measures are well known. For example, the language Ada is currently being used in the UK defence scene. Ada has been adopted by the US Department of Defense and is a standard for both NATO (North Atlantic Treaty Organisation) and the UK MOD. The language was designed in order to provide cost-effective reliability, maintainability and portability for 'embedded' systems. Other standards and recommendations are emerging for 'Open' systems where hardware and software from different vendors can be designed to function together in an integrated environment. At EASAMS, individual projects are sometimes required to conform to design methods such as SSADM (Structured System Analysis and Design Method) and software components such as Ada, SQL (Structured Query Language) and GKS (Graphical Kernel System).

Currently, efforts by the UK Military Survey and NATO partners are being directed towards producing a range of standard digital geographic data products (Military Survey 1988). However, true interoperability lies in the future, in spite of commercial and military pressures to achieve such standardization.

21.4 Conclusion

This chapter describes some military applications of digital terrain models. Some of the problems associated with the applications are identified along with proposed solutions against a background of advancing technology and changing requirements. The three case studies outlined are felt to mirror these advancing technologies and the shifting problems of definition and analysis over the last two decades. Work at EASAMS has reflected the problems and benefits of the terrain modelling environment during this time.

Acknowledgements

This unclassified work is published by kind permission of EASAMS Limited. The opinions expressed are those of the author alone. The author would like to express his thanks to Miss J.A. Hopkinson, Brigadier A.W. McKinnon, Mrs C.A. Hose and Mr P.O. Blanchard of EASAMS Limited.

References

Ahnert, F. (1977) Some comments on the quantitative formulation of geomorphological processes in a theoretical model. *Earth Surface Processes* 2, 191–201.
Ahnert, F. (1987) Approaches to dynamic equilibrium in theoretical simulations of slope development. *Earth Surface Processes* 12, 3–16.
Armstrong, A. (1976) A three-dimensional simulation of slope forms. *Zeitschrift für Geomorphologie, N.F. Supplementband* 25, 20–28.
Armstrong, A. (1980) Soils and slopes in a humid temperate environment: a simulation study. *Catena* 7(4), 327–338.
Barnes, J.G.P. (1984) *Programming in Ada.* International Computer Science Series, London.
Blakelock, J.H. (1965) *Automatic Control of Aircraft and Missiles.* John Wiley, New York.
Bowhuis, G. (1985) *Principles of Optical Disc Systems.* Adam Hilger, Bristol.
Britting, K.R. (1974) *Inertial Navigation Systems Analysis.* John Wiley, New York.
Brubaker, D. and Case, D. (1986) Task partitioning eases concurrent programming. *Computer Design,* 83–87.
Craig, R.G. (1980) A computer program for the simulation of landform erosion. *Computer and Geosciences* 6, 111–142.
Fisher, A.J. (1986) A multi-processor implementation of Occam. *Software Practice and Experience* 16(10) 875–892.
Fournier, A. and Fussell, D. (1982) Computer rendering of stochastic models. *Comm. ACM* 25(6), 371–384.
Griffin, M.W. (1987) A rapid method for simulating three-dimensional fluvial terrain. *Earth Surface Processes* 12, 31–38.
Griffin, M.W. and Gilmour, J.N. (1981) Morphological imaging techniques and the presentation of large-scale erosional features of the Land's End Peninsula. *Proc. Ussher Soc.,* 5(2), 222–226.
House of Lords Select Committee on Science and Technology (1983) *Remote Sensing and Digital Mapping,* (Vol. 1). HMSO, London.
Kennie, T.J.M. and McLaren, R.A. (1988) Modelling for digital terrain and landscape visualisation. *Photogrammetric Record* 12(72), 711–745 (October).
Kennie, T.J.M. and Petrie, G. (1987) Terrain modelling in surveying and civil engineering. *Computer-Aided Design* 19(4) 171–187.
Mandelbrot, B.B. (1982) *The Fractal Geometry of Nature.* W.H. Freeman, San Francisco.
Military Survey (1988) Military Survey Digital Product Information Sheets. Military Survey, Ref: STU/40/25/7/2, September 1988.
May, D. and Taylor, R. (1984) Occam—an overview. *Microprocessors and Microsystems* 8(2) 73–79.
Neel, E.E., Stuttz, S.J. and Tysszka, R.J. (1982) Digital terrain analysis data—a critical element in future tactical C3I. *Signal,* March 1982, 51–54.
Ogier-Collin, J. (1988) The application of digitalized geographical information to C3I systems. *Proc. AFCEA Europe Paris Symposium 1988,* 61–63.

Ogorkiewicz, R.M. (1986) Automated unmanned and robotic tanks. *International Defence Review*, 1283–1290.

Smith, G. (1986) War without people. *Soldier Magazine*, 14–16.

Spriet, J.A. and Vansteenkiste, G.C. (1982) *Computer-Aided Modelling and Simulation*. Academic Press, New York.

Yan, J.K. (1985) Advances in computer-generated imagery for flight simulation. *IEEE Computer Graphics and Applications* **8**(6), 37–51.

22 Graphical display and manipulation in two and three dimensions of digital cartographic data

L.W. THORPE

Introduction

Digital cartographic data capture, display and manipulation is expanding within both civil and military mapping organizations, such that the creation of digital geographic databases is becoming a prerequisite for the new generation of civil and military communications, command control and information (C3I) systems. SD-Scicon's work has been committed to the establishment of suitable databases of this material for efficient retrieval, manipulation and display in both two- and three-dimensional form in conjunction with user information of asset deployment. To this end, SD-Scicon have produced the product VIEWFINDER, which this chapter describes.

The generation of the three-dimensional information and derived mappings of slope, aspect and intervisibility become available by the combination of contour or randomly sampled height data of the region, with the associated topographic or cultural features which make up a conventional map. The following sections highlight the processing of digital map data pertinent to terrain shapes from various capture modes, with the databanking of the cultural features for use in a mission planning, communications planning, navigation and asset deployment environment.

22.1 Digital terrain data capture sources

Three sources of digital material are described which form the basic inputs to digital terrain modelling systems.

22.1.1 Discrete samples
This source of digital terrain data is the most important and the remaining sources reduce to this variety during processing. This type of data is represented by a single triplet of numbers which define the location in three- dimensional space of a point on the surface under consideration.

This surface may be any single-valued function which represents a land surface, a sea surface or perhaps an underground surface formed as the interface between two types of rock. To adequately represent this function in three-dimensional space, the location and height/depth of a suitably large number of points on the surface must be measured, or sampled. It is important to define the surface with the correct number of sample points, such that:

(a) the surface is correctly sampled with respect to the sampling theorem;
(b) the minimum number of samples are obtained that are necessary from an economic standpoint;
(c) the total area of the surface under consideration is sampled at a density (sample/unit area) commensurate with both sampling theorem and economic considerations.

The sampling theorem states that the highest frequency, or shortest wavelength components involved in the Fourier transform of the surface must be sampled at least twice per cycle. This theorem can be interpreted as saying that if the surface is made up of long wavelengths, e.g., rolling hillsides, then the samples can be taken, on average, at a separation less than half the distance between the tops of the hills. If, however, the surface is undulating and irregular, with very steep-sided valleys, then the surface must be sampled at a rate at least twice per cycle of the shortest-wavelength undulations and irregularities that are required in the model of the surface. Under the constraints of the sampling theorem and with a knowledge of the detail that is required on the surface, the number of sample points required to define the surface can be estimated.

Similarly, the cost per sample, in terms of survey equipment, mapping or photogrammetric resources, man effort, etc., involved in the data collection, can be divided into the available budget, to indicate the number of samples that are possible. If the budget for data collection does not allow sufficient samples as dictated by the sampling theorem, then the specification of the detail over the whole surface area may have to be relaxed, or the area of coverage reduced. It is clear that these two factors are in conflict, and their implications must be carefully considered prior to undertaking detailed sampling of the surface. The best-known examples of discrete sample points on surfaces are:

(a) Soundings on hydrographic charts, where the depths of the ocean have been measured at a sequence of sample points taken along the path of the survey vessel as it moved over the ocean surface. This path may have been organized on a criss-cross basis to provide the sampling rate determined by the considerations given above. This would be typical of a detailed hydrographic survey of an area, or it may be a routine measurement of the depth of the ocean taken along the course of a vessel for safety and monitoring purposes.
(b) Borehole data obtained by mineral prospecting companies in their investigation of substrata for oil- and mineral-bearing rocks.

In each case, the information representing the surface is formed from random sampling of the surface in the three axes of the coordinate system, and this is recorded as discrete point information represented as (X, Y, Z) triplets.

22.1.2 Photogrammetric sampling
This source of digital terrain data is obtained via the analysis of overlapping photographic images of the surface obtained from aircraft or satellites. The area being investigated is overflown by a survey aircraft, which takes a series of overlapping photographs of the region. The resulting photographs are subsequently analysed in stereo-viewing systems connected via shaft encoders to digital recording equipment. The human operator focuses the stereo viewing system on to the three-dimensional

image created by two of the overlapping photographs. By this procedure, a particular level or height value is selected. Then by skilful manipulation of the 3D image under his view, a series of points of the same height value are identified as a continuous curve defining a contour line. This procedure is continued until the required series of lines are identified covering the area of interest to the required detail. This material is captured, in a digital format, as a string, or series of strings of (X, Y) coordinates, all associated with a particular Z value, representing the contour height. On older photo-grammetric equipment, without digital interfaces, map output will then require digitization.

22.1.3 Existing contour maps
This source of terrain data perhaps represents the largest source, as a very large number of maps already exist in printed form. The contours on these printed maps have often been created by photogrammetric methods, (without the digital interface). Existing maps are converted into a digital format by digitization. This is achieved using a variety of techniques, each of which has associated advantages and disadvantages.

(a) Hand digitizing, whereby a person follows the contour lines by hand on a flat digitizing table, coding each line in turn with its height.
(b) Mechanical line following, whereby a laser-controlled automatic machine follows the shape of the line and associates the code either automatically or by human voice or finger keying.
(c) Scanning, whereby the paper map is scanned in a raster fashion with a flying spot to identify each 'pixel' on the map containing contours. These pixels are subsequently converted to vectors and the corresponding height codes associated with them are defined.

When the contours are in digital format they are again defined as a string or series of strings of (X, Y) coordinates all associated with a particular Z value representing the contour level.

22.1.4 Conversion of contours to discrete samples
Photogrammetric and digital map data generally occur as line strings of (X, Y) coordinates with an associated Z value for the contour level. It is necessary to convert these line strings into discrete samples by associating the particular Z value for the line string with each component point of the line to form (X, Y, Z) triplets. It can be seen that this form of discrete data point input is formed from random (X, Y) data but quantized Z, to produce the triplets required for further analysis.

22.2 Digital terrain model creation

VIEWFINDER provides a comprehensive data reduction facility for the class of problems relating to surfaces in a three-dimensional space. To use these facilities, it is necessary to convert the array of input (X, Y, Z) data points into a more suitable representation of the surface which can be efficiently and effectively processed by computers. The most convenient representation of a surface for such processing is in the form of a set of height values on a uniform square grid distribution. The numerical approximation, from the irregularly distributed points to height values on a regular grid, provides an efficient and accurate transformation which generates a representation of the surface that best fits each of the input data points. This is achieved using a plane-fitting technique utilizing least-squares methods. This is based on the selection of the nearest set of at least eight sample points distributed in the eight quadrants around the matrix node under consideration. Using these points, the best-fit plane is calculated to pass through the input points from which the value of the node is then interpolated. This process is continued for the whole area to produce the digital terrain matrix.

Once the matrix has been calculated, it can be used for a large variety of purposes. The following paragraphs give a number of examples, which are not exhaustive.

(i) Calculation of contours. From the matrix, the shape of the contour figure field can be produced. This can be done by using linear interpolation through the grid cells, or by an alternative method producing an approximation to a smooth surface based on derivative estimates at grid intersections. The latter produces a more desirable result, and the validity of the matrix can be checked by this algorithm whereby the contours interpolated from the matrix can be compared with the input data (see Figs. 22.1,

Figure 22.1 Input contours of 20 km square area of Yorkshire.

Figure 22.2 50 m recomputed contours of 20 km square area of Yorkshire.

22.2). Similarly, the conversion of input contours in feet, to output contours in metres can also be performed.

(ii) Calculation of cross-sections. From the matrix of terrain heights, the height at any desired location in the area of interest can be interpolated. This is done by linear or curve fitting algorithms, but instead of computing the (X, Y) position for a given height, as for a contour, the height is calculated for a set of (X, Y) coordinates which define the vertical cross-section, or profile through the surface. Generally only two sets of (X, Y) coordinates are specified to define a straight line cross section, but a number of straight line cross-sections can be combined together to define a cross-section through the surface, along a curved line.

22.3 Three-dimensional views

From the matrix of terrain heights, representations of the shape of the surface can be generated in three-dimensional space.

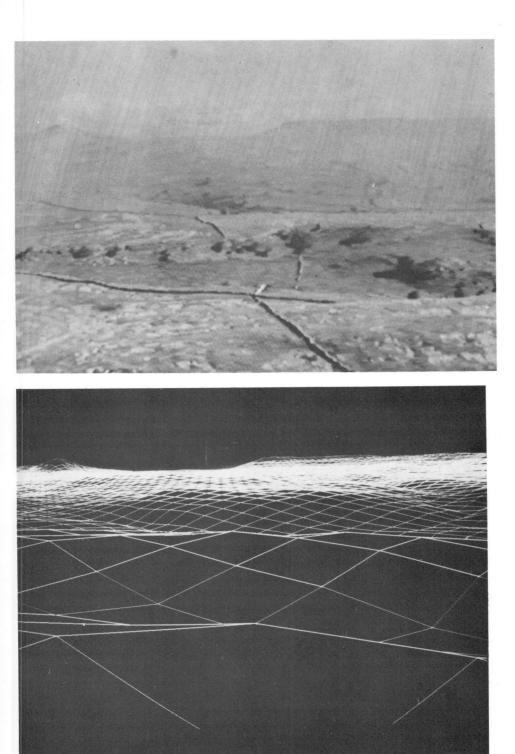

Figure 22.3 (a) Real scene looking NE over area of interest. (b) Computer-generated scene equivalent to (a).

(i) Orthographic projection. An orthographic view represents the surface that contains no distortions due to the distance of the point on the surface from the observation point since the eye point is deemed to be at infinity. Profiles parallel to the X or Y axes are drawn. Each is of constant length (representing the width of the matrix), while the length of the matrix is represented by the set of profiles set back and offset from each other along a line at 45° to the horizontal.

(ii) Perspective projection. A perspective view of the terrain matrix produces a three-dimensional representation that contains a distortion due to the distance of the object, from the observation point, or eye point. A perspective view of the matrix looks very similar to the equivalent orthographic view, particularly if there is no significant difference between the ranges of the set of points on the surface and the eye point.

Perspective views of the surface provide a means of presentation of the matrix information that is particularly easy to comprehend. If the eye point is taken 'on to the matrix', then views of the surface that correspond closely with the actual scenes observed from the equivalent eye point on the ground, can be calculated. Compare Fig. 22.3a with 22.3b which was computed using the SD-Scicon VIEWFINDER Perspective View facility. Having defined the eye point, it is then necessary to define the direction of view and the angle of the cone of vision around this direction.

22.4 Line-of-sight diagrams

Using the digital terrain matrix, diagrams can be generated which display the area of the ground that can be seen from a particular eye point, either on or off the matrix.

Figure 22.4 Intervisibility matrix showing visible and invisible nodes for Fig. 22.3b.

(i) Star diagram. If the location of the eye point is specified in the three-dimensional coordinate system of the matrix, as (X, Y, Z), then a series of cross-sections through the surface can be calculated. For each of these sections, the (X, Y) coordinates of the eye form the start point, while the end point is chosen such that the radial lines in the X, Y plane from the eye point are generated at a set angular separation for the whole $360°$ around the eye point. For each cross-section, the eye point is positioned at the correct height (Z), and lines of sight from this point on to the surface are produced, thereby identifying the points on the cross-section that are visible, and the points that are occluded. The disadvantage with this type of line-of-sight diagram is that, as the range extends, so the resolution of the image degrades.

(ii) Intervisibility matrix. The intervisibility matrix does not exhibit the disadvantage of the star diagram, since every node on the digital terrain matrix is analysed to determine if it is visible or not from the defined eye point. Within the VIEWFINDER system, the intervisibility diagram is computed for each perspective view simultaneously such that a comparison of nodes visible on the 3D view, with those displayed on the intervisibility plot, provides an easy means of establishing the range of features on the surface from the eye. Figure 22.4 shows the intervisibility plot corresponding to Fig. 22.3b and identifies the zone behind the observer's eye point, the zones outside his field of view and the visible and occluded zones.

22.5 Slope and aspect

From the digital terrain matrix, the angle of slope of the ground can be calculated and displayed either as contours of constant slope or by colours.

The VIEWFINDER system calculates the maximum angle of slope at each grid node and the aspect of that slope. The aspect of the slope is the direction (with respect to north) in which a sample of liquid would move under the influence of gravity down the angle of maximum slope. The direction of the colour-coded vector indicates the aspect of the slope as measured from the direction of north.

22.6 Further discussion of the use of perspective views

From the above description, it is clear that there are three important parameters to be defined in order for a 3D image to be calculated:

(a) the eye point;
(b) the direction of view;
(c) the cone of vision around the direction of view.

The manipulation of these parameters to advance the use of digital terrain modelling for military applications is considered prior to a more specific discussion of military applications which follows in the next section.

(i) Change of cone of vision. If the eye point and the direction of view have been defined, then the scene in view is equivalent to that obtained by looking through the viewfinder of a normal camera system (see Fig. 22.3a, 22.3b). If the camera system was fitted with a widely variable focal length lens, or zoom lens, then by changing this focal length an ever more detailed picture would be obtained. The change of focal length directly changes the solid angle viewed through the viewfinder of the camera. By manipulation of the solid angle around the direction of view, the VIEWFINDER system can simulate this widely variable focal length lens, such that the angle of view

L

can been changed from 90° to 1° and a small region of hillside, at a large range, 'fills the field of view' of the camera.

(ii) Change of eye point. The previous section discussed the manipulation of the cone of vision as a smoothly varying parameter. To follow this philosophy with the eye point, as distinct from stochastically changing the observation point and/or field of view to produce random views, the VIEWFINDER system is capable of generating a connected sequence of views as a simulation of an observer flying or moving via some vehicle over the terrain. If the sequence of views are presented as a moving picture on film or television, then about 25 frames must be displayed per second to produce smooth continuous motion. Considerable computing power and other specialized hardware is required to generate such animation.

22.7 Addition of texture and culture

To increase the realism in the computer-generated scene, it is necessary to include both the texture of the surface and other details, generally man-made objects, which make up the set of features on the terrain surface. This requirement necessitates the classification of the various regions of the surface, for example:

Vegetation	Woods	Deciduous
		Coniferous
		Mixed
	Crops	Corn
		Potatoes
		Mixed
Geology	Soil	Clay
		Sand
		Rock
		Silt, etc.

Having established the types of surface cover and their textures, the boundaries of these areas must be digitized, and the facets of the three-dimensional view filled in with the required colour and texture, up to the required boundaries.

The data for culture addition to DTMs is mostly acquired from photogrammetric sources but also from vertical or oblique aerial photographs. This is currently a slow and labour-intensive process, and research is under way to automate the process. Approaches include the use of artificial intelligence (AI) techniques. Only the major simulator companies currently offer any working systems with culture addition, but with the advent of low-cost array processors and high-speed graphics devices, these capabilities become available at significantly lower cost than full mission simulators.

22.8 Defence applications of DTM techniques

VIEWFINDER is a planning tool which will become an essential aid to intelligence and operations staffs, particularly before hostilities, but which will provide increasing assistance during operations as defence forces acquire further ADP capabilities.

22.8.1 Deployment options
The deployment of friendly forces is reactive to the perceived threat. The element of surprise available to an enemy indicate that the plans of friendly forces should be

secure, effective and flexible. This implies the rapid investigation of several deployment options. The previous section explained how the system is able to interact with queries and demands from the user. This provides the facility for preliminary reconnaissance without the need to undertake a cross-country or helicopter mission.

Provided friendly intelligence has acquired data on the deployment of hostile assets (primarily surveillance and target acquisition equipment, and Electronic Warfare (EW) sensors), it is possible to identify whether an area is under surveillance or shielded by natural features or culture. The risk of occupying an area can be estimated and, if necessary, the appropriate responses can be determined: accept the risk of low, eliminate the hostile surveillance, or move to a less vulnerable location.

VIEWFINDER allows the detailed deployment of assets to be made by commanders before a position is occupied. For example, the position of a radar site can be adjusted to provide maximum protection of the main lobe consistent with the best coverage of the threat; artillery can be deployed to avoid crest clearance problems; remotely piloted vehicle (RPV) missions can be planned so that the communications links are established precisely when required. Subordinate commanders can be briefed in detail during a period of tension. Any adjustments to the existing plans can be evaluated, enabling the transition to war to proceed smoothly. SD-Scicon does not consider that any system of this nature can replace the need for a commander, whenever possible, to view the ground and the environmental conditions which apply currently. However it is believed that preplanning using VIEWFINDER will enable a commander to confirm his thinking, and to make whatever tactical adjustments he wishes, in the light of the detailed technical assistance at his disposal. This would provide a more cost- effective utilization of time and resources of the commander and his staff.

During hostilities, VIEWFINDER is available to the planning staffs for preliminary evaluation of other phases of battle, and to operations staffs for real-time mission planning. This would provide quicker reaction, a wider choice of options, and greater threat to an enemy via the increased flexibility which sound planning allows.

Specific examples in which VIEWFINDER may be applied include:

(a) Sensor system siting—where ground surveillance radar is deployed without compromising friendly radar positions. Analytical data provide sufficient information upon which to issue deployment orders with realistic six-figure grid references.
(b) Communications planning—the system is ideal for securing optimum paths where the use of high ground and relay is permitted or for identifying alternative paths which could reduce the probability of hostile intercept.
(c) Fighter Ground Attack (FGA) mission planning—within a mission plan for FGA sorties, the VIEWFINDER system provides the optimum route to weapon release, the time at which AD suppression activity should commence and the escape route post-attack.

22.9 Conclusion

This chapter describes the use of a software package that is independent of host machine and graphics device. It uses basic map data that can and will be used in the printing process of maps, such that the data used for interpretation, planning and decision processing is the same as printed on the conventional paper map. The system then provides the user with an interactive planning and simulation aid for many types of military and civil situations.

23 Applications of Laser-Scan digital terrain modelling systems

D.R. CATLOW, D.J. GUGAN and T.J. HARTNALL

23.1 Introduction

The computerization of cartography began in the 1960s with the processing of vector map data, with the production of paper maps as the end objective. As map data began to be used for analytical purposes, Digital Terrain Models (DTMs) were generated from digital contour data. These were typically used for cut-and-fill computations and terrain visualization, for applications such as highway design. With the growth of Geographical Information Systems (GIS) in the mid-1980s, DTMs were again pushed into the background, since the primary applications of GIS were concerned with linking vector data to relational databases for applications such as land parcel information systems and utilities records systems. Today, DTMs are becoming more widely used products in GIS and digital mapping. The terrain data is being used in conjunction with other information, such as vector data, satellite imagery and attribute data, for an increasingly wide range of applications. This chapter considers three modern applications requiring the analysis of DTMs—a military terrain analysis/C3I (Command, Control, Communications and Intelligence) application, environmental impact analysis, and radio communications planning.

23.2 Processing DTMs

DTMs are generally stored in a raster format. This is a regular array of values where the planimetric position of a data point is inherent in its position in the array (providing the origin and point spacing of the data are known). Data is entered into this format in a number of ways, including:

(a) Conversion from other raster formats, such as US DMA DTED (Defense Mapping Agency Digital Terrain Elevation Data), or OS (GB) NTF (Ordnance Survey National Transfer Format).
(b) Conversion from digital vector data. Data available as, for instance, contours, may be converted to a grid format by an interpolation method. The method used by the Laser-Scan DTMCREATE package is the Delaunay triangulation technique described by McCullagh (1988).
(c) Direct measurement using a photogrammetric stereoplotter. The Laser-Scan/Kern interface allows direct measurement of points on a regular grid. Irregular points (such as breaklines or data measured by progressive sampling) may also be collected and incorporated into the DTM using the triangulation method.

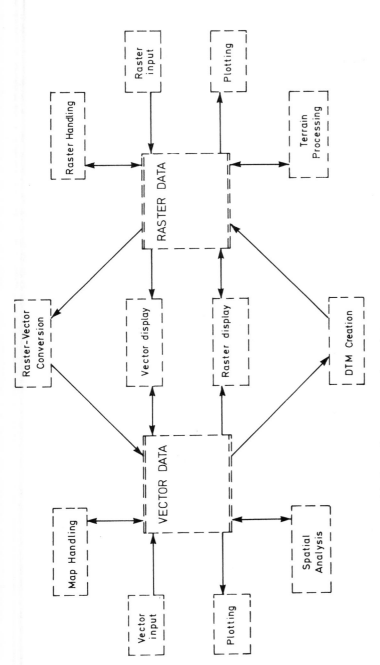

Figure 23.1 Schematic diagram of the Laser-Scan geoprocessing system.

The raster format in which a DTM is stored (which should be able to cope with integer or real data values) allows easy manipulation. Processing utilities for operations on terrain data within the Laser-Scan system (shown schematically in Fig. 23.1) include:

(a) *Slope and aspect generation.* By applying a local operator to the data, (usually a 3 × 3 operator), the gradient in the x and y directions can be computed. The value and direction of maximum slope is derived from these, usually output into user specified classes. An unusual application of this information is possible in countries where the value of land used for vineyards depends on its slope and aspect.

(b) *Shaded overlays.* Combining the slope and aspect information allows a hill-shaded model to be generated. The light source can be placed in any direction and elevation, although the usual position for cartographic purposes is northwest at 45°. The shaded overlay is an effective way of presenting elevation data to a viewer, enabling the perception of form and trends in small-scale surface relief. The use of this technique can be seen on Ordnance Survey 1:63 360 scale tourist series maps.

(c) *Visibility.* Intervisibility may be considered in two ways—the visibility between two points, computed by one of a number of common line-of-sight algorithms, and the cone of terrain visible from a point. The former may be used for the siting of microwave communications transmitters and the latter for the siting of

Figure 23.2 Standard DEC VAXstation computer displaying a perspective view of a DTM of Hawaii with colour-coded height intervals.

observation posts. For many applications, it is necessary to take into account factors such as curvature of the Earth, atmospheric refraction or the characteristics of the transmitter, in the case of radio propagation studies.

(d) *Three-dimensional views*. These are a common product generated from DTM data. Parallel (isometric) or perspective projections can be used to generate a simple profile view or more sophisticated rendered solid models. The advantage of the former is that they can be output to a simple vector plotter, whereas the latter require an image output device, such as an electrostatic plotter. A common use for 3D views is in flight simulation systems.

Examples of these basic processing capabilities are shown in Fig. 23.2 and 23.3.

During the mid 1980s, the processing of terrain model data was carried out largely in isolation from other forms of geographic data. Today, however, it is essential to process and display all forms of data together. The integrated display of data is carried out within the Laser-Scan LITES2 vector display/editing environment. LITES2 was originally developed as a sophisticated vector editing system, and has recently been considerably enhanced by the addition of raster display and manipulation capabilities. Up to nine raster data files can be input and accessed. These are automatically registered using the origin and grid interval data in the file header. An example of using this capability in a data capture mode is when digitizing map data from a 2D geocoded satellite image, with Z values being extracted from a DTM in the

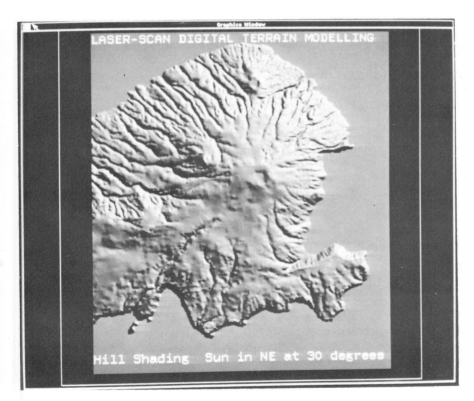

Figure 23.3 Hill-shaded model of the Hawaii DTM.

Figure 23.4 Geometrically corrected SPOT satellite image of South Yorkshire with 1:50 000 vector data overlaid. The LITES2 display/editing environment allows data to be digitized, edited or deleted with reference to the background imagery. Copyright© CNES (1986).

background for each point digitized. A satellite image with vector data overlay is shown in Fig. 23.4.

DTM processing and manipulation is an integral part of a GIS. An essential feature of the system achieved by Laser-Scan is the integration of all GIS functions—data capture, conversion, maintenance, manipulation and output—with in one system. Laser-Scan have standardized on the Digital VAXstation computer and provide the core databases and manipulation capabilities. On-line interfaces are provided to a number of specialist peripheral devices including digitizing tables, Kern photo-grammetric plotters, raster scanners and pen and electrostatic plotters.

23.3 Military applications of DTMs—terrain analysis and C3I

23.3.1 Terrain analysis
Terrain analysis enables the military user to minimize uncertainties in the military decision-making process caused by the ground over which his own planned, or the enemy's predicted movements are to take place (see also Chapters 21 and 22). Marth (1989) identifies six factors of interest for military planning operations: slope, soils, drainage, vegetation, lines of communication and obstacles. Maps, imagery, textual information and any other available sources are used to collect these factors. The terrain analysis takes many forms but the end product is invariably a map, usually presented as an overlay to existing topographic map material.

Terrain analysis is conducted at several levels of detail. At the corps level, the analyst is concerned with generalized thrusts of movement. A typical question might be: 'Will these hills channel the enemy forces into these lowlands?' Or, 'Given the dense deciduous forest in this area, will the minefields be best placed either side of it where the enemy advance will be deflected?' At a divisional level, the commander may be more concerned with assessing where he can conceal his tanks for rest, repair and refuelling or how many of his tanks will become bogged down because of the overnight heavy rain in an area of poor drainage and heavy soils. An assessment of the possible speed of movement is required. For example: 'Will the support vehicles be able to keep pace with the armour given that nearly all the roads are cratered in the area?" Having produced a map overlay to show where movement will be restricted for the tanks, the process has to be repeated for the wheeled support vehicles. An overnight frost renders the predictions obsolete and new map overlays have to be calculated.

Once calculated, the information has to be passed up the chain of command. The geographical information generated from terrain analyses complements the basic topographic mapping over which it is generally overlaid. Boldly portrayed and generalized information is required for presentation to the commander, who in the time of crisis will already be overburdened with information input from other non-geographical sources. The geographical analysis of the terrain forms a backcloth for all the other factors in the decision-making process. Traditionally military terrain analysis has depended on the ready availability of five ingredients:

- Accurate and up-to-date topographic and thematic source mapping
- Accurate and up-to-date non-map data sources, for example bridge loadbearing statistics
- Highly skilled cartographic technicians/terrain analysts
- Abundant chinagraph pencils and rolls of overlay film
- Time

Traditional topographic mapping is available at different scales for different applications, small-scale (1:250 000), for planning at the corps level, medium-scale (1:100 000 and 1:50 000) for the divisional level, etc. The thematic mapping required for terrain analysis, for example slopes, soil types, diggability, vegetation cover, etc., are available either as individual overprints on the topographic map at the required scale or as separate clear film overlays for use with the relevant topographic paper sheet.

Terrain analysis, particularly for prediction of ease of cross-country vehicle movement (CCM), requires simultaneous overlay and assessment of soil character-istics, vegetation cover and type, slope, the results of human activity such as quarry sites and battle information input, for example cratering and minefield location (Fig. 23.5).

It does not overstretch the imagination to see that, after four or five thematic overlays have been placed on the topographic sheet for terrain analysis purposes, the compilation becomes exceedingly cluttered. The thematic polygons often carry quite complex attribute coding which has to be cross-referenced to a legend. The soldier tasked with identifying the best routes through the terrain is usually under pressure and may be working in the back of a box-bodied lorry in the middle of the night. It takes a clear mind to laboriously extract the salient features from the veritable spaghetti of linework and draw them on to yet another overlay with a chinagraph pencil for use at subsequent briefings and planning meetings. If the base topographic mapping to be used at the briefing is at a different scale to that used in the terrain

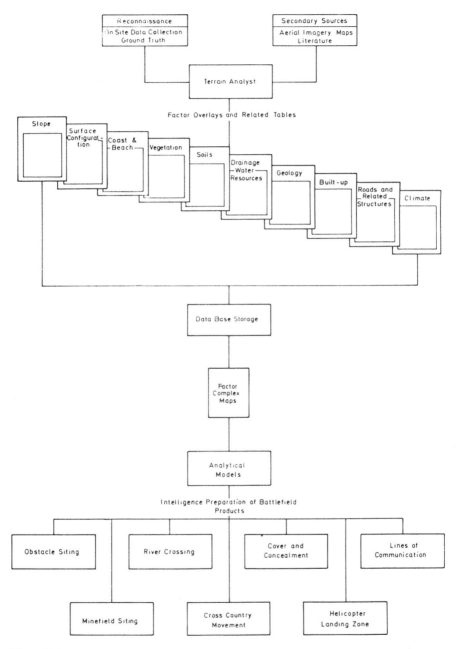

Figure 23.5 Production and use of factor overlays and data tables for manual map overlay (after Falls, 1982).

analysis a laborious manual scale change must be performed. The scene is set for the introduction of digital terrain analysis.

With the ever-increasing availability of digital mapping and DTM data from military mapping establishments, military digitizing programmes have shown a switch of emphasis in the type and manner of data capture. No longer is digital data captured solely to reduce production and revision times of purely topographic map sheets. Design programmes are now in hand for data aimed at multi-role applications, some purely topographic but increasingly topographic and thematic. The wealth of digital terrain models available from DMA (Defense Mapping Agency) are seen to complement vector topographic and thematic digitizing programmes (Fatale and Messmore, 1988).

It is clear from the above that a successful digital terrain analysis system must be capable of simultaneously handling large amounts of both vector and raster data. The Laser-Scan LAMPS system provides such a capability. The developments to LITES2 described above enable the user to simultaneously display and manipulate up to nine vector and nine raster files. Registration between the files is automatic. In this digital system, the files are synonymous with maps and map overlays; the colour video screen with the cartographic technician's light table; the chinagraph pencil is now a screen cursor controlled by a mouse or a puck moved in registration with the source, and the terrain analyst's analogue contour map is replaced by a digital terrain model which may be optionally displayed using coloured altitudinal bands. The altitude for the cursor position may be extracted at any time from the DTM.

The raster data is not restricted to a DTM source, but can be any raster product derived from a DTM (e.g., a slope or aspect map), a scanned image of a paper map source, a thematic product of a matrix combination process or the result of a vector-to-raster conversion. The raster data can be of bit, byte, word, longword or real*4 datatype. Dynamic reclassification of Z values is available for the planimetric display of the raster information. This for example enables the user to selectively display all land above a stated height, or if using a slope map, all slopes which lie in a user specified value range or greater than a threshold value.

Vector data can be digitized directly into LITES2 to complement existing data. The military user lives in an unstable world where topographic features can appear and disappear as a result of battlefield damage. Damage is not restricted to mere bridge demolition or changes to bridge loadbearing capability. Whole forests can be destroyed and open ground rendered impassable by artillery barrage. NBC (Nuclear Biological and Chemical) events create no-go features which have no link with the base topographic mapping but which must be displayed on the terrain analyst's colour video screen. LITES2 graphic and text manipulation facilities enable the user to annotate the screen boldly. The Laser-Scan FRT (Feature Representation Table) lookup table mechanism enables the user to specify exactly how the vector data are to be portrayed. Battlefield obstacles, for example minefields or anti-tank ditches, can be displayed using the standard NATO symbolic conventions. Changes to such obstacles can be rapidly updated on the screen, a far cry from rubbing out chinagraph pencil marks with a wet finger. A raster picture file dump facility is provided to enable the user to take a digital 'snapshot' of the screen contents for direction to a raster plotter of choice or for subsequent redisplay as a raster backdrop file to further data manipulations.

Manipulation of the colour table used in the display is particularly important, for instance if a scanned image of a paper map source is to be displayed in exactly the same colours as used on the paper document with which the user is already familiar.

By using the LITES2 facility to define overlays of screen display planes, it is also possible to independently switch the vector and raster information on and off to alter the rendition of the colours. Many display effects are possible—for example, the user can elect to merge the display of the planes such that background DTM data is visible through the foreground of translucent filled polygons and linework.

The polygon overlay requirement inherent in the terrain analysis process can be tackled in two ways using Laser-Scan LAMPS software. The first route relies on the vector-to-raster and raster-to-vector conversion software within the system. A universal matrix combination utility enables the user to specify a combination expression at run time for mathematical combinations of up to eight separate raster files. DTM or DTM derived data, e.g., slope or aspect information, can be included in the combination process. Run-time classification of the results is available. The matrix resulting from the combination may either be displayed in its raster form or can be vectorized and structured into polygons for input to polygon merging and amalgamation software.

The second, more sophisticated route for polygon overlay involves the use of topologically structured vector data held in LSD (Laser-Scan Structured Data) format. This is an object-orientated structured data format where high-order objects, e.g., road lines and forest areas, are made up using pointers to primitive objects, e.g., edges and faces. The data structure and attribute coding within the data allows rapid vector polygon overlay. Cross-country movement is analysed using the model developed by the Defense Mapping Agency (1983). The results of CCM analysis are displayed as colour-filled vector polygons which underlie vector battlefield, communication and terrain data.

The user interface is one of the most important features of such a system. This must allow easy input of commands, and the display of information in an easily assimilated form. A macro command language enables the user to rapidly issue frequently used commands with a single typed command or a single function button press over a screen menu box. The user can formulate complex data inquiry, display and edit functions using the macro language. Colour tables are themselves defined using the macro language and so in turn are readily manipulated by the user at display time. Colours may be defined to be opaque, transparent, to merge, or to be used additively or subtractively. Analytical display of complex polygonal datasets can often be achieved by manipulation of the colour tables alone. The end result is a display of multiple translucent data sets, with the intersection between polygons clearly visible.

The military terrain analyst's requirement for the simultaneous display of complex topographic and thematic overlay information is met by the integrated raster-vector processing capabilities described above. Laser-Scan FPP (Fast Plotting Program) software enables the user to generate hard copy output at a specified scale. The colour video screen and electrostatic plotter replace the laborious manual registration and transfer of information on to clear film overlays.

So far reference has been made to aspects of terrain analysis which readily transfer themselves from a manual to a digital environment. The production of cross-country movement overlay information can be prepared in advance for different vehicle types and ground moisture conditions.

Standard Laser-Scan software products enable the user to perform tasks which would have incurred a prohibitive time penalty using manual techniques. Identification of reverse slopes for artillery location or vehicle movement shading from SLAR (Side-Looking Airborne Radar) becomes easy, since aspect mapping can be readily generated from the source DTM for the area in question. Intervisibility calculations are straightforward using either fast radial sweep or every-post ray-tracing options using DTM data. The complex intervisibility from multiple observer stations can be calculated and displayed in registration with the planimetric view, which in turn could contain a thematic product overlaying a scanned raster map background. The user gets in tens of seconds what would take several hours to compute manually. The time saved can then be used to experiment with different observer positions, adding limits to the observer's range of view to simulate the effect of smoke, etc.

23.3.2 C3I

Terrain analysis is primarily concerned with the combination of a wide and diverse range of data for the generation of tactical information. In this context, C3I (Command Control Communication and Intelligence) is involved with the application and presentation of this information. DTM data forms an important backdrop to military C3I applications. The terrain analyst's briefing document, so laboriously compiled from multi-overlay sources, comes alive if draped over a perspective or isometric view of the terrain. The commander can immediately 'see' the lie of the ground and identify likely choke points to movement caused by the shape of the terrain. Dead ground, the ground which an observer cannot see, can be visualized either directly from a perspective or isometric view, or can be highlighted by overlaying vector intervisibility cones on to the terrain view. Vector thrust arrows and NATO battlefield symbology are more meaningful if seen in perspective context. The thrust arrows can seen to sweep out of a valley and into the plain in an isometric view. If viewed planimetrically, the commander has either to identify and interpret the contour pattern, or rely on an aide to have accurately generalized the terrain pattern for him using layer-tinted height bands or manual graphic input. Subtle folds of the terrain can be easily missed. A perspective or isometric terrain view provides much more detail and context without cluttering the display with linework. The symbolic representation of a unit is more meaningful if its position on a hilltop or on bluffs beside a river crossing can immediately be seen.

Overlay of the isometric or perspective DTM is not restricted to symbolic representations of field units and battlefield annotation. The views can be overlaid by the results of cross-country movement analysis and topographic data, e.g., forests, drainage and communications. When overlaid with the up-to-date battlefield situation map using NATO symbolic conventions, it provides an even clearer view of the battle situation than could be had from a convenient hilltop. Smoke, poor light and weather conditions no longer obscure a view which, prior to the digital revolution, could only be generated using models made of plaster of Paris or plastic.

Standard Laser-Scan DTM viewing software offers wireframe or solid-surface options for isometric or perspective projections. The solid-surface rendition is fully integrated into the LITES2 display and edit environment. The user can construct perspective views by pointing to the observer's position on the planimetric display and entering the appropriate cone of vision, bearing, observer height and rendering options. The terrain view is then generated in a separate screen window. The perspective screen window can be moved or pushed behind the planimetric window if desired. Current vector data selections from the planimetric view are used to control the content of the vector data overlaid on the perspective view. As in the planimetric view, separate overlays of display bit planes are used to facilitate independent control of the terrain and vector overlay display. Colour tables are user-definable. Work is in hand to enable the user to 'move' vector objects in the perspective view via cursor interaction with the planimetric view. It will then be possible to update the situation map with appropriate NATO symbology in both plan and perspective projection.

Optionally, ground truth (the effect of the height of trees and buildings) can be taken into account as the isometric and perspective views are generated. Thus, if the observer is placed at ground level next to a forest, his view will be appropriately restricted. The provision of more realistic renditions of solid objects is under development. The LITES2 display/edit environment also supports data insertion and update from an external process via mailbox control. It is thus possible for a battlefield situation map to be updated from a remote source while the user performs other display and analysis tasks at the local workstation. All commands are journaled, enabling rapid recovery in the event of a system failure—an important consideration for military applications. Supplementary data files may be read into or exported from the edit/display environment without having to leave the editor. This facility, coupled with the powerful macro command language and colour table manipulation facilities, provides a powerful tool for battlefield situation analysis and briefing. If used in conjunction with a colour screen hardcopy device, up-to-the-minute situation maps can be generated on an overhead transparency in seconds. Figure 23.6 shows an example of data integration for display purposes.

Initial developments of relational database links with the LITES2 display/edit system are now complete. This uses a unique identifier associated with each graphic feature and an entry in a RDBMS index table. Direct linking of LITES2 and the RDBMS ensures fast data retrieval. The user can therefore supplement the spatial display data in LITES2 with tabular data, e.g., the content of logistic depots shown as simple symbols on the screen.

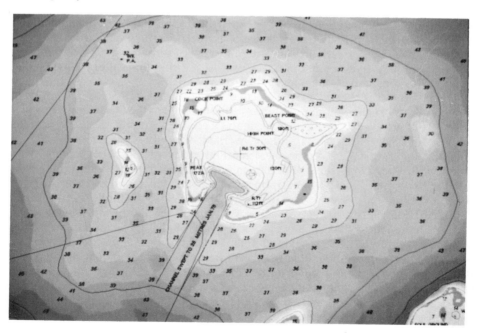

Figure 23.6 The vector data displayed shows an island with depth soundings. A raster DTM has been generated from the depth measurements and is displayed in the background as a depth-shaded model. Display colour table manipulation allows the simulation of tidal states or the switching on/off of layers of vector and/or raster data. External processes can communicate with the display to update the positions of symbols such as ships.

The most exciting extensions of the digital terrain analyst's powers in support of the C3I requirement are in the area of route analysis. DTM-derived data form an important part of the route determination process. Using Laser-Scan LSD data, the user can perform standard on-road or advanced off-road vehicle movement prediction. The user can specify the vehicle characteristics, current ground conditions (e.g., wet, dry or frozen ground) and a start and destination point. Movement can be restricted to roads only, cross-country only, or a combination of the two. Facilities for the display and justification of the chosen route are provided. DTM data can be used to further validate the proposed route against intervisibility criteria. The military application of geographical information, described here, emphasizes the complete and integrated nature of modern analytical applications. DTM data play an important role in these analyses, both as height information (e.g., for perspective views) and for the derivation of other data such as slope, aspect and intervisibility analysis.

23.4 Environmental impact analysis

Environmental impact analysis is concerned with assessing the visual influence of a new development on an existing landscape, and with the prediction of the effect of the development on existing environmental systems and structures. For both applications, a terrain database and associated software, such as for terrain visualization, have a major role to play. The Laser-Scan digital terrain system is now being used by a number of organizations to assess the impact on the landscape of developments such as opencast coal mining, afforestation, transmission line construction and large-scale industrial construction.

Factors that need to be taken into account in the impact analysis are the construction phase (site, plant, etc.), the impact of the associated transportation infrastructure, the removal of existing landscape features, and the addition of new landscape features. In many cases, such as afforestation, the short- and long-term impact of the developments needs to be assessed, while the impact analysis should allow for planned or natural screening from trees and other features in the landscape.

Laser-Scan systems are being used for the interactive design and location of the development, and in representing the development at the public inquiry. Applications that are involved are intervisibility calculation, three-dimensional visualization, and volumetric calculation. For all of these applications, digital techniques have considerable advantages over the traditional manual methods of volume and line-of-sight calculation involving the analysis of contours on paper maps, or the use of an artist and photomontage techniques.

Intervisibility calculations have importance at the planning stage. Questions that need to be answered are whether a development can be seen from a particular viewpoint or number of key viewpoints, and how much of the development can be seen (e.g., the percentage of the feature above the horizon). Using the Laser-Scan coverage prediction software, it is possible to assess the relative merits of a number of different sites quickly and objectively, and to present the information in either a graphical or statistical form depending upon the requirements of different planning departments.

3D perspective views are important in qualitatively demonstrating the visibility of the development in the landscape, and allowing the scale of the development to be assessed. Computer techniques offer the ability to present the images in a flexible and realistic manner to a range of interests involved in the design and planning process. For example, the development can be viewed from many different viewpoints, and under different lighting conditions. The Laser-Scan system offers facilities for the introduction of texture and lighting when visualizing the facets of the terrain surface as a three-dimensional image, and for the modification of the rendered image according to distance from the view point. For example, the visibility of an industrial site at night can be assessed, or the visibility of the construction in mist or fog. For both the visibility and visualization applications, it is necessary to include inform-ation on surface features. For visualization, especially when dealing with a limited area of interest, this may involve the three-dimensional modelling of the feature, and, at the display stage, the removal of surfaces that are obscured by the terrain or by other surface features.

23.5 Radiocommunications planning

In recent years use of DTMs in the area of radiocommunications has increased. Applications in this area include radio and television, and cellular and mobile radio telephones. The principal use for the terrain data, in conjunction with computer analysis, is in the prediction of signal field strength. The signal strength data in turn is used for the planning of communications networks, rapid optimization of radio links, siting of transmitters, identification of complementary areas of coverage and prediction of areas subject to co-channel interference. Signal field strength is calculated by analysing the signal path loss from a transmitter to a receiver. Depending upon the specific application, the receiver may represent a single fixed point, but is more likely to be a circular or conical area surrounding a transmitter. Factors that need to be taken into account in the analysis are the characteristics of the transmitter (e.g., its ground height and height above the terrain); the distance between the transmitter and receiver, and the height and nature of the terrain surrounding and intervening between the transmitter and receiver.

The terrain elevation information may be derived from the DTM; however, a consideration of the terrain surface alone is insufficient, since both natural and man-made surface features have a considerable influence on the modification and distribution of signal strength. In the radiocommunications industry, surface features are referred to as 'clutter', and in many situations, such as featureless geographical areas or areas of large-scale urban development, are more important than the terrain in determining signal strength loss. Characteristics of the clutter that are important include the height and density of the features.

Typically a clutter database for radiopropagation will hold information on the distribution of large areas of water and on the distribution and density of woodland and buildings. These data will normally be stored in a vector database, but in order to allow for the rapid analysis of the clutter information in conjunction with the terrain elevation data, the data in the clutter database is frequently converted to a grid format. The clutter database is generally at a higher resolution than the terrain database, so that information on significant features that have a small spatial dimension is retained.

Using the DTM and the clutter database, signal strength is calculated by sampling both sets of data at a number of points along a single path profile (for point-to-point links), or along a series of path profiles generated radially around the transmitter. In the Laser-Scan system, options are available to construct profiles at a fixed angular interval around the transmitter, or between the transmitter and each node in the DTM. The latter option is generally preferred in order that a signal strength value for areas towards the edge of the transmitter area of influence is calculated directly, rather than being interpolated from adjacent path profiles. The sample locations may be located either at a constant interval along the profiles, or at an interval that varies with distance from the transmitter.

The path profiles hold information about the height of the terrain at each sample point, and about the height and characteristics of the surface clutter. This information, along with data on the transmitter horizontal and vertical angles of departure, transmitter power, and transmitter and receiver antenna gains, is processed through a propagation model in order to predict signal strength path loss. Many different propagation models are in use in the radiocommunications industry. These range from simple 'inverse square' models that relate power loss solely to distance, to models that take into account the diffraction of the signal over obstacles and large areas of open water. As more detailed terrain and clutter databases become available, a trend towards the use of more sophisticated propagation models can be seen. This in turn is leading to the more accurate prediction of signal field strength.

Using the Laser-Scan system, the signal strength or coverage data may be presented in combination with the grid DTM and clutter data, or with vector map data. This allows the coverage information to be presented as a three-dimensional view. Features important to a particular application, such as roads for mobile in-car communication, or urban boundaries for television transmission, can be overlaid on the coverage grid. Alternatively it is possible to contour the coverage grid in order to produce a vector contour map showing signal strength.

Signal strength data relating to a number of transmitters may be combined to determine areas of complementary coverage. This enables major areas of signal interference to be identified, and aids the design of complex radio networks. Similarly, coverage data relating to a single transmitter but modelled with different characteristics (e.g., different height or departure angles) may be combined in order to aid transmitter design.

23.6 Development trends

Developments of terrain model capabilities at Laser-Scan can be split into two areas—the generation of DTMs and the manipulation of these data. Since 1985, Laser-Scan have been involved with the UK government's Alvey programme. The objective of this programme was to encourage links between academia and industry in the development of computer technology to rival the Japanese 'fifth generation project'. Laser-Scan have been collaborating with University College London, Thorn EMI Central Research Laboratories and the Royal Signals and Radar Establishment in a project entitled 'real-time 2.5D vision systems'. The principal objective has been to process stereo imagery to generate depth values. When applied to high-resolution stereo imagery acquired by the SPOT satellite, the data generated is a digital terrain model of the ground. DTM data can be generated with an accuracy of about 10m, sufficient for medium- and small-scale databases. This system replaces manual photogrammetric methods of generating DTM data. The process is quick, generates DTMs covering very large areas, and can be used in areas to which access may normally be restricted (since the data is acquired from satellite).

Laser-Scan are developing this system into a commercial product, using the processing capabilities of the Inmos Transputer chip. The process of stereo matching (identification of the same points on two images) is highly computer-intensive, and a 30 m resolution DTM generated from a SPOT image pair has 4 million points (on a 2000 × 2000 grid). The stereo-matching process takes about 60 ms per point on a 3 MIPS workstation. To generate a complete DTM on the workstation would therefore take around 67 hours of CPU time, but the use of a 32 transputer parallel processor reduces the time required to generate the DTM to a few hours.

The development of manipulation capabilities is concentrated into two areas— graphic and non-graphic data integration. Graphic integration concerns the combination of sophisticated visualization with the planimetric map, with associated vector data overlaid. Non-graphic integration concerns the methods of linking all types of data, including relational databases, raster and vector data, for applications such as cross-country route finding, utilizing a topological data structure.

23.7 Conclusions

The number of applications for digital terrain data, and particularly terrain data held in the form of a regular grid DTM or Triangulated Irregular Network, is increasing rapidly. At the same time, the applications are becoming more varied and more demanding. This chapter has illustrated just three major areas of DTM exploitation within Laser-Scan; many more examples of DTM utilization could have been cited. This situation has important implications for the generation of the terrain data (e.g., the nature of the primary data source, and the algorithms used in the DTM formation); for the data structures used to hold the terrain data; for the resolution at which the data are held, and (importantly) for the design of the terrain exploitation software. The software must offer the flexibility required by users to manipulate the terrain data in conjunction with other data sources, whether those sources be raster in the case of scanned map data and satellite remotely-sensed data, or vector map data.

References

Defense Mapping Agency (1983) Procedural guide for the production of synthesised cross-country movement compilation overlays. DMA Hydro/Topo Center, Washington DC.

Falls, R.A. (1982) Synthesis guide for obstacle siting. Report No. 9 in the ETL series on Guides for Army Terrain Analysis, US Army Corps of Engineers, Engineer Topographic Laboratories, Fort Belvoir, Virginia 22060.

Fatale, L.A. and Messmore, J.A. (1988) The Army's evaluation of Tactical Terrain Data (TTD). *Proc. GIS/LIS'88*, San Antonio, Nov 31–Dec 2, 1988, 791–799.

McCullagh, M.J. (1988) Terrain and surface modelling systems: theory and practice. *Photogrammetric Record* **12**(72), 747–779.

Marth, R.B., Sr. (1989) Combat terrain information systems—terrain analysis and reproduction support in the 1990s. *Proc. ASPRS/ACSM Annual Convention*, Baltimore, Apr 2–7, 1989, **4**, 168–176.

Part F
Accuracy testing, procurement and benchmarking of DTM systems

24 The accuracy of digital terrain models

J.W. SHEARER

24.1 Introduction

Just as there is no such thing as an absolutely accurate map, so the absolutely accurate terrain model does not exist. All digital terrain models will contain inaccuracies, to a greater or lesser extent depending on a number of interrelated factors. As with all mapping operations, the accuracy of the terrain model must be suited to the chosen or intended application. No civil engineer, for example, would contemplate using contours enlarged from standard 1:50 000 scale mapping as a basis for the detailed planning and control of a construction project. It would be equally inadvisable to use a DTM derived from similar contour data for such purposes. The accuracy of digital terrain models and the factors which affect their accuracy are thus important considerations in relation to both the derivation and the application of such models.

24.1.1 Defining and describing accuracy

The only 'true' situation in the context of mapping is the terrain surface itself, and since this condition of 'absolute' accuracy cannot be attained by measurement, the accuracy of any field survey data, photogrammetric measurement or completed map can only be assessed by check comparisons with measurements made to a known higher order of accuracy. Thus when defining accuracy, in most cases this is strictly the relative rather than the absolute accuracy. Two types of error may be recognized:

(a) *Random errors:* as the term suggests, these are errors which are quite random in nature. In terms of height values, random errors would be expressed by values both greater and smaller than the 'true' values.
(b) *Systematic errors:* these have a definite shift or loading. For example, if all height values checked had a positive value (that is, recorded heights were all higher than the correct values) then a positive systematic error would be indicated.

24.1.2 Terminology

The accuracy of a DTM may be assessed by comparison of height values derived from the DTM surface with height values of the corresponding points obtained by measurement of the terrain surface to a known (higher) order of accuracy. For example, a DTM derived from digitized contours could be checked by measurements made by field survey or photogrammetric measurement. The data obtained from such a comparison will consist of height differences (or residuals) at the tested points. These values may be positive or negative in sign, depending on the relative heights of the two surfaces at successive points. This data may then be analysed to yield statistical expressions of the accuracy.

In the formulae given below the height differences are represented by v (residuals at n individual points being $v_1, v_2, \ldots v_n$).

Algebraic mean
The mean, or properly the *algebraic mean*, is computed as follows:

$$\text{Algebraic mean } \bar{X} = \sum_1^n v/n$$

(The standard statistical term $\bar{x}$ is used here for the mean. It is perhaps worth noting that this has no relation to the x-coordinate.)

The expression takes account of the sign of the residuals, and will tend to zero if there are similar magnitudes of positive and negative values. It does not indicate in any way the magnitude of individual values, and indeed, if the differences are truly random, a result of $\bar{x} = 0$ would obviously give no indication of what may be quite large differences in height at individual points. On the other hand, if a significant positive or negative value resulted from this computation, this would indicate that there was a systematic component in the residual values, that is that one surface was systematically higher or lower (related to datum) than the other.

Mean error
The mean error is computed in the same way as for the algebraic mean, but the sign of the residuals is ignored. That is, the absolute values of the height differences are summed discounting the arithmetic sign. Hence:

$$\text{Mean error } ME = \sum_1^n |v|/n$$

(where $|\ |$ denotes the *modulus*, or *absolute value* of the number). Because sign is not considered in this case, the result will always be greater than zero and will reflect the range or distribution of the residuals. 50% of the residual values will lie in the range $-ME$ to $+ME$.

Root mean square error
The problem of having positive and negative values is also obviated through the use of the statistical term 'root mean square error', the computation being as follows:

$$\text{Root mean square error } M = \pm\sqrt{\sum v^2/(n-1)}$$

Properly, $(n-1)$ rather than n should be used in this case, but if the number of points tested is large, then the difference to the computed result through using n will be of little significance. Assuming $\bar{x} = 0$, and that there is a normal distribution, then 68.27% (approximately two-thirds) of the residual values will fall in the range $-M$ to $+M$. The term *standard error* is frequently employed in mapping to describe this expression of accuracy.

Standard deviation
The standard deviation or *deviation from the mean* is the final commonly used statistical expression, and is computed as follows:

$$\text{Standard deviation } S = \pm\sqrt{\sum(v-\bar{X})^2/(n-1)}$$

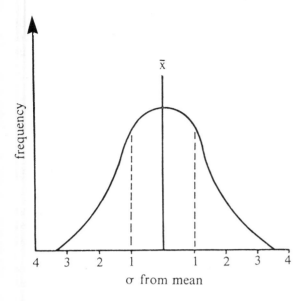

frequency

$\bar{X}$

4 3 2 1 1 2 3 4

σ from mean

Figure 24.1 Normal distribution curve.

Again, while the denominator of this expression strictly should be $(n-1)$, the use of a denominator n makes no practical difference where the number of tested points is large. Thus the above expression can be, and often is, rewritten in the form

$$S = \pm \sqrt{\left(\sum v^2/n\right) - \bar{X}^2}$$

or

$$S = \pm \sqrt{(M^2 - \bar{X}^2)}$$

The distribution of the residuals will then be symmetrical about $\bar{X}$ and again 68.27% of the values will lie in the range $-S$ to $+S$, and 99.73% of the values will fall in the range $-3S$ to $+3S$. This value ($\pm 3S$) is often referred to as the 'maximum error', though it is possible for a few values to exceed this amount (see Fig. 24.1 and Table 24.1).

It should be clear from the above that if $\bar{X} = 0$, then $S = M$, and the root mean square error can replace the standard deviation. In this case, the distribution shown in Fig. 24.1 will still hold true, but $\bar{X}$ will be equal to zero and the deviation from this will

Table 24.1 Probability of errors.

Multiple of S	% of points falling within this multiple	Term
0.6745	50%	Mean error
1.0	68.27%	Standard error
1.5	86.6%	
1.6449	90%	
2.0	95.5%	
2.5	98.8%	
3.0	99.73%	'Maximum' error
4.0	99.99%	

be in terms of multiples of the RMSE (M) instead of the standard deviation S. This perhaps explains the apparent confusion which may arise from the fact that *both* expressions (root mean square error and standard deviation) are often referred to as the *standard error* in the context of mapping accuracy.

24.1.3 The accuracy of height information

With respect to conventional mapping, there are two principal methods of recording and representing height information, both of which may be used as input data for DTMs:

(a) *spot heights* (regularly or irregularly distributed); and
(b) *contours* at a chosen vertical interval related to map purpose, scale and terrain slope).

In both cases, the accuracy has to be considered in terms of both plan position and height value. In simple terms, for example, a spot height may be incorrectly positioned but correct in height, correct in plan position but incorrect in height, or—most commonly—incorrect in both plan and height.

Spot heights are thus normally assessed or specified with respect to accuracy in terms of RMSE values related to plan and height. Contours are linear features, and their accuracy is not quite so simple to define as point-related data. Strictly speaking, errors and hence RMSE values can only be determined at identifiable point locations, yet it is often impossible to clearly establish such positions on contour lines. Though attempts have been made to express contour accuracy in strictly numerical terms (for example Lee, 1985), the common method of defining contour accuracy is to establish a tolerance (RMSE) value for permissible errors in contours, and then to check that points on the map or derived from field or photogrammetric measurement fall within such a tolerance. The accuracy specified is determined with reference to the interrelated factors of scale, terrain slope and vertical interval, and is based on the RMSE values established for points. The normally accepted standard for contours is about 3 times that which can be attained for measured points. It is normally expressed in terms of probability related to the vertical interval, for example, '90% of tested points should be within one-half of the vertical interval of their true value'.

A further point has to be made with respect to contour lines; their morphological correctness, or fidelity in representing the minor (but often critical) irregularities of the

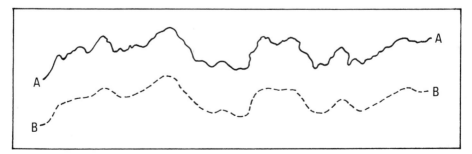

Figure 24.2 Line *A* represents a highly accurate contour, such as might be determined by precise photogrammetric measurement; line *B* represents the much smoother form of the same contour which might be the result of interpolation from a DTM. The 'quantitative' accuracy of *B* might well be quite high, but there is obviously qualitative inaccuracy, and this is difficult to express in numerical terms.

terrain surface. This relates to the detail of the contour rather than the general form (Fig. 24.2).

24.1.4 Specification of height accuracy: mapping standards
The accuracies specified for height representation on maps are, for the most part, based on the work of Koppe carried out in the first years of this century. He demonstrated that the height errors increased in direct relation to terrain slope and established the following expressions, where a is the slope angle:

$$\text{RMSE in height } Mh = \pm(A + B*\tan a)$$
$$\text{RMSE in plan } Mp = \pm(B + A*\cot a)$$

A and B are constants established in relation to the scale (and accuracy requirements) of mapping. It is some 85 years since Koppe first demonstrated this principle, and the validity has been questioned at various times over the years (see Lyytikainen, 1986, for a summary of the arguments). Nevertheless, the formulae still

Table 24.2 Values of A and B for high-quality mapping, where a is slope angle (after Imhof, 1982).

Scale	V.I. (m)	Mh (metres)	Mp (metres)
1:1 000	1	$\pm(0.1 + 0.3 \tan a)$	$\pm(0.3 + 0.1 \cot a)$
1:5 000	5	$\pm(0.4 + 3 \tan a)$	$\pm(3 + 0.4 \cot a)$
1:10 000	10	$\pm(1 + 5 \tan a)$	$\pm(5 + \cot a)$
1:25 000	10	$\pm(1 + 7 \tan a)$	$\pm(7 + \cot a)$
1:50 000	20	$\pm(1.5 + 10 \tan a)$	$\pm(10 + 1.5 \cot a)$

Table 24.3 Values of Mh and Mp for 10° and 30° slopes.

Scale	V.I. (meters)	$Mh(\pm \text{metres})$ $a = 10°$	$a = 30°$	$Mp(\pm \text{metres})$ $a = 10°$	$a = 30°$
1:1 000	1	0.15	0.27	0.86	0.47
1:5 000	5	0.92	2.13	5.26	3.69
1:10 000	10	1.88	3.89	10.67	6.73
1:25 000	10	2.23	5.04	12.67	8.73
1:50 000	20	3.26	7.27	18.51	12.59

Table 24.4 Standards for Mh in current use.

Scale	Country	$Mh(\pm \text{m})$	$(a = 30°)$
1:5 000	W. Germany	$\pm(0.4 + 5 \tan a)$	3.29
1:10 000	Switzerland,	$\pm(1 + 3 \tan a)$	2.73
	UK.	$\pm\sqrt{(1.8^2 + 3^2 \tan a)}$	2.49
1:25 000	W. Germany,	$\pm(0.5 + 5 \tan a)$	3.39
	Switzerland,	$\pm(1 + 7 \tan a)$	5.04
	UK.	$\pm\sqrt{(1.8^2 + 7.8^2 \tan a)}$	4.85
1:50 000	Switzerland,	$\pm(1.5 + 10 \tan a)$	7.27
	USA,	$\pm(1.8 + 15 \tan a)$	10.46
	ISP.	$\pm(1 + 7.5 \tan a)$	5.33

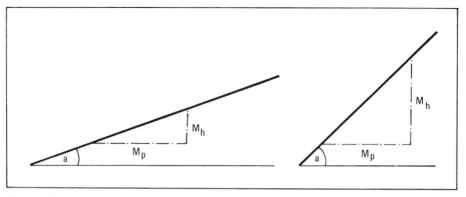

Figure 24.3 On a gentle slope the height error M_h will be small, compared to errors liable to occur on steeper slopes. However, the plan error M_p is liable to be comparatively greater on more gentle slopes.

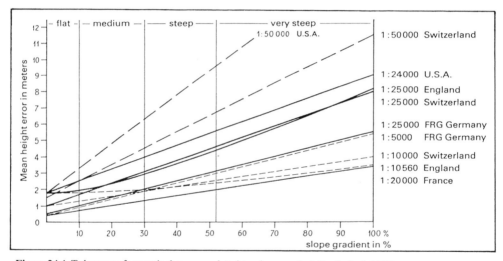

Figure 24.4 Tolerances for vertical errors, related to slope angle (after Imhof, 1982).

form the basis for the accuracy specification of height on most modern topographic mapping. The formulae relate to points, but are also used to specify the accuracy of contour lines, as shown in Tables 24.2, 24.3 and 24.4 and Fig. 24.4.

24.2 The accuracy of digital terrain models

24.2.1 Determination of accuracy

Since digital terrain modelling systems either rely on the production of a regular grid of height points, or can generate such a grid, it is appropriate and relatively simple to check the accuracy of these points by comparison with a control grid of height values obtained by field or photogrammetric measurement. Such a comparison will yield standard error values for the grid nodes and will provide the basis for assessment of the errors liable to occur in interpolated height points and/or contours. In a similar way, the accuracies of different terrain modelling packages can be compared and analysed. The possible situations which may result from the comparison of height values in a DTM with 'true' heights on the terrain surface are illustrated in Fig. 24.5.

Figure 24.5A indicates the situations where one surface is uniformly higher than the other. The value of v at each node is constant, and may be positive or negative relative to a given datum level. This situation reflects a totally systematic shift of one surface relative to the other, and in such a case

$$\bar{X} = ME = v$$

Figure 24.5B illustrates quite random values of v, both in terms of magnitude and arithmetic sign. Over a large sample of points, the value of $\bar{X}$ in such a case would equal or be very close to zero, and the magnitude and distribution of the height difference values could be best expressed by the expression root mean square error.

In the case of Fig. 24.5C, the magnitude of v varies quite considerably but the majority of differences are in one direction. This situation reflects a combination of both random and systematic effects. The value of $\bar{X}$ will indicate the systematic element in such a case, while S will reveal the magnitude and distribution of the residual values. It is worth noting that the value of $\bar{X}$ could be used to 'eliminate' the systematic error from the DTM surface, by appropriately adding or subtracting $\bar{X}$ from each grid node height. The remaining differences at each node would then be random in nature.

In addition to the numerical expression of the grid node accuracy, it is often of value

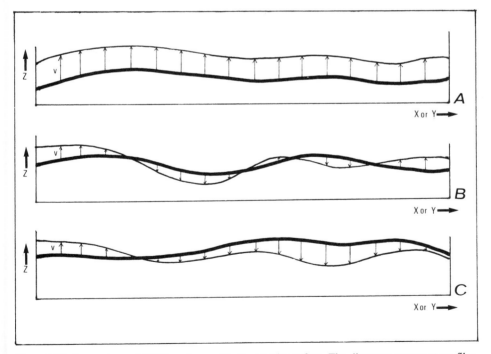

Figure 24.5 Comparison of DTM surfaces with the terrain surface. The diagrams represent profiles through a theoretical DTM surface plotted against the corresponding profile of the terrain.

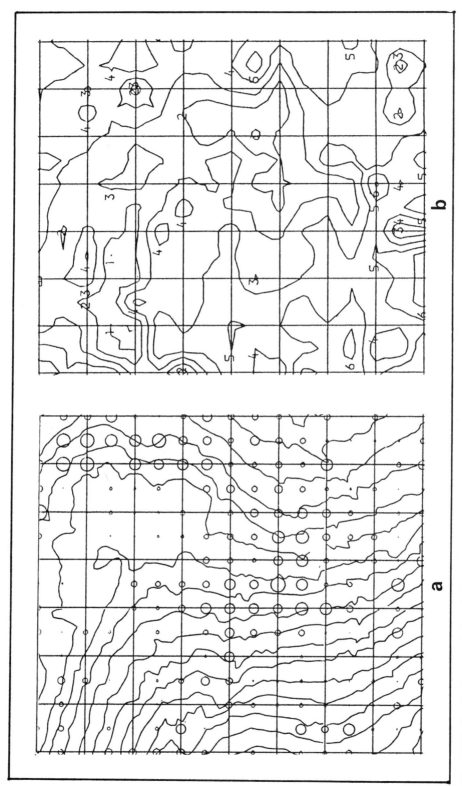

Figure 24.6 Graphical representation of accuracy in digital terrain models. (a) shows errors at grid nodes by means of proportional circles; in (b) the errors are shown by means of contours (after Kassim, 1987).

to have a graphical presentation showing error magnitude and distribution. Figure 24.6a demonstrates one possible method of representation, using proportional circles located at grid nodes. The example shown was generated automatically from residual values obtained by comparison of two digital terrain models derived from the same digitized contour data, processed respectively using the PANACEA DTM package and the 'REGRID' routines of the GHOST general graphics package. In the original output, different colours (red and blue) were employed to differentiate between positive and negative residuals. Such representations have the advantage of clearly indicating grid nodes where serious, and perhaps anomalous errors occur. An alternative and commonly applied method is to plot the errors as contours (Fig. 24.6b).

Comparison of such diagrammatic representations with, for example, a plot of the original input contours, can be extremely informative with respect to the occurrence and magnitude of errors in relation to such factors as the terrain slopes and distribution of input data.

24.2.2 Standards of accuracy for digital terrain models
The standards established for line maps would seem to be an appropriate yardstick for DTMs. One problem, however, is that while map standards are always scale-related, a DTM is, in a sense, independent of scale, as the data is referenced in 'precise' coordinate terms. Ley (1986) suggests that the equivalent of scale in the DTM is the grid node separation, or horizontal resolution, and suggests that node separations of 30 to 100 m are equivalent to map scales of 1:50 000 to 1:250 000. He then considers that accuracy be specified as a function of the grid node distance, for example, the standard error of grid node heights should be one-third of the grid node separation for first-class accuracy. It would appear that this approach is valid only where the grid-based DTM is created by direct measurement of grid height values, for example by photogrammetric or field-survey methods. Where the grid node values are interpolated from originally randomly distributed data, then most DTM packages allow the user to select the node separation. Obviously this has to be done in relation to the density and distribution of the input data. It should be clear that decreasing the node separation does not necessarily lead to an increase in accuracy.

24.3 Case studies of accuracy testing

In recent years, numerous tests have been carried out to ascertain the accuracy of DTMs, comparing the effect of utilizing different input data and processing with different modelling systems. The following examples are representative of the type of work which has been carried out.

24.3.1 The comparative accuracy of DTMs derived from ground-survey and photogrammetric sources
Ackermann (1978) reported on a series of experiments carried out at the Institute of Photogrammetry, University of Stuttgart, to examine the capability of the SCOP DTM package, particularly with respect to contour threading. A test area measuring 0.9 × 1.7km, ranging in relief from 580 to 650m, was measured by ground-survey and photogrammetric techniques to provide input data for SCOP. The resulting contours were compared with one another, with photogrammetrically measured contours, with the existing 1:2 500 scale map coverage of the area and with contours manually

interpolated from the field survey data. The details of the various measurements made are as follows.

A: Ground survey

(S1) *Tacheometric Survey 1.* Carried out using a Zeiss Reg Elta electronic tacheometer. The area was covered with a network of some 6 100 points, on average around 15 m apart. The density of points varied according to slope, with a higher density in steeper areas.

(S2) *Tacheometric Survey 2.* Carried out independently using the same instrument, this time with a reduced number of surveyed points (1 700) with an average separation of 25 m.

(S3) *Plane Table Survey.* Based on some 150 surveyed points at an average separation of 30 m, this survey covered only a small part of the test area.

(S4) *State Base Map.* At the scale 1:2 500. This had been surveyed in 1925, and contours interpolated from some 350 points measured by conventional tacheometry. The average point separation was 45 m.

B:Photogrammetric measurement
Photography at 1:10 000 scale taken with a Zeiss Oberkochen RMK 15/23 wide-angle camera was used for all the measurements. One model covered the whole area. Measurement was carried out in two different instruments—a Zeiss Oberkochen Planimat D2 precision plotter and a Zeiss Oberkochen Planitop F2 topographic plotter. Measurement included direct contouring, and heights measured during scanning—the latter being performed manually (not automatically) at different scan directions and intervals. Because scanning on a regular network inevitably leads to the omission of minor relief features, the scan data was supplemented by the measurement of 1 198 points on breaklines (ridges, etc.). Overall some 11 different sets of measurements were produced, though only three of these were specifically investigated with respect to the accuracy of contour production via the SCOP system. The following data was thus employed.

(P1) *Photogrammetric Profile Scanning.* Using the Planimat, the scan interval of 15 m yielding some 7 580 measured points. The model was scanned separately in x and y directions to give two data sets (P1x and P1y).

(P2) *Photogrammetric Profile Scanning.* Using the Planimat, with a scan interval of 30 m, giving 1 932 points.

(P3) *Photogrammetric Profile Scanning.* With the Planitop, again at 15 m interval, giving 7 114 points.

(P4) *Direct Contouring.* Carried out on the Planimat with a contour interval of 2.5 m. To give a realistic test of contouring accuracy, this plot was edited and scribed to simulate a production situation, hence introducing the minor generalizations and drawing errors inevitable in such a process.

(P5) *Direct Contouring.* Carried out on the Planitop.

The data from S1, S2, P1, P2 and P3 were input to the SCOP program and used to generate a DTM grid with an increased density of 5 m, giving some 70 000 points through which contours were subsequently interpolated. These were rigorously checked for accuracy using some 485 control points obtained by profiling using the Zeiss Reg Elta (estimated to have an accuracy (RMSE) of better than ± 2 cm). The

Table 24.5 Input data and resulting accuracy (Ackermann, 1978).

Method	Grid	RMSE ht (*Mh*)	Notes
S1 (SCOP)	15 m	± 0.25 m	Reg Elta
S1m (manual	15 m	± 0.26 m	Reg Elta
P4 (direct contouring)		± 0.37 m	Planimat, edited and scribed
P1*y* (SCOP)	15 m	± 0.40 m	Planimat, scan in *x* direction
S2 (SCOP)	25 m	± 0.43 m	Reg Elta
P1*x* (SCOP)	15 m	± 0.43 m	Planimat, scan in *x* direction
P5 (direct contouring)		± 0.48 m	Planitop, unedited
P2 (SCOP)	30 m	± 0.59 m	Planimat
P3 (SCOP)	15 m	± 0.64 m	Planitop
S3 (plane table contours)		± 0.75 m	
S4 (state map contours)		± 1.16 m	

photogrammetric plots and the plane table and state map contours were similarly checked, as was a manual interpolation of contours prepared from the first tacheometric survey (S1). The results are summarized (in order of accuracy) in Table 24.5.

The official standard laid down for the state survey mapping at 1:2 500 scale is given as $Mh = \pm (0.4 + 5 \tan a)$. It is clear from Table 24.6 that all the methods tested performed significantly better than the official standard, and this is further emphasized in Figure 24.7.

Four of the contour plots are shown (at reduced scale) in Fig. 24.8. It is clear from these that the contours processed using SCOP (S1 and P1*y*) compare closely in general terms with the photogrammetrically plotted contours (P4). In fact, as Table 24.5 indicates, the RMSE values correspond very closely, P1 and P4 being practically identical, while S1 is somewhat better, reflecting the higher accuracy of survey measurement. However an interesting point emerges in relation to earlier remarks on the qualitative aspect of contours. Both the DTM-derived contour plots are characterized by smooth contour lines; the small irregularities evident in the photogrammetric plot do not appear. This is particularly the case with plot S1. While it is clearly the most accurate in 'numerical' terms, there is some loss with respect to its representation of the minor (but characteristic) undulations of the terrain surface. The fact that plot P1*y* does retain some of these irregularities might suggest that the actual siting of the field surveyed points had at least some influence on the result. A final point may be made with respect to the influence of the photogrammetric scanning direction relative to the terrain. The (marginally) better results from scanning in the *Y*-direction (P1*y*, ± 0.40 m) as opposed to the *X*-direction (P1*x*, ± 0.43 m) might be attributed to the fact that the general trend of the slopes is in the *X*-direction. However, the difference in accuracy in this case is hardly enough to be significant.

Ackermann concluded, both from the accuracy values obtained and by visual comparison of the morphological quality of the contours, that the interpolation of contours from the measured DTM spot height values can be quite acceptable. He makes the point that the nature of the input data is critical, both in terms of accuracy and distribution, and observes, significantly, that the distribution is quite critical and has to be chosen in relation to the terrain rather than on a regular (grid) basis. In the case of photogrammetric scanning, it appears to be essential that the regular grid data is supplemented by profile data from breaklines or by progressive sampling.

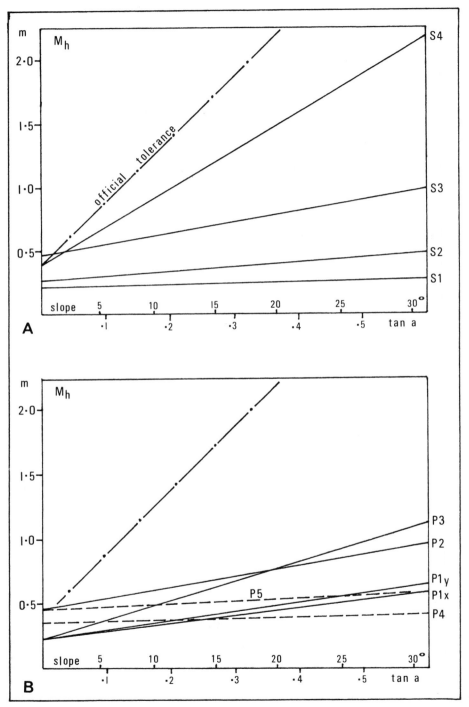

Figure 24.7 Height accuracy of contour lines as functions of slope. (A) Ground survey data. (B) Photogrammetric data. (After Ackermann, 1978.)

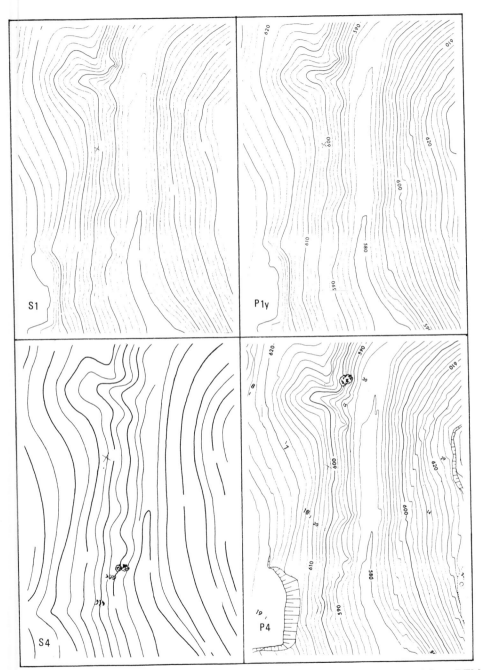

Figure 24.8 Sample contour plots (shown at reduced scale). *S1* and *P1$_y$* are interpolated from DTMs created using SCOP. Plot *P4* shows contours plotted by conventional photogrammetric measurement, while *S4* shows the contours as plotted on the State Base Map. (After Ackermann, 1978.)

M

24.3.2 Comparative tests of DTMs derived from photogrammetrically measured data

Torlegard *et al.* (1987) report on a very detailed series of tests carried out under the auspices of the ISPRS. Measurements were carried out by 15 different organizations using photography from six test areas at scales ranging from 1:4 000 to 1:30 000 (see Table 24.6). All measurements were performed manually, but using different techniques, and included grid measurement, profiling, breaklines, contours, random spot heights and combinations of these. All participating organizations were then asked to derive the heights of some 2 500 control points using their DTMs. The authors then carried out a detailed analysis of these results compared against values obtained for these points by very precise photogrammetric measurement, with a specified accuracy in the order of ± 0.3 to ± 0.7‰ (per mil) of the flying height.

Figure 24.8 shows the results of this analysis for what the authors term 'unfiltered errors', that is direct comparison of the DTM-derived values with the 'true' values. The RMSE figures have been normalized to the image scale by dividing them by the appropriate flying height, hence allowing direct comparison of the different DTMs. The accuracy can thus be related to terrain type, density of measured points, method of measurement and DTM interpolation method. The production organizations had also been asked to designate their various DTMs as intended for low- or high-accuracy applications, and the diagram also shows this distinction. The actual constraints of high and low accuracy are represented as functions of one-third of the normal contour interval for such applications.

It is clear from the diagram that the majority of organizations overestimated the accuracies of their products, particularly with respect to intended high-accuracy applications. The worst results came from area 'D' which is described as 'steep and rugged mountains', though area 'F' ('smooth terrain') also seems to yield rather low-quality results which can be attributed to the difficulty of accurate height measurement in forested areas.

The analysis of the mass of data resulting from these tests has been extremely detailed, and indeed work still proceeds in this area. Obviously, this chapter does not permit detailed examination of the results, but some of the main conclusions are seen as follows.

(a) The standard error in photogrammetrically measured DTMs is in the range ± 0.2 to 0.4‰ of the flying height. Given a flying height range of between 1 000 and 15 000 m, this represents attainable accuracy extremes of between $\pm (0.2$ to $0.4)$ m and $\pm (3.0$ to $6.0)$ m.

(b) In very steep areas, these figures rise dramatically to between ± 1.0 and 2.0‰ of

Table 24.6 Test areas—aerial photography (Torlegard *et al.*, 1987).

Area	Photo-scale	B/H ratio	Terrain type
A	1:30 000	0.63	Farmland and forest
B	1:30 000	0.65	Rugged granite bedrock without soil cover
C	1:20 000	0.61	Urban areas
D	1:17 000	0.59	Steep, rugged mountain area
E	1:10 000	0.67	Moderate hills
F	1:4 000	0.63	Smooth, forested terrain

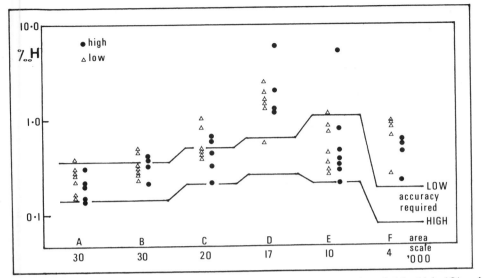

Figure 24.9 RMSE values as ‰ (per mil) of flying height (log scale). DTMs intended for high- (●) and low- (△) accuracy applications from different sources and organizations (Torlegard *et al.*, 1987).

the flying height, hence yielding comparable extremes of ± 1.0 to 2.0 m and ± 15.0 to 30.0 m.

(c) The maximum errors may be as large as 4 to 8 times the standard errors.
(d) Static sampling (spot height measurement) yields better results than dynamic methods of measurement (profiling, contouring).
(e) The accuracy of the DTM increases in relation to the density of the measured points.

As mentioned above, work still proceeds on analysis of the results, and further reports on the findings will continue to be published.

24.3.3 The effect of varying the type and density of input data

A further test employing photogrammetric data as input was reported by Ebner and Reiss (1984). They principally examined the effect of the type and density of input data on accuracy, using the HIFI program to generate the DTMs.

The test area measured 1.2 × 1.2 km, and exhibited steep slopes and marked breaklines. Photography at 1:14 400 scale, taken with a Zeiss oberkochen RMK A 15/23 camera, was used, one model covering the test area. Measurement in this case was carried out in automatic mode using a Planicomp C 100 analytical plotter. The following sets of data were recorded:

(a) a regular grid of height points at 10 m separation;
(b) height profiles at 20 m separation;
(c) contours at a vertical interval of 2.5 m; and
(d) breaklines and spot height data.

Repeated measurements and check measurements established the accuracy of the grid and spot heights as ± 0.25 m (standard error), while the contours and profiles had an accuracy of ± 0.5 m.

Table 24.7 RMSE values of differences between interpolated and measured height points. (Ebner and Reiss, 1984)

Input data set	DTM grid node separation		
	20 m *Mh*	40 m *Mh*	60 m *Mh*
20 m grid	± 0.35	± 0.35	± 0.41
40 m grid	± 0.38	± 0.44	± 0.46
60 m grid	± 0.46	± 0.49	± 0.61
20 m profiles	± 0.47	± 0.49	± 0.61
40 m profiles	± 0.49	± 0.51	± 0.51
40 m profiles	± 0.52	± 0.55	± 0.59
Contours (2.5 m)	± 0.51	± 0.53	± 0.57

Subsets of the measured data were used as input, namely grid and profile data at 20, 40 and 60 m spacing (supplemented by the breakline data). The contour data was also used as input, giving a total of seven test data sets. Each of the data sets was used to generate three separate DTM grids, with node separations of respectively 20, 40 and 60 m. The accuracy was then tested by interpolating grids at 10 m interval from each DTM and checking the accuracy of these interpolated points against the original measured data set. The results are presented in Table 24.7.

The results are rather predictable, in that it is to be expected that the grid inputs would provide more accurate results than comparably spaced profile input. It is significant, however, that even the coarsest grid (60 m) can allow generation of more accurate DTM heights than much denser profile and contour data. This has considerable implications in relation to the time required for data collection and processing. The authors did not, however, report on the accuracy (numerical or morphological) of contours interpolated through the various DTM grids or on the distribution of errors in relation to slope or terrain characteristics.

24.3.4 Comparative accuracy of different DTMs generated using various photogrammetrically derived input data sets with different distributions

Grassie (1982) reported on a series of tests performed on photogrammetrically derived data sets processed using a number of commercially available surface modelling packages and programs which are mainly designed for general or thematic map graphics rather than specifically for DTM generation. Nevertheless, the results obtained are still of considerable relevance, particularly with respect to the effect of the distribution of input data points.

The data used in the tests were digitized photogrammetrically using a Santoni Stereosimplex IIc fitted with digital encoders. Wide-angle photography at 1:21 000 scale was used, and the test area covered an area of 1.5 × 1.5 km, with relief variation in the order of 150 m. Two sets of data were measured to provide control data for analysis of the results:

(a) A regular grid of height points. These were manually measured at a grid interval of 30 m, giving some 2 500 points.
(b) Contours, measured at 10 m interval.

Five separate data sets were measured to provide the test data, chosen 'to represent different selections of points that might be typically acquired from or while generating topographic maps'. Each of these data sets comprised around 300 points. They were:

Table 24.8 Systems tested (after Grassie, 1982).

System	Random to grid interpolation	Contour interpolation method
CONSYS (grid)		Triangle*
SYMAP	Pointwise	
GPCP (1)	Pointwise	Polynomial
GINOSURF	Pointwise	Triangle*
GHOST	Pointwise	Rectangle
KRIGE (S II G(2))	Pointwise	Rectangle
SURFACE II	Pointwise	Rectangle
SACM	Patchwise	Rectangle
TREND (GPCP(2))	Global	Polynomial

(* Triangles formed by subdivision of grid squares)

(i) sampled points on contours;
(ii) breaklines;
(iii) rivers;
(iv) a regular grid of points (85 m interval); and
(v) an irregularly distributed set of height points.

The various DTM forming systems which were tested are listed in Table 24.8.

The actual tests involved separate processing of each data set in turn, generating (where appropriate) grid nodes at different intervals and interpolating contours. A major part of the report is concerned with the visual comparison of the various contour plots, comparison being made relative to the control contours which had been generated photogrammetrically. Two major conclusions were established.

Firstly, all systems exhibited problems in some area and, for the most part, were poor at capturing the minor irregularities present in the measured contours. Considerable variation in quality was found in the methods involving a random-to-grid interpolation followed by contour threading. Patchwise, and especially global, methods of interpolation yielded the poorest results, compared with systems employing pointwise search techniques.

Attention is drawn to a common phenomenon associated with such pointwise search techniques in flat or near-flat areas—the generation of angular contours which in fact follow the grid nodes. This is particularly true in models derived from digitized contours, and is readily explained by the fact that a point search around grid nodes in such areas will normally include a number of points with the same contour value, hence the grid nodes will tend to equal the contour value, with obvious results on the subsequent threading of contours (see Fig. 24.10).

Secondly, the distribution of the input data points is absolutely critical. All systems performed poorly with the 'river' data, which had a very uneven distribution with quite major blank areas. This was particularly evident in the global fit system. Conversely, the best overall results were obtained using the evenly distributed grid data though other distributions performed better in some areas of detail.

The heights at generated grid nodes were also compared with the measured grid points to give comparative standard error values. These are shown in Table 24.9. Again, the importance of data distribution is revealed, with the evenly distributed data giving the lowest standard error values in the great majority of systems, followed by the digitized contours. Predictably, the 'river' data set yielded the highest error values.

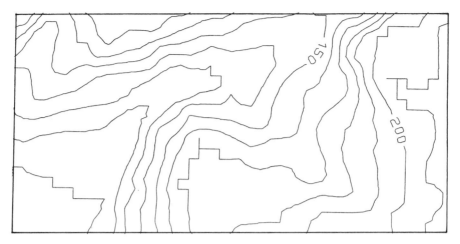

Figure 24.10 Angularity of contours interpolated through grid nodes.

Table 24.9 *Mh* values (± m) from different data sets processed by different packages (after Grassie, 1982).

| System | Breakline | Data sets | | | |
		Contour	Grid	River	Scattered
SYMAP	7.50	6.27	6.54	12.97	6.80
GPCP	7.12	5.19	3.31	12.74	5.81
GINO	5.38	4.55	3.32	11.10	5.00
GHOST	5.43	3.49	3.33	11.93	4.67
KRIGE	4.36	3.85	3.49	11.05	4.09
SACM	4.81	4.27	3.37	10.46	4.86
TREND	14.52	12.49	17.53	21.35	12.13

24.3.5 Testing of digitized contour input, derived from 1:50 000 scale mapping

A further test carried out on the same area at the University of Glasgow (Lowthian, 1986) was aimed at investigating the accuracy of a DTM generated from digitized contours, using data derived from Ordnance Survey 1:50 000 scale mapping. In view of the current establishment of a National Terrain Data Base derived from this source, this preliminary investigation is perhaps of some significance relative to possible applications of such a terrain model. The 1:50 000 scale contour pattern (at 10 m vertical interval) was digitized and the data processed using the PANACEA package. Resultant interpolated grid node values were then compared with the values originally measured photogrammetrically by Grassie. This comparison of 2 500 points yielded results as follows:

$$M = +6.75\,\text{m}$$
$$\bar{X} = -2.21\,\text{m}$$
$$S = \pm 6.44\,\text{m}$$

A significant systematic error might in part be explained by discrepancies in the plan coordinate systems of the data sets. The control data had been digitized on a local grid system, whereas the contour data was, of course, in National Grid coordinates. The reconciliation of these two systems to one common base could only be done approximately. Nevertheless, it is felt that the results do have some validity, and on the face of it, a standard error of about ± 6.5 m does appear to restrict the use of such a DTM to applications where high absolute accuracy is not a requirement. Related to mapping accuracies, such a value would certainly be acceptable for 1:250 000 scale topographic mapping, and might just be acceptable for 1:50 000 scale.

Subsequently, a further series of tests was carried out by Gar El Nabi (1988) who investigated the comparative accuracy of a series of DTMs derived from contours digitized from Ordnance Survey maps at 1:10 000, 1:25 000 and 1:50 000 scales. The area examined was part of the Island of Rhum (one of the Inner Hebrides). The three map scales all had the same contour interval of 10 m. Points on the contours were digitized at approximately the same density for all three maps and DTMs were created with a common 50 m grid interval using PANACEA. The height values at grid nodes were then compared, and standard errors computed for the two smaller scales in relation to the 1:10 000 scale data. The results of this test are shown in Table 24.10.

It has to be remembered that the values of Table 24.10 reflect comparative accuracy of the DTMs rather than absolute accuracy. It is clear, however, that accuracy does reduce with the scale of digitizing and the number of points captured, and that the standard error (compared to the actual terrain surface) for the 1:50 000 scale derived DTM would be greater than the value of ± 2.89 m obtained here.

As mentioned earlier, the Ordnance Survey is currently involved in the creation of a nationwide DTM derived from digitized 1:50 000 scale contours supplemented by breaklines and spot heights. Tests performed by the Ordnance Survey (OS, 1988) suggest that the accuracy of the DTMs so far produced lies between ± 1.7 and ± 2.5 m—significantly better than that suggested by the tests carried out at the University of Glasgow. A major factor in this apparent difference lies in the fact that the Ordnance Survey tests have so far been confined to areas in the south of England where the extremes of slope found in the two areas mentioned above are generally not present. The contour interval of the basic mapping from which the 1:50 000 contours are derived is 5 as opposed to 10 m in such areas, and the standard error specified for basic contouring is consequently lower (± 1.25 as opposed to ± 2.5 m). This alone would lead to a considerable improvement in accuracy.

The method of digitizing employed may also be a contributory factor. At Glasgow, manual line following was employed, while semi-automatic line following (Laser-Scan) using stable film originals has been used by the Ordnance Survey. Nevertheless, it seems unlikely that the standard so far achieved can be maintained over the whole country (if only because 5 m contouring is not nationally available), and it will be

Table 24.10 Comparison of DTMs derived from 1:25 000 and 1:50 000 contour data with that derived from 1:10 000 scale mapping.

Map scale	Number of points digitized	Standard error (S)
1:10 000	11 800	—
1:25 000	5 347	± 0.92 m
1:50 000	1 643	± 2.89 m

interesting to hear of future results for rugged highland areas with extreme slope variations.

24.4 Conclusions

The accuracy of any DTM is a function of a complex interrelation of a number of variable factors. These can be summarized as:

- the methods of data acquisition
- the nature (density and distribution) of the input data
- the methods employed in creating the DTM (the system used).

24.4.1 Data acquisition

Kennie and Petrie have discussed elsewhere in this publication (Chapters 2, 4, 5 and 7) the various possibilities for data acquisition by field survey, photogrammetric measurement and cartographic digitizing. Of the three methods, field survey can yield the most accurate results, while cartographic digitizing is unquestionably the least accurate.

(i) Field survey. Accuracies of ± 3 cm or less can be readily attained at measured points. Against this, however, has to be set the comparatively lengthy time (and hence cost) of data acquisition, which makes this method quite impractical for large areas. For example, Ackermann, in the project described above, quotes a figure of 13 days to survey 6 000 points as compared to 5.5 h to measure some 8 500 points photogrammetrically. The higher accuracy of the survey data is reflected in the RMSE values quoted for respective processing of the two data sets by SCOP (± 0.25 m as opposed to ± 0.4 m). Though the accuracy attained is numerically significant, in practical terms, both data sets would allow the generation of contours at a 1-m vertical interval within normally acceptable accuracy standards.

(ii) Photogrammetry. The factors which will affect the accuracy of the measured data are (see section 4.2 for more detail):

- photo scale and flying height
- base–height ratio
- instrumental accuracy
- the method employed in measurement
  - —static, at discrete points (random or selected)
  - —dynamic, profile scanning or contouring.

With this number of variables, it is not possible to make any overall observations on the attainable accuracy. However, for a given set of parameters—scale, base–height ratio and instrument, it is possible to predict the probable accuracy of different measurement techniques. Conversely, as might be the case in practice, if a particular accuracy level was required, and the camera type and plotting instrument were restricted to those available, then the photo scale and method of measurement could be chosen to achieve the required accuracy. Dynamic modes of measurement consistently yield significantly lower accuracies than static point measurement; a figure of one-third is generally assumed.

However, as the various tests quoted above indicate photogrammetric measurement can yield data of acceptable accuracy, and in practical terms of time and cost of production, photogrammetry is really the only feasible method of providing high-accuracy data for DTM creation over comparatively large areas.

(iii) Cartographic digitizing. This is the least accurate of the methods of providing input data. Obviously, the accuracy of the source map data is a critical factor and it must be realized that the process of digitizing will inevitably lead to further reductions in accuracy due to the planimetric errors which will ensue. This additional error will of course vary according to the instrumental accuracy of the digitizer and the actual procedure of digitizing employed, but for example, with manual digitizing of contours on a typical tablet digitizing table, planimetric errors of up to ± 0.5 mm at map scale may well result. Automatic line following and raster scan digitizing eliminate human errors due to positioning and may thus yield data of a somewhat higher order of accuracy. Nevertheless, while cartographic digitizing is widely used, the accuracy of such input data must be regarded with caution. This is not to say such methods should not be employed, but rather that the applications of DTMs derived from such sources must be carefully chosen. Some loss of accuracy with respect to that of the source contours must be expected.

24.4.2 The density and distribution of input data

These factors obviously have implications related to both the method of data acquisition and also the formation and accuracy of the DTM itself.

Regularly distributed data, for example on a grid or as a series of parallel profiles, has the advantage of complete overall coverage without any serious gaps. However, such a regular sampling of the terrain surface may well result in the omission of minor details, ridges, gullies, breaks of slope, etc. One solution is to reduce the grid interval, but this may greatly increase measurement and processing times. Alternatives of progressive sampling, or supplementing the regularly spaced data with selected height points at key locations and data strings for breaklines, would appear to be more appropriate solutions.

Contour data can yield quite acceptable results in terms of distribution, but it must be remembered that the accuracy of contour data (whether obtained from photogrammetric or cartographic digitizing) will inevitably be lower than that of discrete height points. The distribution of points derived from contour data will be related to the terrain slopes, and will normally result in more data points in steeper areas where the contours are closer together on the source map. The main, or rather the most noticeable, problems will tend to be revealed in flatter areas with subsequent contour interpolation from gridded DTM data (see Fig. 24.9). Again, it seems appropriate to supplement contour data with spot heights and breaklines.

24.4.3 The digital terrain modelling process

This introduces a further set of variables concerning the ways in which the input data is processed to actually create the final digital terrain model. It is difficult to make general statements, and certainly impossible to isolate one system as 'the best' in terms of accuracy. There does seem to be general agreement that methods employing 'global' and 'patchwise' fitting are less satisfactory than those employing 'pointwise' search procedures, and the 'global' methods in particular can produce some quite unpredictable and wildly inaccurate results from given data sets.

Numerous tests on different packages, utilizing different data input have been performed, only a few being indicated above. It should be noted that the majority of tests discussed here deal with programs employing a 'random-to-grid' interpolation as a means of producing the DTM as a regular grid of height points. Contours, or discrete height points, can then be derived by a further interpolation from the grid. The alternative method (such as used by CIP, Wild System 9, etc.) is to triangulate the input data points and to form the DTM using the input values directly. Hence the accuracy of the DTM height values will be *directly* conditioned by the accuracy of the input data. The accuracy of subsequently interpolated grid nodes, discrete heights or contours will again, however, be conditioned by the density and distribution of the data points. The accuracies attainable by such systems will nevertheless at least equal, and in most cases exceed those obtained by grid-based programs. In *both* cases, it appears to be critical that breakline data can be input and recognized.

Ideally, a 'benchmark' of DTM systems should include a performance rating based on the processing of a standard series of input data so that the potential customer or user can readily assess the accuracy performance of the system in relation to different types and accuracies of input data. Likewise, it is desirable that existing DTMs or databases have an accuracy rating, to allow potential users to judge whether they meet the requirements of any proposed application.

References

Ackermann, F. (1978) Experimental investigation into the accuracy of contouring through DTM. *Proceedings of Digital Terrain Modelling Symposium*, St. Louis, 165–192.

Ebner, H. and Reiss, P. (1984) Experience with height interpolation by finite elements. *Photogrammetric Engineering and Remote Sensing* 50(2) 177–182.

Grassie, D.N.D. (1982) Contouring by computer: some observations. *British Cartographic Society Special Publication* 2, 93–116.

Harley, J.B. (1975) *Ordnance Survey Maps: a Descriptive Manual.* Ordnance Survey, Southampton (HMSO, London).

Imhof, E. (1982) *Cartographic Relief Presentation.* Walter de Gruyter, Berlin.

Kassim, M.M. (1987) Accuracy comparison of DTMs created using GHOST, GINO and PANACEA. Unpublished diploma project, University of Glasgow, Dept. of Geography and Topographic Science.

Gar El Nabi, I.M. (1988) Comparative testing of digital terrain models derived from various map sources. Unpublished M.Appl.Sc. thesis, University of Glasgow, Dept. of Geography and Topographic Science.

Lee, Y.G. (1985) Comparison of planimetric and height accuracy of digital maps. *Surveying and Mapping* 45(4) 330–340.

Ley, R.G. (1986) Accuracy assessment of digital terrain models. *Autocarto London* 1, 455–464.

Li, M. (1987) A comparative test of DEM measurement using Kern DSR 11 Analytical Plotter. *Transactions of the Royal Institute of Technology, Sweden, Photogrammetric Reports* Nr 53.

Lowthian, B. (1986) The accuracy of a DTM derived from O.S. 1:50 000 scale mapping. Unpublished diploma project, University of Glasgow, Dept. of Geography and Topographic Science.

Lyytikainen, H.E. (1986) Stereological criteria for primary topographic contours. *Finnish Society for the Surveying Sciences, Publication*, Nr 2, 156.

Ostman, A. (1987) Quality control and accuracy estimation of digital elevation models. *Transactions of the Royal Institute of Technology, Sweden, Photogrammetric Reports* Nr 53.

Tham, P. (1968) Aerial map accuracy in photogrammetry. *Sartryck ur Svensk Lantmateritidskrift* Nr 2, 161–172.

Torlegard, K., Ostman, A., Lindgren, R. (1987) A comparative test of photogrammetrically sampled digital elevation models. *Transactions of the Royal Institute of Technology, Sweden, Photogrammetric Reports* Nr 53.

Anon (1982) Standards for the quality evaluation of digital topographic data. *Canadian Council on Surveying and Mapping, National Standards for the Exchange of Digital Topographic Data*, Vol 2.

Anon (1988) *Digital terrain models and digital contours.* Digital Marketing Pamphlet, Ordnance Survey, Southampton, (HMSO, London).

25 The procurement and benchmarking of a terrain modelling package

R.A. McLAREN

25.1 Introduction

The selection of a Terrain Modelling Package (TMP) is a complex task that can prove both challenging and intimidating, especially to those entering the market for the first time. The success in choosing the correct TMP for a specific application is principally dependent upon the use of a structured procurement strategy. Within this procurement framework, the User Requirement Statement (URS) and Benchmarking phases are the crucial elements that allow the diverse approaches and facilities provided by the variety of commercially available TMPs to be analysed accurately and compared objectively.

25.2 Procurement process overview

The steps involved in a typical procurement process are shown in Fig. 25.1. There are four major phases of activities involved in the process. The first concerns the education and the exposure of the potential user to the technology, allowing the feasibility for investment to be formulated. This is followed by a definition phase where the user requirements describing the proposed product are documented. The next phase involves the evaluation of available products and procurement of the selected package. The final phase concerns the design and implementation of the TMP.

It is recommended that, independently of the scale of investment, all the component steps of the process are observed. Omissions will inevitably lead to the procured product being incompatible with targeted applications or the users of the package being insufficiently skilled to take full advantage of the technology. There are no benefits in taking short cuts. For those entering the market for the first time, it is also preferable initially to 'think small', lowering the risk of the original investment. Once experience with the technology has been gained, the package can be replaced or extended with relative ease.

25.2.1 User requirement statement

The procurement process is primarily guided by the User Requirement Statement (URS). This documents the expectations of the buyer and describes exactly what the TMP must achieve within the context of the user's specific application environment. The document is written from the perspective of the user, and normally contains a detailed description of both mandatory and optional user functionality, expected

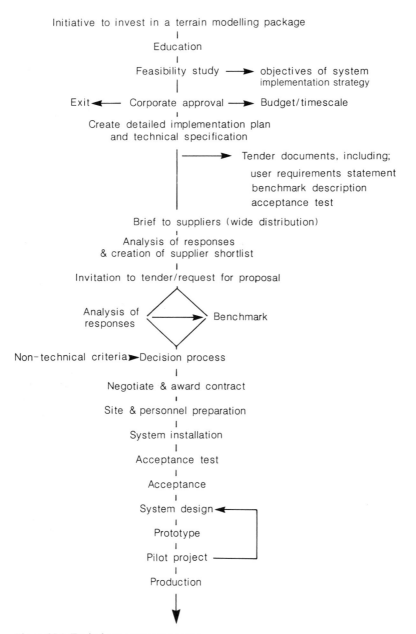

Figure 25.1 Typical procurement process.

performance levels, system architecture constraints, data sources, future needs, and financial guidelines. The document is used as a reference to create the Request for Proposal (RFP) document, distributed to all interested suppliers, and to derive the associated benchmark specification.

25.2.2 Purpose of a benchmark

From a technical perspective, the primary role of a benchmark is to provide an unbiased mechanism to measure the suitability and efficiency of a supplier's proposed TMP within the context of the buyer's application and environment. This allows alternative solutions to be checked for adherence to the User Requirement Statement and their varying approaches to be compared objectively.

In replying to a Request for Proposal, some vendors are tempted to imply that specified software functionality exists in their product, knowing that it is either planned in their product strategy or that it will be included in the system if they are awarded the contract. This is especially the case in the launch of new products. The benchmark presents the opportunity to expose this 'vapour-ware'.

During the benchmark process, the buyer is exposed to many different personnel levels within the suppliers' organizations, ranging from management through to support and demonstration personnel. This allows the company to be evaluated along with their product and provides an ideal opportunity to judge the supplier's ability to support the tendered solution based on the level of professionalism and expertise encountered during the benchmark. Throughout the benchmark, the buyer is continually gaining knowledge through exposure to the state of the art in terrain modelling technology. This knowledge should be used to adapt the weighting of criteria in the decision process and to re-appraise future strategy, as a clearer insight into the trends in terrain modelling is gained.

Apart from being a technical evaluation mechanism, the benchmark is also a useful tool in the internal political contests that prevail in all organizations. Through the shrewd involvement of personnel from a variety of departments, motivation and commitment to the eventual choice can be achieved. This tactic reduces opposition to the investment, provides necessary education to less well informed or less convinced groups, and substantially lowers the risk associated with the project.

25.3 Benchmark content

The benchmark should contain a series of functionality and performance tests that thoroughly analyse the key aspects of the system from the perspective of the terrain modelling application. The scope and emphasis of the tests will vary depending on the buyer's knowledge of TMPs, the scale of the investment and the application's characteristics. Benchmark designs may be guided by strategies to include the following optional contents:

- All mandatory and desirable functions defined in the User Requirement Statement
- Only those aspects judged critical to an organization's needs
- Evaluation of performance aspects allowing the forecasting of workloads and resources
- A focus on the least familiar aspects of the solution
- An emphasis on known weaknesses of the solution.

The tests should be designed to evaluate the existence, efficiency (performance), effectiveness (productivity), limitations and constraints, consistency, integration, adherence to standards, correctness (accuracy), and flexibility of the functionality of the proposed solutions. A TMP benchmark should include tests to evaluate the following aspects of a TMP.

(i) Accuracy. The normal way of evaluating the accuracy of fidelity of a DTM is through the use of an independent set of height control points that have an accuracy equal to or preferably greater than the original dataset used to derive the DTM. The heights of the control dataset are compared with their corresponding interpolated DTM values to obtain a root mean square error (RMSE) or standard deviation value, providing the basis for evaluation.

Ideally, the control dataset should be obtained from field-surveying or photogrammetric techniques. However, in the case where the original data was obtained by digitizing elevation data from maps, the test dataset could be created by digitizing elevations from a larger scale map. In the worst case, the original input dataset, e.g., digitized contours, could be used.

Throughout these accuracy tests it is essential that the characteristics of the test datasets represent those under which the TMP will be applied in an operational environment. Datasets with a variety of data acquisition methods, densities, volumes, structural constraints (breaklines), spatial distributions, accuracies, and terrain types should be used.

(ii) Performance. Elapsed and CPU times should be clocked for operations such as:

- The formation of the DTM from the raw elevation data
- The interpolation of contours
- The display of data (initial and subsequent displays)
- Database retrieval of selected datasets.

A variety of dataset volumes should be tested to allow the relationship between dataset size and performance to be established.

(iii) Limitations. An example is the maximum number of raw data points and breaklines which can be accepted as input to the creation of the DTM.

(iv) Editing facilities. Apart from evaluating the richness of the editing facilities, tests should establish the ability of the system to automatically re-compute higher-order elevation entities. For example, if the raw data is edited, then the system should automatically update the DTM and any derived products such as contours within the area of influence.

(v) Product generation. Aids in the generation of cartographic quality products such as the omission of contour sections based on slope criteria, automatic placement of contour labels, choice of contour smoothing algorithms, and computer graphic visualization techniques, should be evaluated.

(vi) Integration of applications. All the applications supported by the TMP should be closely coupled and form a cohesive package, exhibiting consistency in the user interface and the DBMS used to define and control the environment.

(vii) External data interfaces. The support of interfaces to data sources and GIS/Digital Mapping/CAD products.

(viii) User interface. The acceptance of computer-based technology into conventional environments is frequently based upon the new users' judgement of the 'man–machine interface.' If the functionality is presented in a consistent and efficient manner, using the local language, then users become motivated in their work,

reducing the risk of user rejection. Therefore, the benchmark must evaluate this crucial aspect of a TMP.

(ix) Flexibility. All terrain modelling applications vary sufficiently to render 'off-the-shelf' type products inadequate. Therefore, all buyers are faced with the challenge of customizing the system to merge with their specific environment. The benchmark must be designed to force the supplier to implement buyer-specific functions, thus allowing the evaluation of the available customizing tools. Ideally, the adaptations should be performed using a high-level macro language rather than a low-level programming level.

To fully test the supplier, the benchmark test should be organized into three categories:

- Tasks to be completed prior to the benchmark, e.g., definition of data model, loading of buyer's digital data and creation of buyer-specific functionality
- Predefined tasks to be performed live during the benchmark, e.g., editing of DTM
- Surprise tasks to be performed during the benchmark. These can be either premeditated, or spontaneous 'What-Ifs.'

The most demanding and illuminating tests are the spontaneous 'What-Ifs.' These stretch the skills and ingenuity of the demonstrator as well as the flexibility and power of the TMP.

25.4 Guidelines for conducting benchmarks

The decision to invite certain suppliers to participate in the benchmarking phase is based upon their response to the RFP. The RFP is distributed to a wide audience of potential suppliers, of which a small, shortlisted group is invited to benchmark. Since the benchmarking process is resource-intensive, this is usually limited to a group of three suppliers.

The supplier can only create a good benchmark if the buyer's application is well understood. Therefore, it is essential that the supplier is provided with extensive background information to the project. It is even recommended that the buyer visit the suppliers during their benchmark preparation phases, to ensure that there are no misunderstandings in their interpretation of the benchmark document.

The location of the benchmark is also important. Although a benchmark at the corporate headquarters of the supplier can be informative in judging the health of the company, there is more to be gained by forcing the benchmark to be performed at the local support centre. This exposes the strength of the hardware and software support that the supplier can provide at the local level and gives the buyer a feel for what can be expected in after-sales support.

The benchmark evaluation team should be composed of personnel with varying but complementary interests and skills. Ideally, the team should consist of computer scientists and information technologists, managers of the departments impacted by the new technology, operators of the TMP, and a terrain modelling consultant. The terrain modelling consultant is an essential member of the team, bringing cohesion to the multidisciplinary team and in-depth knowledge about the benchmarked products. The size of the team is normally 4–8 people.

The time between releasing the benchmark specification to the suppliers and the start of the benchmark should be at least six weeks. There is nothing to be gained by pressurizing the suppliers into unreasonable response times. The duration of the benchmark will depend on the complexity and scale of the contract. Most can be accomplished within three days, but some may involve a series of component benchmarks. When conducting the benchmark it is essential to ensure that software and hardware environments are clearly known. The hardware should be exactly what has been tendered for, allowing performance to be clearly evaluated. For multiuser tests, one must ensure that a simulated load is created to more realistically measure performance. The supplier should also be forced to state what software is installed: the release numbers of all packages; a description of all non-standard software and how it was created, i.e., through high-level user tools or a low-level programming language.

The cost of the benchmark is normally met by the potential suppliers (absorbed into the contract cost). However, if the benchmark entails the development of extensive buyer-specific functionality and tailoring, then the suppliers may ask for compensation.

25.4.1 What to look out for
All salesmen play tricks to impress potential buyers, and TMP suppliers are no exception. Most of the deception is aimed at demonstrating performance beyond that achievable through normal operation of the system. The normal tactic is to use small datasets. All standard demonstrations inherently use minimal dataset sizes to enhance the perceived performance. Therefore, it is imperative that realistic dataset sizes are used to exercise the TMP during all phases of the benchmark.

The same illusion can be achieved by using special hardware configurations with fully configured memory or specialized graphics boards. It is essential that the benchmark hardware configuration matches exactly what will be delivered. In an attempt to produce elegant solutions to requested buyer-specific extensions, suppliers sometimes create 'hardwired' software solutions designed for the narrow context of a benchmark example. They are not a general-purpose function and fall apart when applied to a broader, production environment. This is probably the hardest form of deception to identify.

If the supplier cannot demonstrate a requested function, a standard reply is that this is in the next release. The buyer must make sure that it is in the contract document, or it will never be there.

Suppliers will try to restrict the scope of the benchmark to 'safe territory' through strict adherence to a demonstration strategy. *Ad hoc* benchmark requests are a true insight into the system's and demonstrator's capabilities. Again, the buyer must not be initimidated! Throughout the benchmark phase it should be remembered that the objective is to evaluate the system's ability to address the specific applications' requirements. The purchaser should not be blinded by technology.

25.5 Decision process

Although the benchmark activity is designed to identify the most suitable product from a technical perspective, this is only one of many criteria involved in the decision process. The selection process is a convoluted one, with many non-technical objectives and constraints of both the supplier and buyer influencing the transaction. The transaction equation is summarized in Fig. 25.2.

In many cases, the most technically effective solution may have to be compromised by these constraints. This is the dilemma facing potential investors. Many of the constraints, such as the availability of support, existing levels of internal technical

SUPPLIER   ◄── TRANSACTION ──►   BUYER

● PROFITS
● MINIMISE SALES EFFORT
● EXISTING PRODUCT SPECIFICATION
● MARKETING STRATEGY
● PRODUCT DEVELOPMENT PLANS
● RE–USABILITY OF SW. EXTENSIONS
● SUPPORT IMPLICATIONS
● SOFTWARE DEVELOP. IMPLICATIONS
● SUPPORTED HARDWARE
● LOCATION OF BUYER
● PRESTIGE VALUE OF BUYER
● MARKET SIZE OF APPLICATION

● BUDGET
● USER REQUIREMENTS STATEMENT
● LEVEL OF EXPERTISE
● DEPARTMENTAL POLITICS
● ORGANISATIONAL POLITICS
● EXISTING EQUIPMENT & DATA
● INDUSTRY TRENDS
● CORPORATE IMAGE OF SUPPLIER
● INDUSTRY'S OPINION OF SUPPLIER
● RESULT OF BENCHMARK
● STANDARD OF SUPPORT BY SUPPLIER
● OPINION OF SUPPLIER'S
   CUSTOMER BASE
● CORPORATE COMPUTING STRATEGY

Figure 25.2 Objectives and constraints influencing the transaction.

expertise, finance, risk factors in investing in new technology, and a corporate strategy for preferred hardware suppliers, often distort the logical decision process. However, the fundamental objectives of the TMP defined in the User Requirement Statement should not be compromised.

References

Astrahan, M.M. *et al.* (1985) A measure of transaction processing power. *Datamation* **31**(7), April (1983).

Bitton, D., DeWitt, D.J. and Turbyfil, C. (1983) Benchmarking database systems: a systematic approach. *Proc. VLDB Conference,* October, 1983 [Expanded and revised version available as Wisconsin Computer Science TR No. 526.]

Goodchild, M.F. and Rizzo, B.R. (1987) Performance evaluation and work-load estimation for Geographic Information Systems. *International Journal of Geographical Information Systems* **1**(1), January–March 1987.

Raice, J. and Ricotta, J. (1986) *SUN-3 Benchmarks. A SUN Technical Report.* SUN Microsystems, Inc.

Smith, J.E. (1988) Characterizing computer performance with a single number. *Communications of the ACM,* **31**(10), October.

Index